FLOODPLAIN WETLAND BIOTA

IN THE

MURRAY-DARLING BASIN

FLOODPLAIN WETLAND BIOTA IN THE MURRAY-DARLING BASIN

Water and Habitat Requirements

Kerrylee Rogers and Timothy J Ralph

PUBLISHING

National Library of Australia Cataloguing-in-Publication entry

Floodplain wetland biota in the Murray-Darling basin: water and habitat requirements/edited by Kerrylee Rogers and Timothy J Ralph.

9780643096288 (pbk.)

Includes bibliographical references and index.

Floodplain management – Darling River (Qld. and N.S.W.)
Floodplain management – Murray River (N.S.W.–S. Aust.)
Wetland management – Darling River Watershed (Qld. and N.S.W.)
Wetland management – Murray River Watershed (N.S.W.–S. Aust.)
Darling River Watershed (Qld. and N.S.W.) – Environmental aspects.
Murray River Watershed (N.S.W.–S. Aust.) – Environmental aspects.

Rogers, Kerrylee.
Ralph, Timothy J.

333.73160994

Published by
CSIRO PUBLISHING
36 Gardiner Road, Clayton VIC 3168
Private Bag 10, Clayton South VIC 3169
Australia

Telephone: [+613] 9545 8555
Local call: 1300 788 000 (Australia only)
Fax: +61 3 9662 7555
Email: csiropublishing@csiro.au
Web site: www.publishing.csiro.au

Front cover: Channel breakdown in a floodplain wetland, Willancorah Swamp, Macquarie Marshes. Image: Tim Ralph.

Back cover: Floodplain wetland biota of the Murray-Darling Basin: bony bream, *Nematalosa erebi* (Gunther Schmida); desert tree frog, *Litoria rubella* (Jody Rowley); river mussel, *Alathyria jacksoni* (Hugh Jones); Australian shelduck, *Tadorna tadornoides* (Chris Herbert).

Set in Adobe Minion Pro 10/12 and Stone Sans
Edited by Adrienne de Kretser, Righting Writing
Cover design by Alicia Freile, Tango Media
Text design by James Kelly
Typeset by Desktop Concepts Pty Ltd, Melbourne
Index by Russell Brooks
Printed by Ingram Lightning Source

CSIRO PUBLISHING publishes and distributes scientific, technical and health science books and journals from Australia to a worldwide audience and conducts these activities autonomously from the research activities of the Commonwealth Scientific and Industrial Research Organisation (CSIRO).

Feb26_RP_ILS

Contents

Preface

The Murray-Darling Basin is one of the largest inland drainage basins in Australia, encompassing approximately 1 061 000 square kilometres of prime agricultural land. The slopes and plains of the Basin support a range of ecosystems, including more than 30 000 wetlands. Fifteen of these wetlands are recognised for their inherent and international significance under the Convention on Wetlands of International Importance (Ramsar Convention, Iran 1971). All the wetlands play a significant ecological function by providing critical aquatic habitats in otherwise well-drained or typically dry landscapes, often in association with major river courses. They provide numerous services and benefits to Australian society.

Wetlands may be defined in simple terms as areas that are permanently or temporarily covered by fresh, brackish or saline water. The Ramsar Convention defines wetlands as 'areas of marsh, fen, peatland or water, whether natural or artificial, permanent or temporary, with water that is static or flowing, fresh, brackish or saline, including areas of marine water the depth of which at low tide does not exceed six metres'. Clearly, water is an intrinsic and essential requirement for wetlands. The definitions also imply that there are numerous types of wetlands that may be characterised on the basis of their hydrology, geomorphology and/or ecology. In fact, the Ramsar Convention identifies 42 different wetland types, including 19 inland wetland types, and acknowledges that the classification is a broad framework and not an exhaustive list of wetland types.

For this book, our focus is on inland *floodplain wetlands* – wetlands that occur on low-lying, depositional areas of floodplains and are reliant on a supply of freshwater from inland rivers and creeks. The term floodplain wetland broadly refers to the collection of landscape and ecological components subject to inundation around a river, and may incorporate a range of wetland types. In the Murray-Darling Basin, this includes freshwater and saline lagoons and lakes, marshes, distributary channels, anabranches, billabongs, overflows, swamps and waterholes, along with the riparian forests, woodlands and grasslands that intersperse them. These wetlands are regarded as critical habitat for numerous species of flora and fauna, including trees and reeds, waterbirds, fish, frogs, crustaceans, molluscs and other invertebrates within the Murray-Darling Basin.

The ecology of floodplain wetlands of the Murray-Darling Basin is threatened by a range of environmental issues and the systems are regularly at the forefront of debate, managerial concern and research in fields related to environmental management, science and industry. Water resource development within the Basin, which has been marked by an increase in the predictability and consistency of river flows to floodplain wetlands, is implicated in the decline of these environmental assets. This impact is occurring at a time when future water availability for floodplain wetlands is tenuous – climate change due to global warming is expected to cause an overall decline in surface water availability across the Murray-Darling Basin in the 21st century. Floodplain wetlands, which under natural conditions exhibit variable and often unpredictable flood regimes, have been particularly affected by changes to the natural flood regime caused by water resource development. Flood regime changes are projected to continue in the 21st century, exacerbating the dramatic decline of these assets.

Given the importance of floodplain wetlands and the projected further decline in surface water availability, it is timely that consideration be given to the relationship between hydrology and floodplain wetland flora and fauna (collectively termed *biota*) within the Murray-Darling Basin. Scientific literature pertaining to the water requirements of biota within the Murray-Darling Basin has generally been site-specific, rather than focused on the water requirements of species across the broader geographical and ecological range of the Basin. In addition, ecological studies on a site-by-site basis tend to obscure the knowledge gaps that exist for some species, with attention largely given to sentinel and iconic species, such as the river red gum and colonial nesting waterbirds, while the water requirements of other species are foreshadowed. For example, there is a surprising dearth of literature on the water requirements of frogs and invertebrates that occur in floodplain wetlands of the Murray-Darling Basin. Until recently the shrub lignum was largely ignored in the ecohydrological literature despite it being an abundant and essential part of colonial nesting waterbird habitat throughout the Basin wetlands.

This book addresses these issues by presenting profiles of the water requirements of species based on data from the biological and ecological literature for the geographic range of the species. This approach also highlights the ecohydrological knowledge gaps for some species. While acknowledging that a range of biotic and abiotic factors influence the health, condition and/or abundance of biota, due to the fundamental relationship between hydrology and floodplain wetland biota, this book focuses on the influence of the water regime on species survival, health and condition. We have specifically focused on the relationships between species condition/health and aspects of the water regime such as flood frequency, duration, depth, timing/seasonality, inter-flood dry-period, and the rate of rise and fall of floodwaters. We have incorporated biotic factors that may be associated with species performance and flooding, such as germination timing, stimuli/triggers for reproduction, reproductive season and reproduction lag time (see Glossary). Habitat and dietary needs were also incorporated for heterotrophs and were synthesised in terms of the position (or trophic level) of a species in the ecosystem food chain.

The floodplain wetlands of the Murray-Darling Basin support abundant biota. This is particularly evident following a flood when a wetland is alive with the chirping of frogs and insects, the flight of waterbirds and the flurry of fish, and is flush with new vegetative growth. While it is exciting to consider this abundant life, it is virtually impossible to consider the water requirements of all the biota within floodplain wetlands. For this reason a selection of species was included in this book. The species were largely selected on the basis of their occurrence within floodplain wetlands of the Murray-Darling Basin, consideration of their significance in terms of the structure and function of the floodplain wetlands, their relatively widespread distribution throughout the Murray-Darling Basin, and the availability of literature pertaining to their water and habitat requirements. We have divided the biota into five groups: vegetation, waterbirds, fish, frogs, and crustaceans and molluscs. We readily admit that this is far from exhaustive and that it fails to incorporate ecologically significant phyla, such as insects. We hope to address these deficiencies in the future.

The opening chapter of this book gives consideration to the geographic context of floodplain wetlands in the Murray-Darling Basin as well as the ecological responses of their biota to flooding. We summarise the key geographic, climatic, hydrological, geomorphological and ecological variability within the floodplain wetlands of the Basin so as to establish an understanding of the complexity in the landscapes. Since floodplain wetland biota exist in complex landscapes, the response of biota to flooding of these landscapes is also complex. It is largely dependent on the adaptive capacity of biota to respond to flooding (and drought) stress.

Chapters 2 to 6 contain profiles of the water requirements of individual species of floodplain wetland biota that occur in the Murray-Darling Basin. In Chapter 2, Kerrylee Rogers considers aspects of the water regime that are crucial for the survival, maintenance, reproduction and regeneration of 35 key vegetation species. Kerrylee Rogers addresses the water requirements of waterbirds in Chapter 3, highlighting their richness and diversity with 48 species profiles. In Chapter 4, Tim Ralph, Jennifer Spencer and Tom Rayner address the water requirements of 17 native and four alien fish species. Despite a relative scarcity of literature about the water requirements of frogs within the Murray-Darling Basin, Skye Wassens applies her extensive knowledge of the ecology of frogs to document the water requirements of 15 key frog species in Chapter 5. Hugh Jones applies his extensive field knowledge of crustaceans and molluscs in eastern Australia to document the water requirements of 16 species in Chapter 6.

The final two chapters provide an overview of the impacts of hydrological change on biota and the implications of hydrological change for the management of biota. Chapter 7 considers the causes and effects of hydrological changes on biota, with particular reference to the likely impacts of a projected decline in surface water availability in the Murray-Darling Basin. Chapter 8 addresses the needs and challenges for the development and integration of scientific knowledge of the water requirements of biota for management plans and actions. Neil Saintilan discusses the integration of scientific, policy and management information. He highlights the difficulties in applying this knowledge to floodplain wetlands of the Murray-Darling Basin and summarises how best to maximise the use of environmental water.

The main purpose of this book is to synthesise information about the water requirements of biota, to provide a baseline for further education and research and to better inform environmental managers who work with water allocations in floodplain wetlands of the Murray-Darling Basin. In doing so, the book highlights the valuable ecohydrological research that is being undertaken within the Murray-Darling Basin. However, it also underscores knowledge gaps regarding the response of biota to flooding. The authors hope that opportunities to investigate the relationships between flooding and biota within the Murray-Darling Basin are expanded and that this book can inform and inspire further ecohydrological research within the Basin.

The Aquatic Ecosystems Climate Change Adaptation Research Project which preceded this book was proposed by Tim Pritchard (Manager Water and Coastal Science Section, DECCW) and funded through the NSW Greenhouse Office and NSW Department of Environment, Climate Change and Water (DECCW). A database of the water requirements of biota in the Murray-Darling Basin was developed as part of the project and, in conjunction with XIION, complements this book. The database can be accessed through the NSW Department of Environment, Climate Change and Water.

There are a number of people who deserve special acknowledgement. Our thanks go to Jeff Kelleway and Dr Liza Miller, previously of the DECCW, who contributed to the Aquatic Ecosystems Climate Change Adaptation Research Project. Dr Bruce Chessman provided insightful and comprehensive comments on the final report for this project and we thank him for this contribution. We wish to thank Dr Neil Saintilan and Dr Joanne Ling, who were great advocates for our book proposal and the preceding project. Many people provided comments on and contributed to reviewing and editing this book. We acknowledge that this is often a tedious task and we extend our gratitude to Dr Neil Saintilan, Dr Mike Maher, Dr Bruce Chessman, Dr Joanne Ling, Dr Jennifer Spencer and Dr Bob Creese for their willing assistance. Our gratitude is also extended to the people who supplied photography for this book. Finally, we thank the publishing team at CSIRO, namely Briana Melideo and John Manger, for their assistance and support.

Kerrylee Rogers and Timothy J Ralph
April 2010

List of contributors

Hugh A Jones
Landscape Modelling and Decision Support Section, NSW Department of Environment, Climate Change and Water

Timothy J Ralph
Department of Environment and Geography, Macquarie University

Thomas S Rayner
Australian Wetlands and Rivers Centre, School of Biological, Earth and Environmental Sciences, University of New South Wales

Kerrylee Rogers
Rivers and Wetlands Unit, NSW Department of Environment, Climate Change and Water

Neil Saintilan
Rivers and Wetlands Unit, NSW Department of Environment, Climate Change and Water

Jennifer A Spencer
Rivers and Wetlands Unit, NSW Department of Environment, Climate Change and Water

Skye Wassens
School of Environmental Sciences, Charles Sturt University

Chapter 1

Floodplain wetlands of the Murray-Darling Basin and their freshwater biota

Timothy J Ralph and Kerrylee Rogers

Introduction

In inland Australia, remarkable wetlands occur in low-lying and often extensive areas of floodplain that are subject to inundation by freshwater from rivers and creeks. These floodplain wetlands provide critical aquatic and riparian habitat for flood-reliant and flood-tolerant flora and fauna, collectively termed *biota*, in otherwise semiarid or arid landscapes. This diverse range of plants, animals and microscopic organisms, including endemic and threatened species, occupy habitats and ecological niches that are created and maintained by the flows and flood regimes of the floodplain wetland systems in which they survive and flourish. These ecosystems are naturally variable and are characterised by complex interrelationships between their flood patterns, landforms and soils, and ecological communities. The flow regimes of inland Australian rivers are driven by weather and climate variability, and so inland floodplain wetlands experience changes in the frequency, magnitude and duration of flooding in response to cycles and extreme events of rainfall and runoff in their catchments. Like the impacts of land use and water resource development in our river catchments today, future climate change related to human-induced global warming is likely to compound the effects of natural climate and hydrological variability, potentially altering the balance of biophysical and ecological processes in many of Australia's rivers and iconic floodplain wetlands.

Understanding the water requirements of freshwater biota in inland floodplain wetlands is critical for the survival, regeneration, maintenance and management of these ecosystems and their ecological communities. This book focuses on key floodplain wetland biota and their water requirements in one of Australia's largest and most environmentally, economically and culturally significant catchments – the Murray-Darling Basin. The floodplain wetlands of the Murray-Darling Basin are made up of freshwater lagoons and lakes, distributary channels, anabranches, billabongs, marshes, swamps, waterholes and overflow areas, as well as the riparian forests, woodlands and grasslands that intersperse them. These different components provide essential habitat and energy sources for many species of plants, waterbirds, fish, frogs, molluscs, crustaceans and invertebrates. This chapter considers the geographic context of the floodplain wetlands in the Murray-Darling Basin, as well as general ecological responses of the biota to flooding. It synthesises key environmental and ecological factors to establish an introduction to the complexity of flow–ecology relationships and processes in the rivers and floodplain wetlands of the Murray-Darling Basin.

Rivers and floodplain wetlands of the Murray-Darling Basin

Geography and climate

The Murray-Darling Basin has a catchment area of approximately 1 061 000 km^2. Its 17 large and complex floodplain wetland systems are associated with relatively large perennial or intermittent rivers (Figure 1.1, Table 1.1). Most of these rivers run from temperate uplands on the southern and eastern margins of the Basin (>500 m above mean sea level) and drain inland towards increasingly low elevation (<300 m above mean sea level) and low relief (*lowland*),

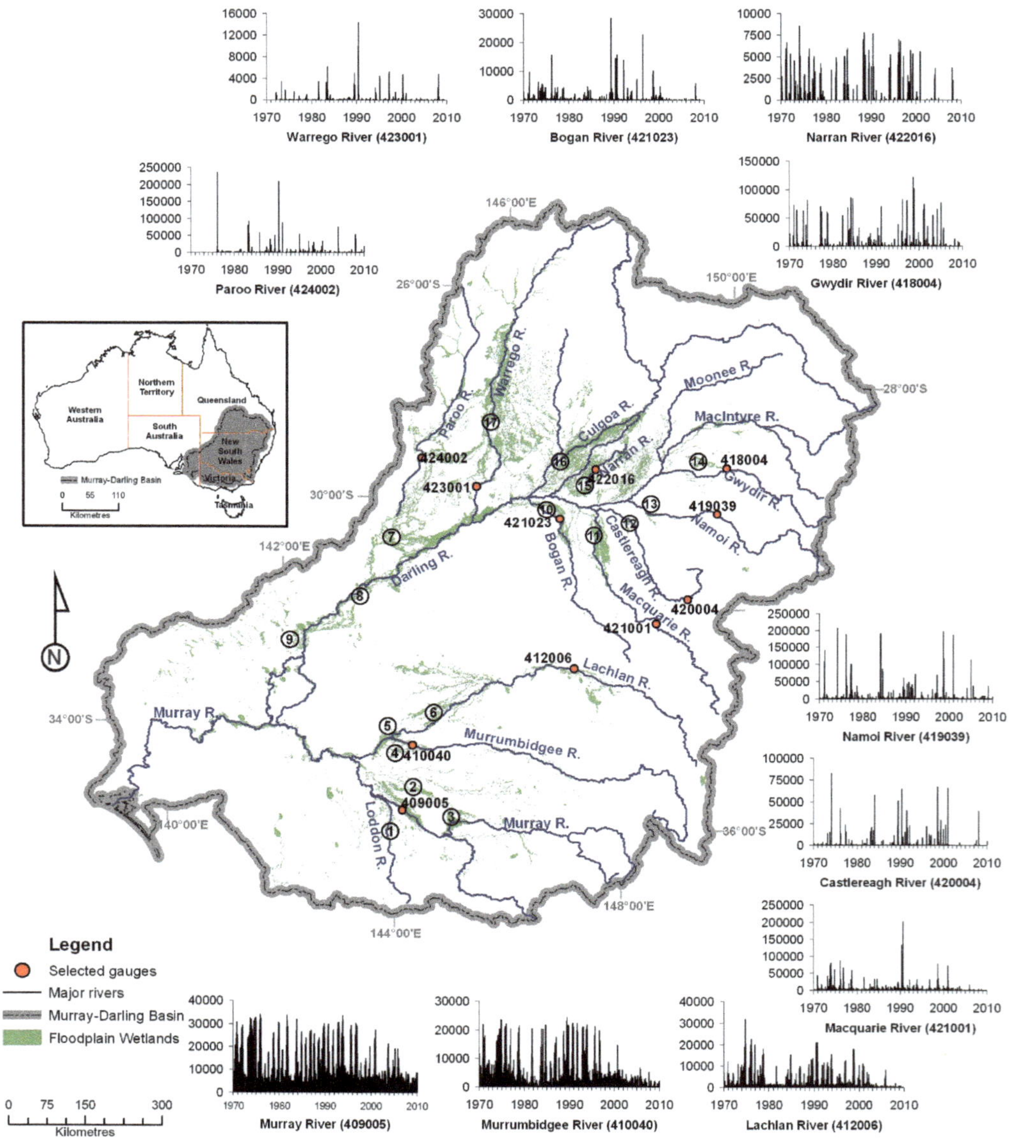

Figure 1.1: Map of key rivers and floodplain wetlands in the Murray-Darling Basin. Adapted from Ralph (2008). Hydrological data sourced from NSW Government (2009).

Table 1.1: Overview of selected key floodplain wetlands in the Murray-Darling Basin

Floodplain wetland	Number in Fig. 1.1	River and reach location	Geomorphic setting	Approximate extent (ha)
Kerang Lakes	1	Lower Loddon River and Kerang district, VIC	Riverine plain, deflation lakes	9172
Werai Forest	2	Middle Edward and Neimur rivers, NSW	Riverine plain	11 234
Millewa Forest	3	Middle Murray River between Tocumwal and Barmah, NSW/VIC	Riverine plain	33 636
Lowbidgee Floodplain	4	Lower Murrumbidgee River and Yanga, NSW	Riverine plain, distributaries	200 000
Booligal Wetlands	5	Middle Lachlan River, NSW	Riverine plain, distributaries	5000
Great Cumbung Swamp	6	Lower Lachlan River, NSW	Riverine plain, terminal swamp	50 000
Paroo Overflow	7	Lower Paroo-Warrego alluvial plains, NSW	Riverine plain, distributaries, floodouts	720 000
Talyawalka Anabranch and Teryawynia Creek	8	Mid-lower Darling River between Wilcannia and Menindee, NSW	Riverine plain, anabranches	Variable
Menindee Lakes and Darling Anabranch Lakes	9	Mid–lower Darling River downstream of Menindee, NSW	Riverine plain, anabranches, deflation lakes	45 000–269 000
Bogan Marshes	10	Lower Bogan River, NSW	Riverine plain	Variable
Macquarie Marshes	11	Lower Macquarie River, NSW	Riverine plain, distributaries and floodouts	40 000–250 000
Castlereagh Marshes	12	Lower Castlereagh River, NSW	Riverine plain, distributaries	Variable
Namoi Floodplain	13	Mid-lower Namoi River, NSW	Riverine plain, distributaries	Variable
Lower Gwydir Wetlands	14	Lower Gwydir River and Gingham watercourse, NSW	Riverine plain, distributaries	102 120
Narran Lakes	15	Lower Narran River and Lakes, NSW	Riverine/lacustrine plain, distributaries and terminal lake	10 000
Culgoa-Briari Floodplain	16	Mid-lower Culgoa and Briari Creek alluvial plains, NSW	Riverine plain, distributaries	Variable
Warrego Floodplain	17	Mid-lower Warrego alluvial plains, QLD/NSW	Riverine plain, distributaries	Variable

Source: MDBC (2006).

semiarid to arid (*dryland*) depositional settings (Thoms and Sheldon 2000; Ward *et al.* 2002; Warner 1986). This means that typically these lowland–dryland rivers are allogenic, fed from their headwater catchments and rarely receive tributary inflows along their middle and lower reaches to compensate for the evaporation, infiltration and distributary losses that occur on the dryland plains through which they flow.

Despite the dryland environments surrounding the lower reaches of these inland rivers, intermittent and semi-permanent floodplain wetlands including lotic (flowing water) marshes, lentic (standing water) swamps, and riparian woodlands and grasslands are maintained by flooding from the rivers, their anabranches and their outflowing distributary channels and floodouts (places where channels break down on the floodplain and water flows over the ground surface). For example, the Murray River feeds the Barmah-Millewa Forest, the Lachlan River feeds the Booligal Wetlands, the Lachlan Swamp and the Great Cumbung Swamp, the Murrumbidgee River feeds the Lowbidgee Floodplain, the Macquarie River feeds the Macquarie Marshes and the Gwydir River feeds the Gwydir Wetlands (Kingsford 2003; Kingsford and Thomas 2004; MDBC 2006) (Figure 1.1). Altogether, the floodplain wetlands are very extensive, making up around 6% of the Murray-Darling Basin (the total wetland component, including lakes, is ~6.5%) and accounting for >95% of all wetland areas in the inland catchments (Kingsford *et al.* 2004).

The reliance of the lowland–dryland rivers and floodplain wetlands of the Murray-Darling Basin on flows from their upper and middle catchments means that these systems are particularly susceptible to changes in climate and water supply. Regional climate variability – natural deviations from the prevailing climatic conditions in a region over years to decades – has impacts on these rivers and wetlands including changes to direct rainfall and maximum/minimum temperatures. It also affects their catchment hydrology. This climate variability is driven by large-scale ocean-atmosphere fluctuations in the Pacific, Indian and Southern oceans that influence regional air pressure and circulation patterns, and weather and rainfall. As a result, the rivers, floodplain wetlands and aquatic ecosystems of the Murray-Darling Basin have adapted to cope with natural environmental variability, which makes it difficult to generalise or simplify their preferences and requirements in terms of flow and flood regimes.

Hydrology

In the past, changes in climate have greatly influenced changes in the flow regimes of inland Australian rivers. Today, the characteristics of rivers are maintained by the current climate and the hydrology of their catchments, but rivers are also subjected to immense pressure from water resource developments, river regulation and water extraction by humans. The lowland–dryland rivers of the Murray-Darling Basin have either perennial, seasonal, intermittent or ephemeral hydrological regimes, and their flows tend to be highly variable over yearly, decadal and centennial time-scales (Finlayson and McMahon 1988). For example, monthly maximum hydrological data for some of these rivers show brief periods of very high flow interspersed by periods with moderate or very little flow, or no flow at all (Figure 1.1). Several of the rivers also experience general downstream declines in river discharge and valley slope, leading to lower energy conditions and a propensity for reductions in stream capability and efficiency. Since lower energy flows typically transport less sediment than higher energy flows, this downstream decline in discharge and stream power tends to lead to greater sediment deposition compared with upstream reaches. This in turn promotes a greater proportion of overbank flows during floods along the lower reaches of the lowland–dryland rivers compared with upstream reaches, and greater interconnection between the main river channels and the surrounding floodplain wetlands.

In general, five groups of flow regime variables are critical for floodplain wetland ecosystems and the biota that rely on water from lowland–dryland rivers of the Murray-Darling Basin (Young 1999). The first, flow magnitude, describes the total or maximum discharge volumes and associated water levels (or area inundated) and duration of flooding in a river or wetland

Figure 1.2: Some key features of inland rivers and floodplain wetlands in the Murray-Darling Basin. (a) Dams and river regulation. (b) Areas of channel breakdown and floodout. (c) Aquatic vegetation. (d) Colonial waterbirds. (e) Mosaics of flood-reliant floodplain wetland vegetation. Photographs: Tim Ralph, Macquarie University.

over daily, monthly or yearly periods. The second, flow variability, describes the frequency and periodicity of certain flood volumes and water levels that occur during a certain period of time. The third, magnitude and frequency of extreme events, describes the volume or size and length of time between severe or prolonged floods and droughts. The fourth, rates of flow changes, describes the speed at which water levels rise and fall. The fifth, flow seasonality, describes the timing of flows for a series of months or for a season in a year. These flow variables combine to influence patterns of inundation and the duration of flooding in floodplain wetlands, to which the biota will typically respond (discussed below). Extreme floods and droughts cause disturbances to normal flow regimes, and strongly influence the structure and function of wetland ecosystems (Figure 1.2).

Geomorphology and soils

Contemporary lowland–dryland river floodplains represent important sinks for the storage of sediment, nutrients and contaminants which have been mobilised from the upstream catchments. Floodplain wetlands in these systems develop by processes of vertical accretion, particularly where channels feed well-vegetated wetlands which filter sediments out of suspension and promote in-channel and near-channel sedimentation. The topography and geomorphic features (e.g. levees and floodbasins) formed by this sediment deposition provide the underlying form of the floodplain wetlands and the structure of the shallow aquatic habitats.

Many floodplain wetlands, such as billabongs (wetlands formed in meander bends that have been partly or wholly cut off or isolated from regular river flows), occur along the long meandering middle and lower reaches of lowland–dryland rivers of the Murray-Darling Basin. These rivers also tend to have large multi-channelled anabranching and distributary networks in these reaches, where interconnected and divergent channels occur on broad flat floodplains made up of fine-grained cohesive sediments (clay, silt and sand). In some cases, the main rivers lose their capacity to transport sediment and maintain channels, due to loss of stream power and discharge or due to blockage by bedrock or other barriers to flow (see reviews in Nanson *et al.* 2002; Tooth 2000). This can lead to channels that divide and eventually break down in floodplain wetlands. Channel discontinuity causes greater overland flooding on floodplains and can lead to outright river termination on land (O'Brien and Burne 1994; Tooth 1999). This phenomenon is termed *channel breakdown*: it is a fluvial state where the combined effects of hydrological and geomorphic factors lead to the partial or whole disintegration of a drainage network with sediment accumulation, channel diminution and discontinuity at the downstream end of lowland–dryland alluvial rivers, and a dominance of non-channelised flows and overland flooding (Tooth 1999).

Although floodplain wetlands can exist in areas where rivers do not break down, wetlands including riparian forests, lagoons, swamps and marshes often occur in distributary zones and zones of channel breakdown in lowland–dryland rivers of the Murray-Darling Basin (Hesse *et al.* 2005; Kingsford 2003; O'Brien and Burne 1994; Ralph and Hesse 2010; Yonge and Hesse 2009). These wetlands are periodically or continuously inundated areas of floodplain with a range of lotic and lentic environments (Thoms and Sheldon 2000; Ward *et al.* 2002). The biological components and ecology of these systems are adapted to temporarily or permanently flooded conditions (Paijmans *et al.* 1985; Tooth *et al.* 2002, 2007). Due to these relationships, the geomorphology and ecology of the floodplain wetlands often take the form of complex mosaics (Semeniuk and Semeniuk 1995). The floodplain wetlands tend to experience periods of geomorphic and ecological adjustment in response to floods, droughts and changes in sediment supply associated with the parent and subsidiary streams (Ward 1998; Ward *et al.* 2002). The floodplain wetlands can respond rapidly to processes of new channel formation, channel abandonment and associated changes in flood pattern.

The soils of floodplain wetlands in the Murray-Darling Basin are related to the water, sediment, nutrients and organic matter supplied from their rivers, as well as the vegetation growing on the floodplain surface, the inter-flood dry-periods that allow soil oxidation and the biotic activity within the soil profile (e.g. bioturbation – sediment/soil turnover by ants, earthworms etc.). The modern, low-energy alluvial systems are usually dominated by silt- and clay-sized sediments that slowly accumulate on top of older, and often coarser, sediments that were deposited by much earlier river systems on the alluvial plains. In these inundated areas, heavy-textured grey-brown soils are common. These are characterised by high clay and silt content, fairly uniform texture and colour profiles and cracks when dry. In contrast, surrounding soils developed on palaeochannels (old or abandoned river courses) and sediments that are less regularly flooded tend to have red-brown earths that are weakly structured, or massive texture-contrast soils (e.g. soils that have a grey-brown to red-brown loamy near-surface layer and a brighter brown to red clayey underlying layer). Modern and ancient sediment deposits and their related soils coalesce and overlap in the floodplain wetlands, providing sources of energy and habitats for the overlying biota.

Human impacts

Humans have been part of the environment in Australia for more than 40 000 years, and Australian landscapes, ecosystems, people and other biota have had to adapt to a range of environmental pressures and changes. Recent human impacts in the form of major water resource developments and intensive land-uses have had the most rapid and significant impacts upon the hydrology and ecology of lowland–dryland rivers in the Murray-Darling Basin (Frazier *et al.* 2005; Kingsford 2000a, 2000b; Sheldon *et al.* 2000; Thoms and Sheldon 2000; Thoms *et al.* 2005). River regulation by dams and water extraction for domestic, industrial and agricultural uses have generally led to a reduction in the frequency, magnitude and duration of flood events in the lower reaches of many systems (Jolly 1996; Reid and Brooks 2000). Flow seasonality is also affected, while floods can be less intense and tend to recede at a faster rate (Jolly 1996). There is some evidence to suggest that geomorphic changes (e.g. erosion and sedimentation in channels) are related to hydrological changes due to river regulation and abstractions (Thoms and Walker 1992). Catchment land-use changes, such as vegetation clearing for agricultural cropping, have led to altered runoff and sediment supply regimes in lowland–dryland river systems in the Murray-Darling Basin.

Human impacts, particularly water resource developments, have led to changes in the natural drought and flood cycles in floodplain wetlands of the Murray-Darling Basin (Kingsford 2000a, 2000b). Ecological fragmentation has occurred due to a reduction in the lateral connection between river channels and floodplains (Thoms 2003). It is widely known that floodplain wetlands have been adversely affected by river regulation and water extraction in terms of their spatial extent (i.e. reduced flood coverage), ecological health and biodiversity (Frazier and Page 2006; Kingsford 2000a, 2000b; Kingsford and Thomas 2004). For example, many wetland vegetation communities (e.g. reedbeds and eucalyptus forests) and wildlife populations (e.g. waterbirds and fish) have suffered declines due to altered flood regimes, lower flood levels and shorter flood durations (Lemly *et al.* 2000). Severe and long natural droughts are not uncommon in the Murray-Darling Basin despite the development of large off-river water storages, canals and artificial levee banks; these have also altered the wetland areas on the floodplains in many lowland–dryland rivers (Thoms 2003). Clearly, all the above mentioned factors have the potential to alter the short-, medium- and long-term ecological components of the floodplain wetlands.

Flow–ecology relationships and the response of biota to hydrological variability

According to the flood pulse concept, biota respond to characteristics of the flood pulse, including flood timing, duration and rate of rise and fall. However, the response of biota will vary depending on their adaptations, the characteristics of the flood or drought, and whether a flood is regarded as a subsidy or stress (Odum *et al.* 1979). The subsidy–stress hypothesis is based on the concept that too much of a 'good thing' or perturbation may be detrimental to the performance and ultimately to the survival of a species, community or ecosystem. The theory relies on a perturbation being an alteration or deviation from what is usual or expected. A perturbation may cause a subsidy, defined as a favourable deflection from the expected, or a stress, defined as an unfavourable deflection. The effect of perturbations on performance may be expressed on a curve to indicate peak performance or decline in performance as the perturbation increases (Figure 1.3). The transition of a perturbation from a subsidy or stress is dependent on characteristics of the perturbation.

For floodplain wetlands, both flood and drought may be regarded as perturbations that may be beneficial or detrimental to the performance of biota. Peak performance of a species in

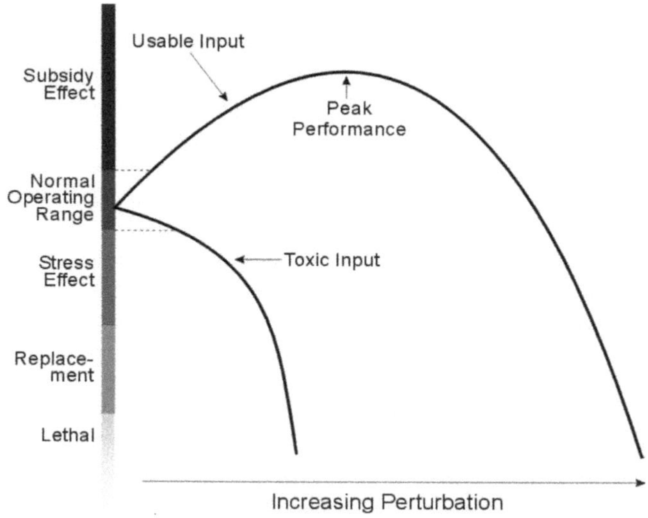

Figure 1.3: Hypothetical performance curves for a perturbed ecosystem subjected to a stress (lower curve: toxic input) or a subsidy (upper curve: usable input). Source: Odum *et al.* (1979).

response to a perturbation depends on its unique adaptations to flood and drought as well as on the characteristics of the perturbation, such as duration and intensity. Some of the characteristics of flood and drought perturbations that may influence the performance of biota include flood frequency, duration, depth, timing/seasonality, inter-flood dry-period, rate of flood rise and fall, and antecedent flood conditions.

Response of flora to flooding

Unlike terrestrial vegetation, floodplain wetland vegetation exhibits varying degrees of adaptation to flooding and, to some extent, drought. Due to these adaptations, vegetation performance in response to flooding is unlikely to exhibit an initial decline (Figure 1.3, lower curve); rather, vegetation is likely to respond with an initial increase in performance then a subsequent decline when a flood perturbation continues for longer than a species' adaptations can withstand (Odum *et al.* 1979). The likelihood of a flood perturbation becoming a stress for vegetation depends on the characteristics of the flood and the initial condition of the vegetation. However, a stressful perturbation may not necessarily be lethal or limit the viability of a species. Plants exhibit different phases in response to stress, and recovery may occur once the stressful perturbation has ceased.

The stress concept of plants provided by Lichtenthaler (1996) acknowledges the potential for vegetation regeneration after the removal of a stressful perturbation, provided the damage is not too severe (Figure 1.4). This concept differentiates the response of plants to stressful perturbations according to four phases:

- response phase or alarm reaction;
- restitution phase or stage of resistance;
- end phase or stage of exhaustion;
- regeneration phase.

The response phase is characterised by a decline in physiological function, such as photosynthesis, which causes a deviation from the normal physiological performance and a decline

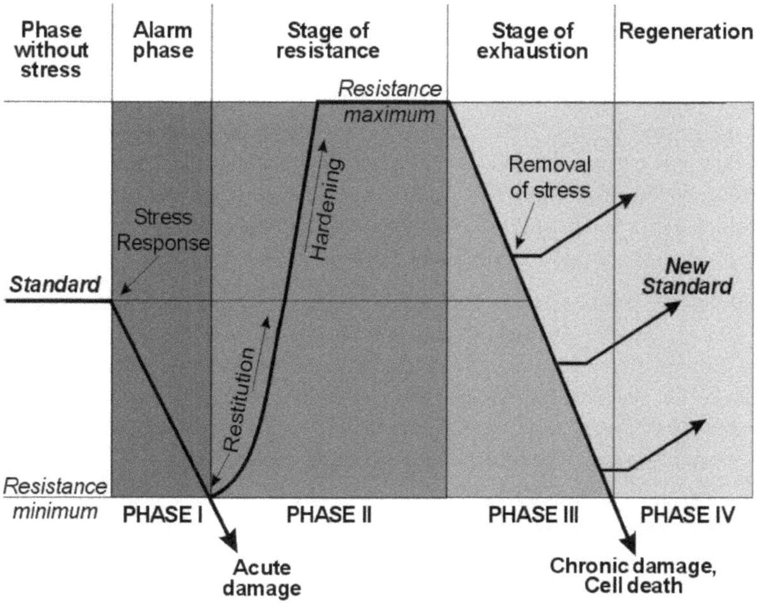

Figure 1.4: Plant stress phases induced by exposure to stressful perturbations. Source: Lichtenthaler (1996).

in vigour. In the absence of adaptations or 'tolerance mechanisms', acute damage may occur. The restitution phase is characterised by repair of damage caused by the stressful perturbation, but also a hardening of their physiological function according to the conditions of the stressful perturbation. This phase occurs only in species with adaptations enabling restitution and hardening. When the stressful perturbation continues and a plant's tolerance mechanisms have been exceeded, the plant is said to be in the end phase or stage of exhaustion. This phase is characterised by a progressive loss of vigour and vitality. If the stressful perturbation continues, chronic damage, cell death and finally plant death will result. The regeneration phase, which is unique to the concept presented by Lichtenthaler (1996), provides for the establishment of a new standard in physiological function provided that the stressful perturbation is removed before senescence dominates.

Application of this concept to floodplain wetland vegetation recognises that species' adaptation to optimal flooding conditions facilitates continued function at a physiological standard. However, it also recognises that due to tolerance mechanisms there is a range of flooding conditions that will enable plants to survive under prolonged flooding or drought, perhaps at a new physiological standard.

Numerous studies have classified wetland vegetation on the basis of functional strategies for coping with stress. The purpose of these classification systems is to categorise the response of wetland plant species to stress and to predict the composition and zonation of an ecosystem in the presence of stress. The CSR model of Grime (2001) incorporates species' physiological adaptations and competitive strategies in classifying the response of plants to stress and disturbance, where stress comprises phenomena that restrict photosynthetic production (e.g. limited water availability) and disturbance comprises phenomena attributable to the destruction of plant biomass (e.g. grazing or fire). The model is based on the concept that plants have evolved strategies that enable them to exploit conditions of:

- low stress and low disturbance – plants evolved for these conditions are referred to as 'competitors' (Grime 2001) and are characterised as productive plants occurring in undisturbed habitats (Menges and Waller 1983);
- high stress and low disturbance – plants evolved for these conditions are referred to as 'stress-tolerators' (Grime 2001) and are characterised as slow-growing stress-tolerant plants due to the unproductive environment they inhabit (Menges and Waller 1983);
- low stress and high disturbance – plants evolved for these conditions are referred to as 'ruderals' (Grime 2001) and are characterised by high growth rates, short life-spans and high reproductive ability (Menges and Waller 1983).

When applying the CSR model to floodplain wetlands, flooding may be regarded as both a disturbance and stress, depending on the morphological and physiological adaptations of plants (Menges and Waller 1983). Menges and Waller (1983) identify ruderals as annuals (short life-span), which due to their life-history and reproductive ability are able to avoid flooding disturbance and stress. Stress-tolerators are identified as perennials that have developed physiological and morphological adaptations to cope with the stress and disturbance of flooding. Menges and Waller (1983) identify a two-axis gradient between competitors, stress-tolerators and ruderals: physiological and morphological adaptations differentiate between stress-tolerators and ruderals, and disturbance frequency separates competitors (Figure 1.5). According to this model, competitors are species that have limited adaptations to disturbance and stress (in this case, flooding) and that generally establish at higher elevations where flood frequency is low. These species may even be regarded as fully terrestrial with limited tolerance to waterlogged soils.

An alternative classification is the 'environmental sieve' approach applied by van der Valk (1981) to model wetland vegetation dynamics. The model recognises 12 functional groups and classifies species on the basis of life-span, seed longevity and seed establishment requirements. The model was developed only on the basis of species response to flooding and drought, rather than incorporating other interactions between species, such as competition. In terms of life-span, species are classified as annuals (A), perennials (P) or vegetatively reproducing perennials (V). They are also classified either as seed bank species (S), with long-lived seeds stored

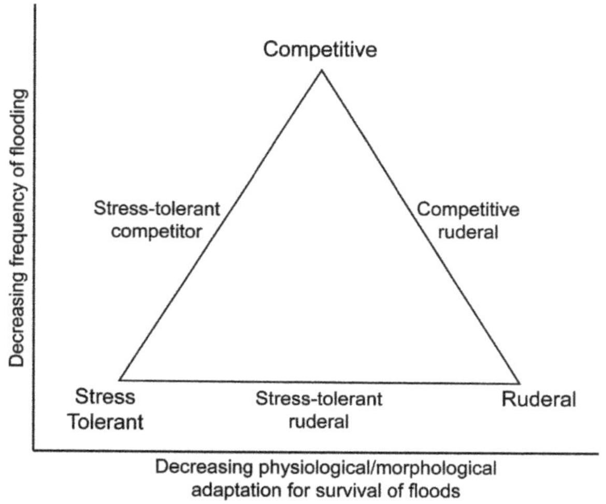

Figure 1.5: The CSR model (Grime 2001) relative to frequency of flooding and plant adaptations. Source: Menges and Waller (1983).

Table 1.2: The wetting and drying model (Brock and Casanova 1997), with classification of wetland species based on responses to wetting and drying patterns

Primary category	Secondary category	Description
Terrestrial	Dry species: Tdr	Species which germinate, grow and reproduce where there is no surface water and the watertable is below the soil surface
Terrestrial	Damp species: Tda	Species which germinate, grow and reproduce on saturated soil
Amphibious fluctuation-tolerators	Emergent species: ATe	Species which germinate in damp or flooded conditions, tolerate variation in water level, and grow with their basal portions under water and reproduce out of the water
Amphibious fluctuation-tolerators	Low-growing species: ATl	Species which germinate in damp or flooded conditions, tolerate variation in water level, are low-growing and tolerate complete submersion when water levels rise
Amphibious fluctuation-responders	Morphologically plastic species: ARp	Species which germinate in flooded conditions, grow in both flooded and damp conditions, reproduce above the surface of the water, and have morphological plasticity (e.g. heterophylly) in response to water level variation
Amphibious fluctuation-responders	Species with floating leaves: ARf	Species which germinate in flooded conditions, grow in both flooded and damp conditions, reproduce above the surface of the water, and have floating leaves when inundated
Submerged	Submerged: S	Species which germinate, grow and reproduce underwater

Source: Casanova and Brock (2000).

within seed banks enabling germination and establishment whenever conditions are suitable, or as dispersal-dependent species (D), with short-lived seeds that can only germinate and establish when environmental conditions and seed availability coincide. Two types of seed establishment requirements are recognised: those requiring no standing water for establishment (Type I) and species that can establish in standing water (Type II).

Van der Valk (1981) applied the model to a theoretical wetland to predict potential species transitions between drawdown and flooded conditions. Figure 1.6 illustrates that flooded conditions may result in the loss from established vegetation of annual and perennial species that

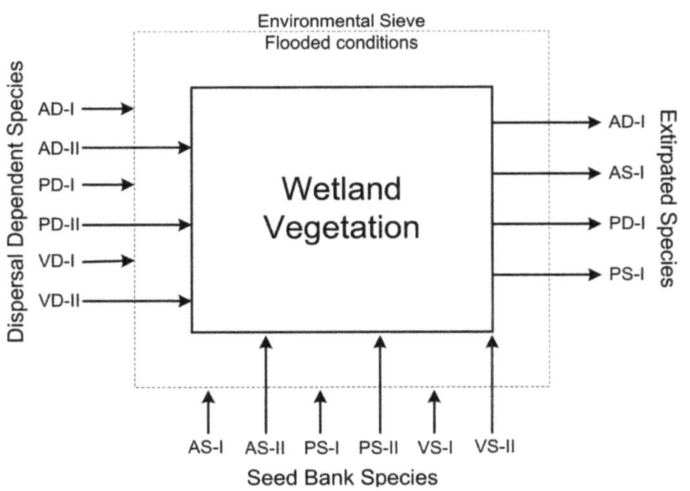

Figure 1.6: The environmental sieve model (van der Valk 1981), which illustrates the potential loss of species from established vegetation in response to flooded conditions. The species that may be lost are those characterised by seed establishment in drawdown conditions (Type I species).

rely on drawdown conditions for establishment. According to the model, seed bank species (S) are virtually impossible to eliminate from a wetland once they have established, while dispersal-dependent species (D) may be entirely eliminated once the adult population has reached the end of its life-span. Re-establishment of D-type species would require a nearby source of seed that may be redispersed to the site, as well as suitable conditions for establishment.

A classification of plants occupying edges of wetlands has been based on experience from Australian wetlands (Brock and Casanova 1997). According to this 'wetting and drying' model, plants occupying the upper edge of wetlands are regarded as 'terrestrial', those on the lower edge are 'submerged', and plants located between those zones are referred to as 'amphibious'. Similar to the environmental sieve scheme (van der Valk 1981), these groupings are further classified according to life-cycle, including water conditions for germination from a seed bank, growth response to water and reproduction in response to water. This classification scheme identified seven functional groups (Table 1.2) on the basis of wetting and drying patterns. It highlights the impact that changes in water regime may have on species richness and composition of a wetland.

Response of fauna to flooding

Floodplain wetland fauna exhibit adaptations that enhance performance in response to flooding. Reproduction is the most prominent response of many floodplain wetland fauna species to the flood pulse. For example, as the flood pulse stimulates productivity throughout a wetland, prey items of waterbirds become abundant, thereby enabling waterbirds to store fat for sustenance throughout the breeding season and to stimulate gonadal development and egg formation (see Chapter 3). Flooding acts as a stimulus for breeding in most waterbirds within the Murray-Darling Basin; only two species, the musk duck and blue-billed duck, are identified as purely seasonal breeders (Briggs 1990). In some fish species, such as silver perch and golden perch, a high-flow event or flood may stimulate spawning or trigger migration to suitable breeding habitat. However, for the majority of fish flooding is one of a suite of factors required for fish recruitment and their use of floodplain wetlands appears to be mainly opportunistic (see Chapter 4). Several species of frogs do require flooding to coincide with breeding activity if their recruitment is to be successful (Chapter 5). Crustaceans and gastropods generally require variable flow regimes, with species that favour permanent flowing waters (e.g. pea shells, basket shells and river mussels) becoming more abundant since river regulation reduced flow variability. There has been a corresponding decline in species reliant on lotic habitats (e.g. river snails and pond snails; Chapter 6).

The performance of fauna is also indirectly linked to the flood pulse via the response of habitats and food items to the flood pulse. Theories on the role of flooding in waterbird reproduction now largely depend on the link between waterbird condition, their trophic position as top-order consumers and the productivity of biota at lower trophic levels (Maher 1991; Kingsford and Norman 2002). Due to the trophic link between waterbirds and their ecosystem, poor reproductive performance may signify long-term environmental change related to reduced ecosystem productivity at lower trophic levels (Kushlan 1993). Fish, on the other hand, require permanently wet aquatic habitats in rivers and wetlands, and will move in and out of suitable habitats in search of more abundant food sources during their life-cycle. When considering the response of fauna to flooding, it is therefore essential to consider the indirect links between flooding, faunal habitat and dietary needs. There have been no classifications of the response of wetland fauna to flooding; this may be due to the combination of direct and indirect links to flooding and the associated complex nature of faunal response to flooding.

In contrast to the response of vegetation to flooding, some fauna species exhibit adaptations that enable them to disperse, to take advantage of the subsidising effect of flooding. Due to the nomadic nature of Australian waterbirds, they may paradoxically be abundant even within a landscape with a variable and often unpredictable flood regime (Roshier *et al.* 2001). Fish may take advantage of high flow conditions and utilise the floodplain and wetlands for spawning, they may use the floodwaters to disperse eggs and juveniles, or they may migrate upstream behind the flood front to access better habitats.

Not all faunal species may be regarded as highly dispersive or able to migrate to suitable habitats where water needs are met. These species may exhibit restitution and hardening adaptations that enable them to survive the stressful periods between flood events. Williams (1985) established that many species are unable to survive dry phases in the adult state. Waterbirds, some insects and fish are able to migrate to refuge areas during dry phases, while some crustaceans (*Coxiella striata, Haloniscus searlei*) have impermeable shells. Others, such as frogs of the *Cyclorana* and *Limnodynastes* genera, exhibit a range of adaptations to cope with dry conditions. These include impermeable cocoons, subcutaneous sacs and bladders bloated with water, burrowing behaviour and physiological adaptations to cope with water loss. Alternatively, many insect and crustacean species survive dry periods in embryonic states that resist drying (e.g. resistant eggs) or that limit the effects of drying (e.g. aestivation or dormancy) (Williams 1985). In either case, the performance subsidy imposed by flooding results in population booms for floodplain wetland fauna (Balcombe *et al.* 2005; Jenkins and Boulton 2003; Kingsford *et al.* 1999).

Knowledge of water and habitat requirements of floodplain wetland biota

Clearly, there is a diverse range of plants, animals and microscopic organisms that occupy the ecosystems and habitats created and maintained by the flows and flood regimes of the floodplain wetlands of the Murray-Darling Basin. However, we have limited knowledge of the water requirements of the freshwater biota, despite water being critical for the survival, regeneration, maintenance and management of the ecological communities and ecosystems. The biota have adapted to cope with natural flow and climatic variability, which makes it difficult to simplify their preferences and requirements in terms of general flow and flood regimes.

The following chapters provide a fundamental synthesis of the knowledge of water requirements of key flora and fauna in freshwater floodplain wetlands in the Murray-Darling Basin. The authors integrate aspects of the biological and ecological requirements of the aquatic biota in the context of their life-cycles, trophic linkages, preferred habitats and wetland water regimes. The aim is to provide a versatile reference and a platform for educational, academic and managerial endeavours related to understanding and maintaining the functional and ecological characteristics of these types of aquatic biota.

References

Balcombe SR, Bunn SE, McKenzie-Smith FJ and Davies PM (2005) Variability of fish diets between dry and flood periods in an arid zone floodplain river. *Journal of Fish Biology* **67**, 1552–1567.

Briggs SV (1990) Waterbirds. In *The Murray.* (Eds N MacKay and D Eastburn) pp. 337–334. Murray-Darling Basin Commission: Canberra.

Brock MA and Casanova MT (1997) Plant life at the edge of wetlands: ecological responses to wetting and drying patterns. In *Frontiers in Ecology: Building the Links.* (Eds NK Klomp and I Lunt) pp. 181–192. Elsevier Science: Oxford, UK.

Casanova MT and Brock MA (2000) How do depth, duration and frequency of flooding influence the establishment of wetland plant communities? *Plant Ecology* **147**, 237–250.

Finlayson BL and McMahon TA (1988) Australia v the world: a comparative analysis of streamflow characteristics. In *Fluvial Geomorphology of Australia*. (Ed. RF Warner) pp. 17–40. Academic Press: Sydney.

Frazier P and Page K (2006) The effect of river regulation on floodplain wetland inundation, Murrumbidgee River, Australia. *Marine and Freshwater Research* **57**, 133–141.

Frazier P, Page K and Read A (2005) Effects of flow regulation on flow regime in the Murrumbidgee River, south-eastern Australia: an assessment using a daily estimation hydrological model. *Australian Geographer* **36**, 301–314.

Grime JP (2001) *Plant Strategies, Vegetation Processes and Ecosystems Properties*. John Wiley and Sons: Chichester, UK.

Hesse PP, Ralph TJ and Yonge D (2005) Dryland wetlands: the Holocene response of inland rivers in Australia. In *Joint BGRG-BSRG International Conference: Drylands – Linking Landscape Processes to Sedimentary Environments*. London. Geological Society: London.

Jenkins KM and Boulton AJ (2003) Connectivity in a dryland river: short-term aquatic microinvertebrate recruitment following floodplain inundation. *Ecology* **84**, 2708–2723.

Jolly ID (1996) The effects of river management on the hydrology and hydroecology of arid and semi-arid floodplains. In *Floodplain Processes*. (Eds MG Anderson, DE Walling and PD Bates) pp. 577–609. John Wiley and Sons: New York.

Kingsford RT (2000a) Ecological impacts of dams, water diversions and river management on floodplain wetlands in Australia. *Austral Ecology* **25**, 109–127.

Kingsford RT (2000b) Protecting rivers in arid regions or pumping them dry? *Hydrobiologia* **427**, 1–11.

Kingsford RT (2003) Ecological impacts and institutional and economic drivers for water resource development: a case study of the Murrumbidgee River, Australia. *Aquatic Ecosystem Health and Management* **6**, 69–79.

Kingsford RT and Norman FI (2002) Australian waterbirds – products of the continent's ecology. *Emu* **102**, 47–69.

Kingsford RT and Thomas RF (2004) Destruction of wetlands and waterbird populations by dams and irrigation on the Murrumbidgee River in arid Australia. *Environmental Management* **34**, 383–396.

Kingsford RT, Curtin AL and Porter J (1999) Water flows on Cooper Creek in arid Australia determine 'boom' and 'bust' periods for waterbirds. *Biological Conservation* **88**, 231–248.

Kingsford RT, Brandis K, Thomas RF, Crighton P, Knowles E and Gale E (2004) Classifying landform at broad spatial scales: the distribution and conservation of wetlands in New South Wales, Australia. *Marine and Freshwater Research* **55**, 17–31.

Kushlan J (1993) Colonial waterbirds as bioindicators of environmental change. *Colonial Waterbirds* **16**, 223–251.

Lemly AD, Kingsford RT and Thompson JR (2000) Irrigated agriculture and wildlife conservation: conflict on a global scale. *Environmental Management* **25**, 485–512.

Lichtenthaler HK (1996) Vegetation stress: an introduction to the stress concept in plants. *Journal of Plant Physiology* **148**, 4–14.

Maher M (1991) Waterbirds back o'Bourke: an inland perspective on the conservation of Australian waterbirds. PhD thesis. University of New England.

MDBC (2006) *Wetlands – Murray-Darling Basin Authority*. Murray-Darling Basin Authority, Canberra. http://www2.mdbc.gov.au/nrm/water_issues/wetlands/.

Menges ES and Waller DM (1983) Plant strategies in relation to elevation and light in floodplain herbs. *American Naturalist* **122**, 454–473.

Nanson GC, Tooth S and Knighton AD (2002) A global perspective on dryland rivers: perceptions, misconceptions and distinctions. In *Dryland Rivers: Hydrology and Geomorphology of Semi-arid Channels.* (Eds LJ Bull and MJ Kirkby) pp. 17–54. John Wiley and Sons: Chichester, UK.

NSW Government (2009) Pineena CM – version 9.3. NSW Office of Water.

O'Brien PE and Burne RV (1994) The Great Cumbung Swamp: terminus of the low-gradient Lachlan River, eastern Australia. *AGSO Journal of Australian Geology and Geophysics* **15**, 223–233.

Odum ET, Finn JT and Franz EH (1979) Perturbation theory and the subsidy-stress gradient. *BioScience* **29**, 349–352.

Paijmans K, Galloway RW, Faith DP, Fleming PM, Haantjens HA, Heyligers PC, Kalma JD and Loffler E (1985) 'Aspects of Australian wetlands'. CSIRO Division of Water and Land Resources: Canberra.

Ralph TJ (2008) Channel breakdown and floodplain wetland morphodynamics in the Macquarie Marshes, south-eastern Australia. PhD thesis. Department of Physical Geography, Macquarie University.

Ralph TJ and Hesse PP (2010) Downstream hydrogeomorphic changes along the Macquarie River, southeastern Australia, leading to channel breakdown and floodplain wetlands. *Geomorphology* **118**, 48–64.

Reid MA and Brooks JJ (2000) Detecting effects of environmental water allocations in wetlands of the Murray-Darling Basin, Australia. *Regulated Rivers: Research and Management* **16**, 479–496.

Roshier DA, Robertson AI, Kingsford RT and Green DG (2001) Continental-scale interactions with temporary resources may explain the paradox of large populations of desert waterbirds in Australia. *Landscape Ecology* **16**, 547–556.

Semeniuk CA and Semeniuk V (1995) A geomorphic approach to global classification for inland wetlands. *Vegetatio* **118**, 103–124.

Sheldon F, Thoms MC, Berry O and Puckridge JT (2000) Using disaster to prevent catastrophe: referencing the impacts of flow changes in large dryland rivers. *Regulated Rivers: Research and Management* **16**, 403–420.

Thoms MC (2003) Floodplain-river ecosystems: lateral connections and the implications of human interference. *Geomorphology* **56**, 335–349.

Thoms MC and Sheldon F (2000) Water resource development and hydrological change in a large dryland river: the Barwon-Darling River, Australia. *Journal of Hydrology* **228**, 10–21.

Thoms MC and Walker KF (1992) Channel changes related to low-level weirs on the River Murray, South Australia. In *Lowland Floodplain Rivers: Geomorphological Perspectives.* (Eds PA Carling and GE Petts) pp. 235–249. John Wiley and Sons: Chichester, UK.

Thoms MC, Southwell M and McGinness HM (2005) Floodplain-river ecosystems: fragmentation and water resources development. *Geomorphology* **71**, 126–138.

Tooth S (1999) Floodouts in central Australia. In *Varieties of Fluvial Form.* (Eds AJ Miller and A Gupta) pp. 219–247. John Wiley and Sons: Chichester, UK.

Tooth S (2000) Process, form and change in dryland rivers: a review of recent research. *Earth Science Reviews* **51**, 67–107.

Tooth S, McCarthy TS, Hancox PJ, Brandt D, Buckley K, Nortje E and McQuade S (2002) The geomorphology of the Nyl River and floodplain in the semi-arid Northern Province, South Africa. *South African Geographical Journal* **84**, 226–237.

Tooth S, Rodnight H, Duller GAT, McCarthy TS, Marren PM and Brandt D (2007) Chronology and controls of avulsion along a mixed bedrock-alluvial river. *Geological Society of America Bulletin* **119**, 452–461.

van der Valk AG (1981) Succession in wetlands: a Gleasonian approach. *Ecology* **62**, 688–696.

Ward JV (1998) Riverine landscapes: biodiversity patterns, disturbance regimes, and aquatic conservation. *Biological Conservation* **83**, 269–278.

Ward JV, Tockner K, Arscott DB and Claret C (2002) Riverine landscape diversity. *Freshwater Biology* **47**, 517–539.

Warner RF (1986) Hydrology. In *Australia: A Geography.* (Ed. DN Jeans) pp. 49–79. Sydney University Press: Sydney.

Williams WD (1985) Biotic adaptations in temporary lentic waters, with special reference to those in semi-arid and arid regions. *Hydrobiologia* **125**, 85–110.

Yonge D and Hesse PP (2009) Geomorphic environments, drainage breakdown and channel and floodplain evolution on the Lower Macquarie River, central western New South Wales. *Australian Journal of Earth Sciences* **56**, S35–S53.

Young WJ (1999) Hydrologic descriptions of semi-arid rivers: an ecological perspective. In *A Free-flowing River: The Ecology of the Paroo River.* (Ed. RT Kingsford) pp. 77–96. NSW National Parks and Wildlife Service: Sydney.

Chapter 2

Vegetation

Kerrylee Rogers

Introduction

According to the flood pulse concept (Junk *et al.* 1989), vegetation responds to characteristics of the flood pulse, including flood timing, duration and rate of rise and fall. However, the response of vegetation will vary depending on plant adaptations, characteristics of the flood or drought, and whether the flood is regarded as a subsidy or a stress (see Chapter 1). This chapter explores the response of vegetation to water and, more specifically, the water requirements that will promote the growth of vegetation within floodplain wetlands. Water requirement profiles for individual species are provided.

Particular emphasis is given to flood and drought perturbations, due to the overarching importance of water availability for the performance of floodplain wetland vegetation. Biotic factors (e.g. competition, herbivory and grazing) and abiotic factors (e.g. light, temperature, nutrient availability and soil conditions) may influence the performance or survival of floodplain wetland vegetation, but this chapter generally does not discuss the response of vegetation to these factors. Excluding these components from analyses of the response of vegetation to flooding may be simplistic, but consideration of the response of floodplain wetland vegetation to water is an essential first step when examining the performance of floodplain wetland vegetation.

It is acknowledged that other factors linked with surface water availability may influence the performance and survival of floodplain wetland vegetation. For example, groundwater availability may subsidise surface water contributions and enhance plant growth, and salinity may limit access to available water. Where possible, consideration is given to these factors, but the primary focus of this chapter is the influence of surface water availability, with particular reference to flooding and lack of flooding.

Species included in this chapter have been selected for a number of reasons, the most important being that they are regarded as floodplain and/or wetland species and exhibit a distinct reliance on flooding. This excludes vegetation species that may be widespread throughout wetlands within the Murray-Darling Basin yet are regarded as intolerant of flooding, as marginally tolerant or as competitors according to the CSR model (Grime 2001; see also Chapter 1). Some of these species include the poplar box (*Eucalyptus populnea*), belah (*Casuarina cristata*) and wilga (*Geijera parviflora*) that may populate wetland areas in response to rainfall and tolerate infrequent small flood events. Species have also been included if they are relatively widespread or dominant within the floodplain wetlands of the Murray-Darling

Basin. Information about the water requirements for some species is severely lacking and those species have therefore been excluded from the species profiles; as much information as possible about water requirements of species has been incorporated.

In undertaking the species profiles, a specific methodology was followed. First, plants were categorised into the major plant groups of trees, shrubs, grasses, sedges and rushes, herbs and forbs and submerged aquatic macrophytes as undertaken previously by Roberts and Marston (2000). Further definitions of these major plant groups are provided in the Glossary.

Second, specific aspects of the water regime were considered important and incorporated into the species profiles. It is acknowledged that, while the response of plants to a flood pulse are dependent on characteristics of the flood pulse, the distribution of plant species within a floodplain wetland may largely reflect the water regime (e.g. Blanch *et al.* 1999b, 2000; Casanova and Brock 2000), defined as the prevailing pattern of flood pulses over a period of time. A description of the water requirements of vegetation should therefore consider specific aspects of the water regime. Aspects of the water regime considered important for the species profiles included flood frequency, duration, depth, timing and inter-flood dry-period (see Glossary). Consideration was also given to the biotic aspect of germination timing. Due to a lack of supporting literature, some aspects of the water regime were excluded, including the rate of rise and fall of floodwaters and the antecedent flood conditions. Further research is required to ascertain the importance of these aspects.

Third, consideration was given to the water requirements of plants at their various life-stages. For simplicity and to limit redundancy, two main life-cycle stages were considered: the established phase when a plant is at maturity and able to reproduce, and the regenerative phase when plants are germinating and/or establishing (Grime 2001). For the species profiles within this chapter, these life-cycle stages are referred to as 'survival and maintenance' and 'reproduction and regeneration', respectively. Recognition was given to the different life-histories of annual and perennial plants, particularly perennials that are able to reproduce both sexually from seed and vegetatively. It is emphasised that regeneration is not considered complete until plants have matured and are able to reproduce (Figure 2.1).

Finally, the chapter applies the plant stress concept (Lichtenthaler 1996) to floodplain wetlands by recognising that, while there is an optimal water regime for the maintenance of plants, there is also a range of water regimes that will support the survival of plant species, perhaps at limited reproductive capacity. For the purposes of the species profiles, 'maintenance' refers to the water regime required to ensure growth, flowering and survival of established plants at a standard state or heightened levels of productivity (Lichtenthaler 1996). 'Survival' refers to the water regime required to enable established plants to survive, perhaps in a state of stress but not at the point of no recovery, chronic damage or cell stress. Within species profiles, water regime values for survival are presented as maximum or minimum values for specific aspects. Maintenance values are described as 'ideal'.

Trees

River red gum: *Eucalyptus camaldulensis*

The river red gum is among the most widespread eucalypt tree in Australia, occupying watercourses and wetlands throughout mainland Australia (Brooker *et al.* 2002). *Eucalyptus camaldulensis* var. *camaldulensis* is the most abundant variety in south-eastern Australia and dominates the Murray-Darling Basin (Brooker *et al.* 2002). The river red gum is a perennial,

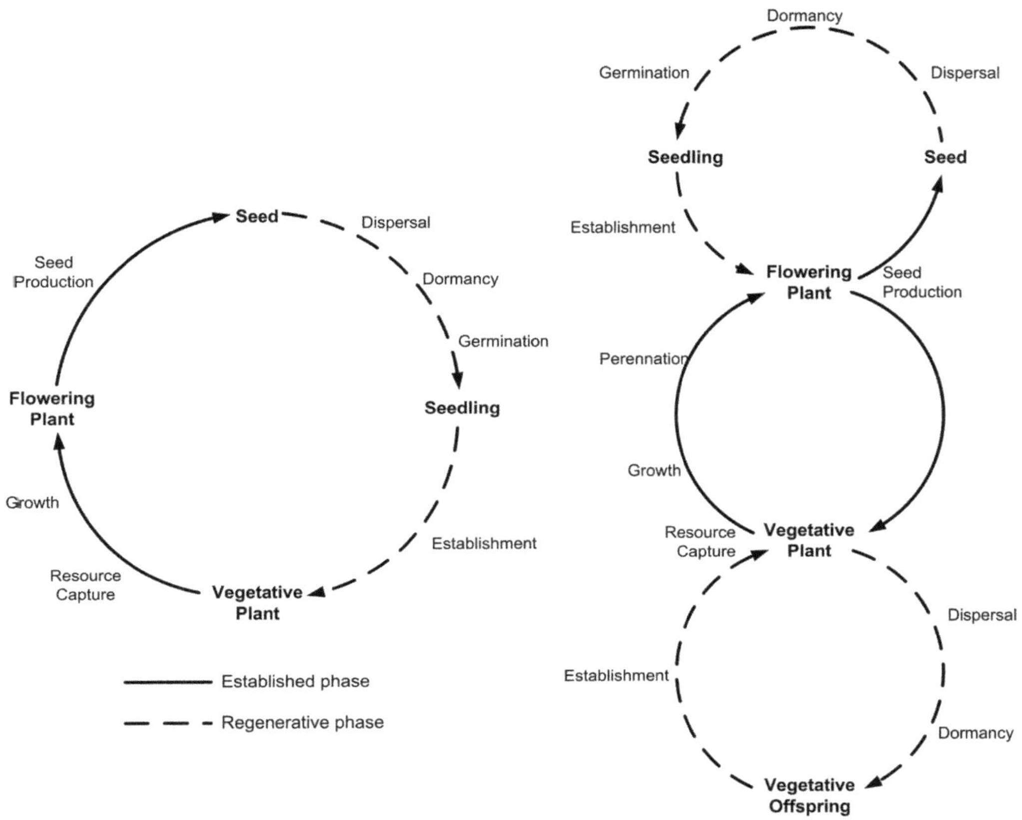

Figure 2.1: Schematic of life-cycles of (a) an annual flowering plant and (b) a perennial plant producing both sexually or asexually. Source: Grime (2001, p. xxii).

single-stemmed, large-boled, medium to tall tree of up to 45 m (Brooker *et al.* 2002; Figure 2.2). It is a relatively long-lived tree, reaching ages of 500 to 1000 years (Jacobs 1955). Flowering generally occurs in the warmer months of late spring and summer, but has been recorded as early as June (Brooker *et al.* 2002).

Survival and maintenance

The survival and maintenance of the river red gum is dependent on the availability of water. The river red gum acquires water from four sources: direct rainfall, surface flooding from floodwaters, stream water and groundwater. The river red gum may utilise all water sources, or any combination of the sources. Since the river red gum has a wide distribution throughout semiarid Australia, rainfall is sparse and intermittent, and is therefore not a primary water source. However, rainfall does act to recharge soil water.

Floodwaters, derived from rainfall within the catchment, enter the floodplain via ephemeral creeks and overbanking of inundated rivers and streams. Floodwaters are a primary water source for the river red gum, and numerous studies indicate that healthy river red gums are situated in areas receiving adequate surface flooding (Bren and Gibbs 1986; Robertson *et al.* 2001). Adequate surface flooding depends substantially upon topographic position (Bren and

Figure 2.2: River red gum: *Eucalyptus camaldulensis*. Photograph: Tim Ralph (Macquarie University).

Gibbs 1986; Chisholm and Stone 2003) and distance from the water source (Stone and Bacon 1994, 1995a). Tree densities also affect the watering requirements of a site, with river red gum forests requiring more water than woodlands (Paul *et al.* 2003).

Flood frequency has a significant influence on the growth of the river red gum (Robertson *et al.* 2001). Numerous studies have found accelerated growth in association with abundant surface waters (Kidson *et al.* 2000a; Robertson *et al.* 2001). In fact, Robertson *et al.* (2001) found that one large inundation every three years appeared to have the same effect on the productivity of river red gums as regular, short, spring flooding over a number of years. Flooding reportedly occurs every six to eight years in every 10 years at Barmah Forest (Bren and Gibbs 1986), while George (2004) indicated that river red gums require flooding one to four times every five years, and Bacon *et al.* (1993) reported increases in leaf area with increased flood frequency from zero to two years. However, depending on access to other water sources, river red gum forests and woodlands are maintained with flooding at different frequencies and may survive extended drought periods (MDBC 2003; White *et al.* 2000). This may occur at the expense of regeneration and limit the viability of river red gum populations (George *et al.* 2005b).

Ideal inter-flood drying appears to be in the order of five to 15 months for river red gum forests at Barmah, but longer dry-periods in the order of three to four years may be sustained within river red gum woodlands (Young *et al.* 2003). Individuals reportedly start to show evidence of a decline in health with inter-flood periods greater than two years, with death usually occurring after about five years (Johns *et al.* 2009). The duration of inter-flood drying tolerated may also depend on tree condition prior to the initial flooding, with stressed trees showing a more rapid decline in condition (Johns *et al.* 2009).

Duration of flooding appears to be as important as frequency, particularly since river regulation has resulted in frequent smaller floods (Roberts and Marston 2000). Small frequent floods may have the same result on productivity as less-frequent large floods (Robertson *et al.* 2001). In addition, due to the ability of flooding to recharge soil moisture at some distance from the edge of floodwaters, small-duration floods may significantly improve tree growth in large forest areas (Bacon *et al.* 1993). Barmah Forest reportedly has historical flood durations of one to seven months (Roberts and Marston 2000). However, river red gums have survived much longer periods of flooding, such as 24 months at Barmah forest, three to four years behind the Hay Weir (Bren 1987) and three years on the Lachlan River (Briggs and Maher 1983). At Barmah Forest, tree death from permanent inundation appears to occur in excess of four years flood duration (Chesterfield 1986).

Flood timing also significantly influences the growth of river red gums. With controlled flood frequency and timing, higher rates of wood production were observed in the Barmah Forest during summer flooding than in spring flooding or no flooding. This occurred despite spring flooding being the natural flood pattern for the forest. Due to the high temperatures, summer is a time when river red gums try to conserve water by reducing leaf density through litterfall (Robertson *et al.* 2001). Despite higher growth rates reported during summer flooding, unregulated flooding in the majority of the Murray-Darling Basin typically occurs in later winter and spring with low flows in summer and autumn (Dalton 1990). However, flooding at any time can be sustained by river red gums by transpiring heavily when excess water is available (Heinrich 1990).

While observations indicate that reductions in flood frequency reduce river red gum condition, few studies have found a direct relationship between decreases in flood frequency and a decline in the health of river red gums (Bren and Gibbs 1986; MDBC 2003). Due to the ability of established river red gums to access other water sources (Roberts and Marston 2000), it is more likely that small changes to flood regimes would become evident in a change of understorey species (Chesterfield 1986), including regenerating river red gums. River red gums are likely to be able to survive solely on other water sources if available, but there would be a lack of regeneration in the absence of surface flooding (George *et al.* 2005b).

Due to the capacity of river red gums to have deep, extensive root systems (MDBC 2003) they depend greatly on groundwater, provided that salinity is not excessive (Overton and Jolly 2004). They also have the ability to opportunistically select water from available water sources, depending on availability (Mensforth *et al.* 1994). At Chowilla, river red gums that were prevented from accessing stream water were found to rely on groundwater in summer and on a combination of groundwater and rain-derived surface-soil water in winter (Mensforth *et al.* 1994; Thorburn *et al.* 1994). Trees were less affected by changes in creek flow (Thorburn *et al.* 1994), and moderately saline groundwater appeared to be a more important water source than fresh surface water (Thorburn and Walker 1993).

Salinity and depth to groundwater are important factors influencing the ability of river red gums to utilise groundwater sources. Groundwater salinity should be less than 40 dS/m, as river red gums are less salt-tolerant than other floodplain trees (Overton and Jolly 2004). However, tolerance is dependent on the tree provenance (Marcar *et al.* 2003). There is also a critical depth to groundwater; as flood frequency and duration increase, the depth to groundwater and groundwater salinity decrease (Overton *et al.* 2006). Little is known of the critical depth at which river red gums access groundwater but it is likely to vary depending on soil quality (Overton and Jolly 2004) and depth of individual river red gum roots, which may be partly determined by the flood regime in which the individual was established and reached maturity.

Floodwaters also recharge soil moisture, enabling river red gums to be sustained once floodwaters recede. The length of time that soil moisture remains recharged depends on numerous factors, such as soil type, vegetation type and density, and climatic conditions (Bacon *et al.* 1993). Studies from Barmah Forest indicate that recharge of soil moisture lasts approximately one month, after which time trees must obtain water from other sources (Bren 1987). The spatial extent of soil moisture recharge from flooding depends on access by river red gums to shallow aquifers. A study of soil moisture recharge at Barmah found that the depth ranged from 1.3 m to over 6 m below the surface and there was a horizontal movement of 0–38 m from the edge of floodwaters (Bacon *et al.* 1993). In fact, the study reported that even short-term flooding of 15–20% of a river red gum forest floor may increase tree growth in up to 70% of the forest. Alternatively, river red gums located on the Chowilla floodplain at distances greater than about 15 m from the river/stream used no stream water (Mensforth *et al.* 1994). Despite these discrepancies, it is apparent that river red gums are opportunistic in the sources of water used (Mensforth *et al.* 1994). Short floods, even confined to rivers and ephemeral creeks and runners, can provide water for large sections of floodplains (Robertson *et al.* 2001).

The occurrence of stream-bordering river red gums attests to some reliance on stream water and soil moisture from stream infiltration (Mensforth *et al.* 1994; Thorburn *et al.* 1994). Stream-bordering river red gums have been shown to utilise stream water, however, it is suggested that a lack of complete use of stream water may be an adaptation enabling individuals to respond to changes in stream flow by accessing groundwater (Thorburn *et al.* 1994). Stream water use did not change greatly when river red gum individuals had access to stream water for two weeks or 10 months (Thorburn and Walker 1994).

Water availability has been found to have a number of other impacts on river red gum condition. Leaf abscission from insect herbivory is reportedly greater in river red gums under water stress (Stone and Bacon 1995b), and this insect herbivory appears to be related to smaller leaf size rather than to other leaf characteristics such as foliar nitrogen or cineole concentrations (Stone and Bacon 1994). Grazing may affect the condition of river red gum saplings and seedlings during prolonged dry conditions. As other feed sources, such as weeds, become scarce cattle, kangaroos and rabbits will graze on river red gums, destroying the majority of river red gum juveniles and significantly influencing the regeneration of populations (Dexter 1978).

Reproduction and regeneration

Early observations of river red gum seedling establishment and regeneration indicate that flooding is an important process for abundant seedling germination (Jacobs 1955). The observations suggested that flooding induced copious amounts of seed production, created moist conditions suitable for germination and decreased the amount of seed-robbing insects on forest floors, thereby increasing the probability of germination (Jacobs 1955). These observations led to further studies of the influence of water on river red gum seed availability, germination, regeneration and maintenance of viable producing river red gum communities (Dexter 1978; George 2004; George *et al.* 2005a, 2005b; Jensen 2008; Meeson *et al.* 2002; Pettit and Froend 2001).

Seed production occurs in response to pollination and development of inflorescences, usually from late spring to mid-summer. Flowering and seed fall vary geographically (Jensen

2006) and may occur in response to peak flood timing. Buds and fruit may be shed during excessively dry conditions and seed yields may be influenced by water availability in the 24–36 months prior to seed fall (Jensen 2006). Seeds generally take about nine months to develop and seed fall occurs throughout the year, providing year-round potential for seed germination when conditions are favourable (Pettit and Froend 2001). Seed fall appears to be highest in spring when floods recede (Dexter 1978), but peaks are evident in spring and autumn in the Lower Murray (George 2004) and in winter in the Mount Lofty Ranges (Pudney 1998). Adult river red gums have been estimated to produce 600 000 seeds (Jacobs 1955), but seed falls significantly less than this number have been reported (George *et al.* 2005a). The intensity of flowering and seed production is observed to vary widely (Dexter 1978) with healthy individuals yielding significantly more seeds and more viable seeds than river red gums of poor health (George 2004; George *et al.* 2005a). While numerous factors influence the health of river red gums, moisture stress is a significant factor which may reduce the health of river red gums and cause individuals to reduce reproductive effort, resulting in less seed production. Adequate access to water, which enables maintenance of tree health, is essential for trees to produce adequate seed supply for regeneration.

Flooding may also provide secondary benefits for seeds, further enhancing the probability of germination. Flooding may promote the growth of feed for cattle, kangaroos and rabbits, thereby protecting seeds from grazing impacts (Dexter 1978; Meeson *et al.* 2002). Flooding may also reduce post-dispersal predation of seeds by ants (Meeson *et al.* 2002) and aid wider distribution of seeds within flood debris (Pettit and Froend 2001).

Once seed falls, access to water from rainfall or flooding is necessary for seed germination. Seeds may germinate in unflooded areas provided that rainfall is adequate to create moist soil conditions, generally during significant wet periods (Dexter 1978). During winter and early spring, exposed seedlings may be susceptible to freezing temperatures and frosts (Roberts and Marston 2000).

Prolific germination commonly occurs in response to the recession of floodwaters or drawdown, which creates moist soil conditions. However, winter floods with winter recession expose seeds to frosts and freezing conditions, making germination unfavourable (Dexter 1978). Prolonged flooding may destroy seeds, while flood recession in mid- to late summer exposes very young undeveloped germinates to extremely hot temperatures. Germination of seeds is greatest with widespread flooding which recedes in spring or early summer; prolific seed fall in spring is suggested as an adaptation to spring flood recession (Dexter 1978; Pettit and Froend 2001). This creates moist soil conditions that can sustain the growth of seedlings through higher temperatures in summer.

Regeneration also requires seedling establishment, which may be more problematic than germination (Jacobs 1955). Seedlings must adapt to heat and moisture stress and to flooding. The development of adventitious roots and aerenchymatous tissue helps protect seedlings from the anoxic soil conditions caused by flooding (Dexter 1978; Heinrich 1990). Flooding which causes complete immersion of seedlings for several months is reportedly lethal for undeveloped seedlings of less than 25 cm height. Seedlings of greater stature (50–60 cm) appear to have greater survival rates and can survive inundation for four to six months (Dexter 1978).

Seedlings adapt to heat stress by shedding leaves, allocating resources to the development of roots that thus enable them to access soil moisture (Dexter 1978; Roberts and Marston 2000). However, in drought years, seedling growth may be slow, thereby limiting the ability of roots to access deep soil moisture (Dexter 1978). In non-flood years, young germinates are exposed to cold temperatures which may cause significant losses to seedling populations.

When conditions are suitable seedlings may become established and develop into saplings and pole stage trees over a number of years. However, regeneration is not complete until the saplings mature and are able to reproduce (George *et al.* 2005a). Maintenance of access to water is vital to retaining viable, reproducing adult river red gum populations.

Black box: *Eucalyptus largiflorens*

Black box has a limited distribution but is relatively common in New South Wales. It also occurs in Queensland, Victoria and South Australia (Pryor and Briggs 1981). Black box typically occurs in locations with heavy clay soils (Harden 1991) that are seasonally flooded and for this reason is commonly found in association with river red gums and coolibah on floodplain wetlands. Black box are commonly single-stemmed trees growing to heights of up to 20 m. They have a dark fibrous bark, typical of box trees (Figure 2.3). Flowering reportedly occurs between August and January, with seed ripening within a few months of flowering (Boland *et al.* 1986).

Survival and maintenance

The survival and maintenance of black box depends on the availability of water. Similar to the river red gum, black box demonstrates opportunistic water use and is able to access water from four sources: floodwaters, rainfall, stream water and groundwater (Holland *et al.* 2006; Thorburn and Walker 1993). Unlike the river red gum, black box is relatively tolerant of both flood and drought (Roberts and Marston 2000); due to this tolerance the source of water is not important as long as it is of adequate quality. Indeed, some studies at Chowilla have not clearly illustrated black box growth or increased transpiration in response to flooding in the short term (Akeroyd *et al.* 1998; Bramley *et al.* 2003; Jolly and Walker 1996). Infrequent floods play a role in boosting water availability and increasing available water quality where saline soils and groundwater are an issue (Akeroyd *et al.* 1998; Bramley *et al.* 2003).

Jolly and Walker (1996) noted that black box at Chowilla was readily able to access rainfall to meet transpiration needs. It is likely that, where rainfall is regular and in adequate volume, black box may not require flooding or access to other water sources. However, black box is

Figure 2.3: Black box, *Eucalyptus largiflorens*. Photograph: Sharon Bowen (DECCW).

distributed throughout semiarid parts of the Murray-Darling Basin where rainfall is commonly low and irregular, and unlikely to meet maintenance watering needs. Thus, black box commonly requires that available water supplies be boosted by flooding.

Since black box is less commercially valuable than river red gums, less is known of the flood requirements (in terms of frequency, duration, timing and inter-flood dry-period) for black box in semiarid Australia. There appears to be a great range in the required flood frequency for population maintenance at different sites. The range in flooding requirements may reflect the ability of black box to opportunistically access other water sources of adequate quality, such as groundwater. For example, healthy black box have been reported at sites in south-western New South Wales at flood frequencies of one in every four to five years (Shepheard 1992) and at the Chowilla floodplain at sites with flood frequencies of one in two to five years (Sharley and Huggan 1995). Roberts and Marston (2000) suggested that black box can tolerate flood frequencies of one in seven to 10 years provided flood duration is adequate. George (2004) suggested that adequate inundation should be received 10–15% of the time to retain species viability and maintain populations, thereby implying a frequency of one flood in eight to 10 years.

Flood duration appears to be as important as flood frequency for the maintenance of black box populations. Duration should be long enough to boost water availability, but not exceed the capacity of black box to transpire. A reduction in transpiration indicates anoxic conditions, which may be detrimental to plant maintenance and survival. Field measures of transpiration of black box at Chowilla showed that stomatal closure did not occur until after 32 days of flooding, implying that soil oxygen was adequate over this period (Jolly and Walker 1996). Similarly, no relationship was reported between flood length and increases in transpiration for black box flooded for up to 78 days (Akeroyd *et al.* 1998). These results may be largely dependent on site-specific soil properties, which influence water and oxygen availability (Akeroyd *et al.* 1998; Bramley *et al.* 2003; Slavich *et al.* 1999). Floods in excess of four months may result in stomatal closure and increased transpiration as an adaptation to flooding, thereby causing loss of vigour and limiting the capacity of individuals to reproduce. Acute stress has been observed in black box flooded in excess of 13 months (Briggs and Townsend 1993).

Since black box is relatively drought-tolerant (Jolly and Walker 1996; Thorburn and Walker 1993), does not exhibit a relationship between growth and temperature (Young *et al.* 2003) and can opportunistically access water from different sources (Akeroyd *et al.* 1998), the timing of flooding does not appear to be critical (Roberts and Marston 2000). This is reportedly a common characteristic of opportunistic arid-zone species (Young *et al.* 2003). For similar reasons, the inter-flood dry-period is not a critical aspect for the growth and survival of black box.

Studies of water use by black box at Chowilla have indicated that it is able to utilise groundwater when rain-derived soil water and flood-derived soil water are depleted (Jolly and Walker 1996). Depth to groundwater and the salinity of the groundwater were determining factors in the ability of black box to utilise groundwater (Jolly 1996; Thorburn *et al.* 1993). Black box is reportedly able to utilise moderately saline groundwater of less than 40 dS m^{-1} when flood- and rain water were limited (Thorburn *et al.* 1993). Salinities greater than this appear to be detrimental; the extensive dieback of black box at Chowilla has been attributed to the contribution of saline groundwater to floodplain salinisation (Jolly *et al.* 1993).

Black box overlying highly saline groundwater has reportedly been able to maintain adequate health by accessing lower-salinity deep soil water (Holland *et al.* 2006). The source of this water is suggested to be vertical infiltration from rainfall and horizontal bank recharge from nearby surface waters such as lakes, creeks and rivers, which were able to overlie the saline groundwater. Bank recharge was common within approximately 50 m of the river, creek or lake (Holland *et al.* 2006).

Like river red gum, water availability may have other impacts on black box health. Miller *et al.* (2003) found that black box is more vulnerable to infection by the mistletoe *Amyema miquelii*, with decreasing water and/or salinity stress. The form of black box under drier conditions is twisted, with dead limbs and hollows that provide refuge and breeding holes for fauna, while very erect forms occur in well-watered sites (Roberts and Marston 2000).

Reproduction and regeneration

Black box appears to have a stronger flood requirement for reproduction and regeneration than for survival and maintenance. Early observations of regeneration indicate that it largely occurs after flooding and seldom occurs after rainfall (Treloar 1959). Seedlings are reportedly observed after flood recession (Treloar 1959), possibly indicating a link between seed release and flooding.

Phenological studies of black box indicate that buds may be retained on trees up to 12 months prior to flowering, with buds shed in response to poor condition. Flowering is therefore dependent on water availability in the year prior to flowering (Jensen 2008). The timing of flowering varies geographically, primarily occurring between August and January (Boland *et al.* 1986) but it may occur from May to October (George *et al.* 2005b; Jensen 2008; Roberts and Marston 2000). The name *Eucalyptus largiflorens* relates to flower structures and means 'abundant flowers' (George 2004). Black box reportedly produces abundant flowers (Cunningham *et al.* 1992), but little is known of seed and seedling yields in relation to flowering. However, it is apparent that seed matures in a few months (Boland *et al.* 1986), thereby coinciding with peak flooding periods throughout the Murray-Darling Basin. Black box exhibits serotiny with fruits retained in the canopy for up to two years before the seeds are released when conditions are suitable (Jensen 2008). The triggers for seed release remain largely unreported.

Once seed is released, flooding provides the primary source of moisture for germination (George 2004; Treloar 1959). Studies indicate that black box germinates at an optimum temperature of 35°C (Grose and Zimmer 1958), but germination may occur at temperatures of 30°C (Turnbull and Doran 1987). Light also appears to be important for germination (unpublished data, cited in Grose and Zimmer 1957).

Since seedlings do not have the well-developed root systems of mature black box, they are not opportunistic water users and sufficient soil moisture must be available for seedlings to be well established by the drier summer months. Roberts and Marston (2000) reported that germination for black box in the northern parts of the Murray-Darling Basin (western New South Wales and the lower Darling) is most effective between May and October, although air temperatures may not be optimal for germination at that time. Germination is most effective in the southern Murray-Darling Basin in spring, thereby giving seedlings moist conditions during summer.

Black box seedlings do not have the physiological adaptations of river red gum seedlings, such as adventitious roots, and are therefore susceptible to water stress from flooding (Akeroyd *et al.* 1998). Two-month-old seedlings can tolerate flooding, but not complete submersion, for approximately one month. Periods longer than one month limit growth and seedlings flooded for up to 70 days showed signs of stress (Heinrich 1990). Since black box is susceptible to complete flooding, it may be inferred that inundation depths of 1–30 cm are optimal for seedling establishment.

Coolibah: *Eucalyptus coolabah*

Coolibah is a eucalypt tree growing up to 20 m tall, with a grey box-like bark (Harden 1991; Figure 2.4). It is found in arid and semiarid Australia. It is structurally similar to *Eucalyptus*

Figure 2.4: Coolibah, *Eucalyptus coolabah*. Photograph: Tim Ralph (Macquarie University).

microtheca and was originally regarded as a subspecies of *E. microtheca*. It is primarily a riparian tree, situated on heavy clay soils near permanent or regular water supplies (Harden 1991). Coolibah has three subspecies: *E. coolabah* subsp. *arida, E. coolabah* subsp. *excerata* and *E. coolabah* subsp. *coolabah* (Harden 1991). *E. coolabah* subsp. *arida* is generally located on sandy or gravelly creek lines typical of north-western New South Wales, while the latter two species are located on heavy clay soils consistent with floodplains of the northern and western parts of the Murray-Darling Basin (Harden 1991).

Survival and maintenance

The review by Roberts and Marston (2000) highlighted that little physiological research had been undertaken into the water requirements and tolerance of coolibah. Disappointingly, little has changed. However, its location near water sources shows that coolibah requires ready access to water at some point in its life-cycle. River flooding is reportedly necessary for restoring soil moisture, which is essential for germination and seedling establishment (Roberts 1993).

Reports indicate that coolibah is able to survive with relatively lower flood frequencies than other floodplain eucalypts, however, there is great variability in flood frequencies between sites. Coolibah on the Gwydir appears to have a flood frequency of one in 10–20 years (Bennett and McCosker 1994), while coolibah at Cooper Creek has reportedly been flooded in the order of one in every five to six years (Roberts 1993). Coolibah associated with floodways has highly variable flood frequencies in the order of two to three flood pulses in five years, which may be followed by another five years without surface flooding (Surrey Jacobs, cited in Roberts and Marston 2000). Differences in flood frequencies may be related to antecedent site conditions or different physiological tolerances between subspecies. The range of results indicates the need for further research into the flood responses and water requirements of coolibah at a range of sites and in all subspecies.

Due to its location throughout arid and semiarid Australia, it is apparent that coolibah is tolerant of a range of inter-flood dry-periods. Roberts and Marston (2000, p. 12) indicated that coolibah is tolerant of 'relatively long dry inter-flood conditions as well as periodic flooding'. Others have reported that coolibah can survive great variability in flooding (Surrey Jacobs, cited in Roberts and Marston 2000).

Like other floodplain eucalypts, flood duration is an important factor – death of coolibah can result from waterlogging of soils (Roberts 1993). The flood duration necessary to promote seed production is likely to be in the order of a few to several weeks, depending on soil type (Roberts and Marston 2000) and flood timing. Flood timing appears to be most beneficial when it coincides with the peak growing season and air temperature peaks. Roberts and Marston (2000) reported that the ideal time for flooding is summer and autumn, but highlighted that the effects of shifting flood timing are unknown.

Measures of soil water electrical conductivity indicate that coolibah is salinity-tolerant to some extent. Soils with electrical conductivity greater than 0.2 dS m^{-1} reportedly supported coolibah with reduced reproductive capacity and with canopy dieback (Roberts 1993). Presumably, more frequent flooding would dampen the effects of saline soils and groundwater on tree condition and promote maintenance of coolibah communities.

Reproduction and regeneration

Similar to black box, coolibah appears to have a stronger flood requirement for reproduction and regeneration than for survival and maintenance. Roberts (1993) associated stand age with previous flooding opportunities, to highlight the correlation between recruitment and flooding. Relatively little is known of the water regime required for recruitment of coolibah.

Coolibah flowers between October and December (Boland *et al.* 1986) and fruits mature relatively quickly during summer and autumn (Doran and Boland 1984). For this reason, it is likely that coolibah does not store seed in the canopy like river red gum. Long-term storage of seed requires temperatures of 3–5°C (Boland *et al.* 1986) and it is likely that seed stored within the soil profile would quickly deteriorate (Doran and Boland 1984). This is further supported by seed bank studies which show an absence of coolibah seeds despite its presence within the floodplain (e.g. Capon and Brock 2006).

Germination experiments indicate that viable seedlings are produced at temperatures of 28–37°C, the optimal temperature being 35°C. These temperatures are consistent with late summer flooding or rainy seasons throughout the distribution range of coolibah (Doran and Boland 1984; Roberts and Marston 2000).

Weeping myall: *Acacia pendula*

Weeping myall, also known as boree, is an erect spreading perennial growing to a height of up to 13 m (Figure 2.5). The botanical and common names refer to the pendulous shape of mature

Figure 2.5: Weeping myall, *Acacia pendula*. Photograph: Tim Ralph (Macquarie University).

individuals. This wattle tree has a widespread distribution in inland areas of New South Wales, Queensland and Victoria and is known to grow on major floodplains in heavy clay soils (Harden 1991). Weeping myall flowers mainly during summer and autumn (Harden 1991).

Survival and maintenance

Little information is available about the flood requirements of weeping myall. While it occurs on floodplains, its distribution appears to relate to its tolerance of flooding (Kidson *et al.* 2000b) rather than a requirement of flooding for survival. In the Macquarie River region, weeping myall most frequently occurs on gilgais (Metcalfe *et al.* 2003).

Weeping myall is regarded as reasonably salt-tolerant (NT Dept of NREA 2006), but there is no known scientific literature of its salinity tolerance range.

Reproduction and regeneration

The distribution of weeping myall on floodplains may reflect the water requirements of weeping myall for germination, rather than water requirements for survival. While no scientific information is available about the germination of weeping myall, nurseries indicate that germination is best when seeds are soaked in near-boiling water for some time. After sowing,

Figure 2.6: River cooba, *Acacia stenophylla*. Photograph: Tim Ralph (Macquarie University).

germination should take two to three weeks at temperatures of approximately 25°C (Faucon 2005; Windmill Outback Nursery 2003).

River cooba: *Acacia stenophylla*

River cooba, also referred to as river myall, is an erect spreading perennial growing to a height of up to 13 m (Figure 2.6). This wattle tree has a widespread distribution throughout central and western New South Wales and all other mainland states (Harden 1991). It is known to grow near watercourses and swampy areas, which is characterised in its common name. Flowering occurs between March and August, but river cooba may flower throughout the year (Cowan 1996, cited in CSIRO 2004a).

Survival and maintenance

Little is known of the water requirements for the survival and maintenance of river cooba. River cooba is regarded as both drought- and flood-tolerant (Marcar *et al.* 1995) and reportedly grows well where groundwater is shallow or rainfall is adequate (Boxshall and Jenkyn 2001; Johns *et al.* 2009). The distribution of river cooba within floodplains commonly lies between zones occupied by river red gum and black box, implying that water requirements for survival and maintenance of populations lies within the ranges established for river red gum and black box (Johns *et al.* 2009).

River cooba is somewhat salt-tolerant, but growth may be reduced at soil groundwater salinities of 10–15 dS m^{-1} and survival is limited at salinities in excess of 15 dS m^{-1} (Marcar *et al.* 1995). However, recent research indicates that river cooba may survive under markedly higher saline conditions of up to 40 dS m^{-1} (Doody, cited in Johns *et al.* 2009).

Reproduction and regeneration

River cooba generally flowers between March and August (Harden 1991) with the seeds maturing between October and December (Marcar *et al.* 1995). However, flowering has been observed at other times, possibly in response to favourable flood conditions (Doody, cited in Johns *et al.* 2009). Seed maturity coincides with peak flood timing within much of the Murray-Darling Basin and prolific germination of seed is commonly observed along the floodline (Cunningham *et al.* 1981). Little is known of the flooding requirements for germination; however, flooding evidently increases the likelihood of germination.

Shrubs

Lignum: *Muehlenbeckia florulenta*

Lignum is a multi-stemmed woody perennial shrub (Sainty and Jacobs 1981, 2003) located in arid and semiarid areas of eastern Australia. It occurs as an understorey in eucalypt woodlands or as shrublands on floodplains. Lignum grows to a height of 2–3 m and may form dense thickets when conditions are favourable (Figure 2.7). Lignum appears dry and lifeless during dry periods, but exhibits prolific growth in response to favourable flood conditions. While lignum occurs in flood-prone areas, it is relatively salt- and drought-tolerant and may survive

Figure 2.7: Lignum, *Muehlenbeckia florulenta*. Photograph: Tim Ralph (Macquarie University).

for some time without rainfall and flooding (Roberts and Marston 2000; Young *et al.* 2003). Despite the wide distribution of lignum throughout the Murray-Darling Basin and its ecological significance as a waterbird breeding habitat, few studies of the water requirements for its survival and maintenance have been conducted.

Survival and maintenance

Flood frequency reportedly does not significantly influence the cover of lignum, but it is apparent that lignum is located in areas with a flood frequency of every three to 10 years (Craig *et al.* 1991). Lignum responds quickly to an increase in available water by developing shoots, leaves and flowers (Chong and Walker 2005; Roberts and Marston 2000) and lignum cover is observed to be greatest in high flood frequency locations compared to lower flood frequency zones (Capon 2005).

Little information is available on the duration of flooding necessary for the survival and maintenance of lignum. However, lignum shrublands are located throughout south-eastern Australia in areas with a flood duration of a few to six months, even 12 months (Roberts and Marston 2000). On the Lower Murray floodplain, the majority of lignum was inundated for 92–228 days over a two-year period (Blanch *et al.* 1999b). While lignum may initially show signs of vigorous growth with flooding, under prolonged flood conditions (>12 months) lignum will die due to the anoxic conditions (Kozlowski 1984). This is supported by observations of significant decreases in shrub cover and mortality of lignum following a major flood at Cooper Creek in 2000 (Capon 2003).

The timing of flooding contributes to the survival and maintenance of lignum shrublands. Since a positive relationship has been established between soil moisture and lignum cover (Craig *et al.* 1991), flooding that occurs at optimal times and acts to maintain soil moisture reportedly produces vigorous growth in lignum shrublands (Young *et al.* 2003). Floods in late spring and early summer increase soil moisture conditions throughout the warmer summer months when productivity is highest and vigorous growth is greatest. The importance of soil moisture is further evident in the occurrence of lignum at infrequently flooded sites, where summer rainfall contributes to soil moisture (Roberts and Marston 2000).

Inter-flood dry-period may also influence the growth of lignum, but little information is available on the maximal inter-flood drying that may limit its ability to reproduce. Craig *et al.* (1991) found a negative relationship between lignum cover and time since last flooded, implying that maximum growth occurs when flooding is an annual event. However, as lignum occurs in regions with flood frequency of three to 10 years, it is implied that lignum can survive inter-flood dry-periods of up to 10 years.

The distribution of lignum is reportedly influenced by the depth and salinity of groundwater. While lignum is relatively salt-tolerant and able to withstand salinities of at least 10 000 mg L^{-1} (Van der Sommen 1980), growth may be limited at higher salinities. Craig *et al.* (1991) suggested that maximal growth occurs at higher salinity locations where floods occur more frequently.

There is little reported information on optimal flood depths for the maintenance and survival of lignum, but the majority of lignum on the Lower Murray River is located on the banks receiving inundation no deeper than 60 cm (Blanch *et al.* 1999b).

Reproduction and regeneration

Lignum is reliant on flooding to reproduce both vegetatively and sexually from seed. Vegetative growth occurs rapidly in response to flooding or sufficient rainfall (Chong and Walker 2005; Jensen *et al.* 2006). Production of shoots, leaves, flowers and seeds also occurs in response to rainfall and flooding. Expansion in response to rainfall appears to be largely limited to shoot extension and arching, with few new plants observed (Jensen *et al.* 2006).

Lignum seeds respond rapidly to floodwaters, with the time between ripening and dispersal being a relatively short 12 days (Chong and Walker 2005). To take advantage of floodwaters, ripening and dispersal may occur in winter or spring. Lignum seeds are buoyant for five to 25 days and are dispersed by floodwaters. Seeds are able to germinate within water, but germination generally occurs within wet mud once floods recede (Chong and Walker 2005). There is no evidence that lignum seeds are dispersed by wind or other dispersal methods (Chong and Walker 2005; Jensen *et al.* 2006).

The opportunistic growth of lignum in response to flooding is also evident in the rapid germination of seeds under favourable growing conditions. Deterioration of the seed perianth occurs within three weeks in moist conditions (Chong and Walker 2005). Lignum seeds have remained viable for up to 15 years when stored in paper bags at room temperature, but within floodplains the moist soil conditions and the rapid deterioration of the perianth mean lignum seeds do not appear to remain viable for extended periods and lignum does not maintain a persistent seed bank.

Flood duration is likely to be important for germination of lignum. Viable seeds are produced 14–30 days after flower development (Chong and Walker 2005). Since seeds quickly disperse after ripening and their deterioration is also rapid, it is inferred that, for germination to occur, flood duration should not exceed approximately eight weeks after flower development.

Timing of seed dispersal may be important, with a laboratory-based experiment indicating that germination is inhibited at constant temperatures of 12°C and 24°C (Chong and Walker 2005). This indicates that temperatures must be fluctuating and not freezing for germination to occur, such as those evident in spring and summer. Chong and Walker (2005) suggested that germination may occur after a particularly cold winter.

Grasses

Water couch: *Paspalum distichum*

Water couch is a perennial grass that is typically located in or near fresh water. It has a wide distribution throughout Australia and internationally. It occurs in all Australian states and is commonly regarded as an invasive weed throughout Europe and the US (Aguiar *et al.* 2005; Huang and Hsiao 1987; Huang *et al.* 1987). It is described as a stoloniferous (prostrate stems) and rhizomatous (horizontal stem-like roots) perennial, growing to a height of 0.5 m and with stolons up to 5 m long (Harden 1993; Figure 2.8). In shallow water conditions, stolons may become erect and extend up to 1 m in height when conditions are optimal (Roberts and Marston 2000). In Australian wetlands, water couch may form dense monospecific stands or it may occur as a mixed grass community. Composition of communities may alter in response to changes in the water regime (Watt *et al.* 2007) and the timing of flooding (McCosker 2001). Water couch grasslands are relatively common on the floodplains of the Macquarie and Gwydir rivers (Roberts and Marston 2000) and are a valuable food source for graziers.

Survival and maintenance

Water couch has relatively high water requirements and for this reason is commonly located in creeks, drainage lines and irrigation channels. It is also common along the margins of rivers and on floodplains where the water regime is adequate to sustain growth. Due to its high reliance on water and wide distribution throughout semiarid Australia, it is unlikely that water

Figure 2.8: Water couch, *Paspalum distichum*. Photograph: Tim Ralph (Macquarie University).

couch pastures can be maintained on intermittent rainfall and rainfall infiltration to the root zone. However, Bennett and Green (1993) suggested that water couch may survive for some years on rainwater alone when not exposed to grazing pressure. Since the root system of water couch is relatively shallow, it is unlikely that deep groundwater sources can sustain water couch. Water couch requires regular inundation from floodwaters or water-level fluctuations along the banks of rivers and streams (Bennett and Green 1993).

Water couch exhibits a relatively wide flood frequency tolerance. Blanch *et al.* (1999b) regarded water couch as a common floodplain species that occurs in three flood frequency zones: water couch was most abundant in areas flooded for a median of 163 days in two years or a maximum of 552 days in two years, it was relatively common in zones flooded for a median of approximately 45 days every two years, and it occurred in zones flooded as little as 11 days in two years. Where the duration of inundation is short, flooding is likely to be most beneficial when floods do not occur consecutively – water couch appears to have a maximum inter-flood dry-period of 290 days in two years (Blanch *et al.* 1999b). Field studies on the Gwydir indicate that water couch requires flooding at least once a year during the summer months (Bennett and Green 1993). Buried rhizomes provide some degree of drought tolerance, but there is loss of vigour. Repeated droughts increases the vulnerability of water couch grasslands to competition from invasive species such as lippia (*Phyla cansecens*) (Bennett and Green 1993).

Flood duration appears to be quite variable. Blanch *et al.* (1999b) suggested that water couch may survive short-duration floods when they do not occur consecutively. Water couch may also survive flooding of up to 513 days in two years, but this is less likely to occur in consecutive years. Distribution of water couch pastures on the Gwydir floodplain indicate that duration is most important during the summer months and that floods should last for four to eight weeks. Prostrate forms establish when inundated for shorter periods (Bennett and Green

1993). Laboratory experiments on water couch growth showed that the effect of flood duration on growth depended on the height of individual plants and on flooding depth (Hsiao and Huang 1989). Taller plants did not exhibit growth limitation when flooded for extended periods at shallow flood depths, while growth was limited earlier when flood depths were greater (Hsiao and Huang 1989).

While water couch is commonly located in frequently flooded sites, it generally does not occur where water depths exceed approximately 60 cm. However, it may occur at depths of up to 200 cm (Blanch *et al.* 1999b). Field studies on the Gwydir floodplain suggest that depths of 10–15 cm are optimal for the maintenance of water couch grasslands (Bennett and Green 1993). Complete submergence for up to three months may reduce plant growth (Hsiao and Huang 1989). It is likely that water couch may survive complete submersion, provided it is not for an extended duration.

Despite water couch being distributed over a range of flood frequencies, depths and durations, it is likely that the timing of flooding is essential for the maintenance of water couch grasslands. Laboratory experiments indicate that growth is greatest at temperatures of 30–40°C, and that little growth is evident at 10°C (Huang *et al.* 1987). Water couch reportedly has a spring to summer growth season (Roberts and Marston 2000) and flowering and seed production primarily occurs during the summer months (Harden 1993). Conditions should be optimal during the main growing season to sustain growth and seed production. Therefore flooding provides greatest benefit when it occurs throughout the warmer months of spring and summer.

Grazing and flooding may have a significant influence on the survival of water couch. In India, simulated grazing experiments using clippings of water couch found that clipping water couch that is completely submerged usually caused death. Under moist conditions (0 cm water depth), water couch usually survived when clipped at fortnightly intervals but it exhibited less total and root mass (Middleton 1990).

Reproduction and regeneration

Water couch reproduces both vegetatively from its numerous rhizomes and aerial stems, and sexually from seed (Huang *et al.* 1987). Studies in India suggest that regeneration of water couch is greater by vegetative expansion than from the seed bank (Middleton 1999). Water regime and its impact on flooding play a significant role in sexual and asexual reproduction.

High temperatures and flooding coincide with maximum growth of water couch at both the Macquarie Marshes and Gwydir Wetlands, where extensive water couch grasslands exist. While studies of vegetative spread of water couch are limited, Huang *et al.* (1987) indicated that expansion occurs both from creeping rhizomes and from aerial stems (stolons) that are able to sprout and root. Growth of rhizome segments and stems was greatest at temperatures of about 30°C and declined at temperatures in excess of 40°C (Huang *et al.* 1987).

High temperatures without moisture expose aerial shoots and rhizomes to desiccation (Huang *et al.* 1987). Flooding experiments indicate that a small degree of root flooding (15% and 35%) promoted shoot growth (Manuel *et al.* 1979, cited in Huang *et al.* 1987). However, the role of flooding in growth apparently becomes limited and it may inhibit growth when the degree and duration of root flooding is increased. For example, root flooding greater than 50% and for a duration of approximately three months caused significant decreases in the number of shoots and root dry weights (Hsiao and Huang 1989).

Water couch flowers in summer (Harden 1993) and produces a large number of seeds, estimated at 100 seeds per panicle or 100 000 per square metre, apparently stored within seed banks (Middleton 1990; Valk *et al.* 1992). However, only a small number of seeds (approximately 5–10%) may be viable (Okuma and Chikura 1984, cited in Huang and Hsiao 1987).

Similarly, Middleton (1999) found that, even under ideal germination conditions, less than one seedling of water couch germinated per square metre.

Ideal conditions for seed germination are similar to those for vegetative expansion. The optimum temperature for germination is reportedly 28–35°C; germination is limited at temperatures below 20°C and above 40°C (Huang and Hsiao 1987). Moist conditions are essential for germination; studies indicate that germination from seed occurs best under moist conditions which simulate drawdown (Middleton 1999; Middleton *et al.* 1991). Conditions for growth are similar to those required for survival and maintenance. However, due to less-developed rhizomes, water couch juveniles may not survive extended drought conditions.

Common reed: *Phragmites australis*

Common reed is a widespread grass species occurring on most continents, except Antarctica (Clevering and Lissner 1999). Common reed was formerly classified as *Phragmites communis* until 1968, when a revision was undertaken (Clayton 1968). In Australia, common reed occurs in all states. It is largely limited to temperate regions within creeks, streams, channels and drains, swamps and areas that are seasonally inundated (Sainty and Jacobs 2003). Common reed is replaced by *Phragmites vallatoria* (also referred to as *P. karka*) throughout tropical regions of Australia (Haslam 1972; Hocking 1989a).

Common reed is a clonal perennial grass growing to 4 m under ideal conditions in Australia (Sainty and Jacobs 2003; Figure 2.9). The seasonal growth of common reed within Australia has been described by Hocking (1989a, 1989b). Budding and generation of young shoots throughout south-eastern Australia appears to be dormant throughout the late winter months, but rapid growth occurs by early October. Growth continues throughout summer with a peak by late summer or early autumn. Dry matter then accumulates until mid winter and coincides with peak below-ground biomass. Flowering commences in late summer during the peak growing season and flowers grow quickly, reaching maturity by March (Hocking 1989a).

Figure 2.9: Common reed, *Phragmites australis*. Photograph: Jeff Kelleway (DECCW).

Common reed has a high water requirement, occurring within permanent standing water or under fluctuating water levels (Blanch *et al.* 1999b). Common reed is described as a well-adapted wetland plant with considerable tolerance to flooding and exposure (Blanch *et al.* 1999b, 2000; Pagter *et al.* 2005). Due to its ability to tolerate extensive drought periods, common reed may occur in wetlands with highly fluctuating hydroperiods (Pagter *et al.* 2005). While common reed is unlikely to survive on rainfall alone, it may survive in areas where runoff can accumulate (Haslam 1970). Common reed has also been reported at sites where the groundwater is at a depth of 4 m (Haslam 1972), and rhizomes may grow to considerable depths when the watertable is below the surface (Frankenberg 1997). Haslam (1970) suggested that competition rather than limited water availability limits the distribution of common reed.

Survival and maintenance

While common reed is able to survive permanent waterlogging (Blanch *et al.* 1999b), studies indicate that this is not the optimal condition for growth (Blanch *et al.* 1999b; Saltmarsh *et al.* 2006). Blanch *et al.* (1999b) found that common reed grew under a range of flood frequency conditions, from permanently wet to high on the floodplain where conditions are near-permanently dry. Common reed survived where flooding occurred for only 33 days over a two-year period, perhaps due to groundwater seepage from the nearby Murray River. Optimum growth occurred with fluctuating water levels creating moderate flood frequency conditions. It is apparent that, for survival of above-ground biomass and in the absence of access to other water sources such as groundwater, flooding should occur near annually or every one to two years (Roberts and Marston 2000).

As common reed occurs under a range of flood frequency conditions, it is likely that flood duration is of little importance for the survival of stands. However, since the physiology of common reed enables it to inhabit sites with fluctuating or static water levels (Blanch *et al.* 1999b), optimal growth takes place where flooding occurs for extended periods. For example, Blanch *et al.* (1999b) found optimal growth when flooded for approximately six months per year (499–351 days per two-year period).

Little information is available about the preferred timing of flooding. Roberts and Marston (2000) suggested that there is no seasonal requirement for flooding. Haslam (1970) observed that common reed stands flooded in late spring performed well, and that spring emergence of shoots was delayed when a stand that was normally flooded became dry. Since flooding commonly occurs from late autumn to summer within the Murray-Darling Basin, it is likely that the biotypes within the Basin are adapted to a variable water regime. This timing coincides with optimal temperatures for growth (Frankenberg 1997).

Common reed occurs in a range of hydroperiods and inter-flood dry-periods (Blanch *et al.* 1999b). Optimal growth on the Lower Murray was reported at sites where the inter-flood dry-period was relatively short, at 125–236 days over a two-year period. However, common reed was also located in zones with no dry-period or with extremely long inter-flood dry-periods (Blanch *et al.* 1999b).

While common reed prefers fluctuating water levels rather than permanently wet conditions, it does occur in areas which are rarely inundated. By maintaining the carbon and water balance of the whole plant, common reed can survive extended inter-flood dry-periods (Saltmarsh *et al.* 2006). To adapt to drier conditions, common reed may initially exhibit a reduction in leaf area and leaf biomass, thereby enabling it to maintain some capacity for photosynthesis (Pagter *et al.* 2005; Saltmarsh *et al.* 2006). It may increase the proportion of water-absorbing root biomass as another means to exploit available water resources (Pagter *et al.* 2005). Under severe drought stress, common reed reportedly reduces osmality in leaves (Pagter *et al.* 2005).

Most studies indicate that common reed is able to survive inundation depths of up to 2 m (Sainty and Jacobs 1981). Optimal flood depth for maintenance of common reed stands depends on the nature of flooding, as common reed occurs in both static and fluctuating water conditions. Under static water conditions, common reed stands generally occur at shallow water depths, ranging from an average of 45 ± 20 cm (Coops *et al.* 1996), or within the range of 50 cm above to 20 cm below the waterline (Haslam 1970). Similarly, biomass is reportedly limited at depths of 80 cm (Coops *et al.* 1996). Like other emergent macrophytes, the ratio of above-ground to below-ground biomass increases in deeper water. The elongation of stems and subsequent increase in above-ground biomass with increasing water depths appear to be adjustments to maintain the biomass of photosynthetically active material above the water level (Coops *et al.* 1996; Hayball and Pearce 2004). Greatest stem lengths reportedly occur at depths of 55 cm (Coops *et al.* 1996).

Numerous studies indicate that flood depth for optimal growth is complex, apparently reliant on the relationship between plant elevation or topography and water fluctuation (Alvarez-Cobelas and Cirujano 2007; Deegan *et al.* 2007; White *et al.* 2007). Deegan *et al.* (2007) found, in a pond experiment over a 14-week period, that the largest biomass for common reed occurred under moderately fluctuating water levels of ±30 cm rather than static, low (±15 cm) or highly (±45 cm) fluctuating conditions. The response of plants was highly dependent not only on the amplitude but on the elevation of plants along the inundation gradient. For example, a negative response was detected only when the amplitude of fluctuations was high and plants were positioned at low elevations. The study was supported by that of White *et al.* (2007), who confirmed that common reed is suited to fluctuating water levels when growing at high elevations. The negative response may reflect the reduced capacity of common reed to photosynthesise and respire in deep water (Coops *et al.* 1996). Similarly, higher stem densities have been associated with shallower water.

While common reed tends to occur where soil conditions are waterlogged for much of the year, it may survive in areas where groundwater is not at great depths (Haslam 1972). Rhizomes tend to occur in the upper 1.5 m of the soil profile (Haslam 1970), but, where common reed relies on groundwater, rhizomes may extend for some depth to gain access to groundwater resources (Frankenberg 1997).

The ability of common reed to survive on groundwater alone also depends on the quality of the groundwater. Common reed exhibits some degree of salt tolerance and can establish in brackish conditions within estuaries (Sainty and Jacobs 2003). It can survive in water with up to 10 ppt total dissolved salt (16 dS m^{-1}) (Sainty and Jacobs 2003), and germination has been reported at salinities of approximately 25 ppt (Greenwood and MacFarlane 2006; Mauchamp *et al.* 2001).

Temperature may influence the growth of common reed, however, due to the species' wide distribution throughout many climatic zones, temperature mainly affects the timing of the growth cycle (Haslam 1975; Soetaert *et al.* 2004). Glasshouse experiments indicate that increases in temperature (to 25°C) and humidity enable the common reed to produce new shoots throughout the year, rather than during the typical spring and summer growing season (Haslam 1969b; Hocking 1989a, 1989b). Higher temperatures increase the length of the growing season (Haslam 1975) and may promote increased stem density and biomass, if optimal access to water is maintained.

Reproduction and regeneration

Common reed can regenerate both from seed and vegetatively. It is more effective when expanding vegetatively, particularly in deeper water (Weisner *et al.* 1993). In fact, vegetative expansion is so effective that stands of common reed may be one entire clone (Frankenberg

1997; Koppitz *et al.* 1997). Recent research suggests that populations may initially develop from seed, but over time become dominated by one or two clones that are well-adapted to the prevailing site conditions (Koppitz and Kühl 2000). Low genetic diversity of common reed stands may partly explain the expansion of common reed in North America (Amsberry *et al.* 2000) and the dieback of populations throughout Europe (Koppitz and Kühl 2000).

Expansion of rhizomes commences in summer and continues until it peaks in spring, in south-eastern Australia (Hocking 1989b). Buds develop on rhizomes year-round but remain dormant near the soil surface (Haslam 1969a). Rapid emergence of buds occurs during spring, enabling new aerial stems to develop (Frankenberg 1997; Haslam 1969a). Where temperature is too low or access to water is limited, emergence of buds is also limited (Haslam 1969a). It is therefore evident that conditions for vegetative expansion are similar to those required for maintenance of common reed stands.

Regeneration of common reed from seed appears to be relatively inefficient in many parts of the world (Haslam 1971; Mauchamp *et al.* 2001), despite there being up to approximately 1000 fertile seeds per panicle or flower head (Haslam 1972; McKee and Richards 1996). The viability of seeds is variable, ranging from 0.1% to 59.6% (Ishii and Kadono 2002) or from 0–100% in locations throughout Europe (McKee and Richards 1996). A number of studies report the existence of sterile clones of common reed (Frankenberg 1997; Haslam 1972).

In Australia, flowering generally occurs rapidly during summer with full height attained in early autumn (Frankenberg 1997; Hocking 1989a). Seed is generally available for germination in spring (Frankenberg 1997) or it may enter the seed bank if conditions are not suitable for germination. Stored seeds remain viable for three to four years (Haslam 1972), but the viability of seeds stored in soil seed banks is unknown.

Once viable seeds are available for regeneration, a number of 'sieves' (Coops and Velde 1995) limit the regeneration of common reed from seed. First, seeds need to be transported to a location appropriate for germination. Common reed seeds rely on hydrochory for transport of seeds by floating on water. The seeds may be small enough to be transported by wind but they still require some degree of flooding, and therefore hydrochory, for establishment (Coops and Velde 1995). Common reed seeds float for a relatively short period of one to three days (Coops and Velde 1995), which may limit distribution to lower elevations within the flood zone.

The second filter for regeneration from seed is germination and seedling survival. Seeds appear to prefer drawdown conditions for germination rather than flooded conditions. However, germination can often be relatively slow (approximately 10 days in a laboratory experiment by Coops and Velde 1995). Haslam (1971) suggested that the soil should be wet but not inundated more than 1 cm, that severe frosts must be absent and that light, temperature and phosphate must be high. Germination is slow (approximately one month) at low temperatures. Under optimal conditions, over 25°C, germination may occur in a number of days (Frankenberg 1997). Based on temperature and water requirements for germination, ideal inundation timing for germination is likely to be in spring and summer (Young *et al.* 2003).

Established seedlings appear to tolerate a narrower range of water levels than adult plants (Haslam 1971), which may relate to the amount of emergent material available to photosynthesise (Mauchamp *et al.* 2001; Weisner *et al.* 1993). Seedlings may survive inundation for extended periods (Coops and Velde 1995; Mauchamp *et al.* 2001), but growth may be slow or absent during flooded conditions (Coops and Velde 1995; Weisner *et al.* 1993). Coops and Velde (1995) reported that week-old seedlings survived inundation for up to seven weeks, but with reduced growth. Alternatively, young seedlings may not survive extended flooding, while older seedlings may have greater tolerance (Haslam 1971; Mauchamp *et al.* 2001). Armstrong *et al.* (1999) indicated that seedling shoots exposed to permanent submergence may never become emergent. Biomass of seedlings grown under flooded conditions may be significantly less than those grown under drained conditions, possibly due to the senescence of older basal leaves and the reallocation of resources to younger leaves (Coops and Velde 1995; Mauchamp *et al.* 2001). Improvement in growth may become evident upon drainage (Coops and Velde 1995).

Due to the dispersal characteristics of common reed, seedling survival may be limited despite water requirements being suitable. Common reed produce abundant seed (Coops and Velde 1995), much of which may not be transported far from the parent plant. Some seeds may germinate under drawdown conditions, but competition between common reed clones and abundant seedlings for resources is high and seedlings are unlikely to survive. Competition between faster-growing plants and common reed seedlings may also limit the survival of common reed (Haslam 1971).

The third filter is the response of seedlings to water levels and other environmental conditions. Once seedlings have established, water requirements must be maintained until seedlings reach maturity so that regeneration is complete. This largely depends on maintenance of water regimes or adequate access to water, and on the location at which seedlings originally established (Coops and Velde 1995). If seedlings establish at higher elevations or flooding is not substantial in subsequent years, access to water may be limited.

Seedlings generally germinate in spring and summer throughout south-eastern Australia, followed by a dormancy period during winter. There is a relatively short period during which seedlings can develop root systems to adequately access water. Spring and summer drought conditions may limit the survival of seedlings to maturity and winter mortality during the dormancy period may be high (Haslam 1971).

Salinity may also limit germination and regeneration of common reed, as salt tolerance appears to increase with age (Haslam 1971). Adult common reed may survive in salinities of up to 3% salt (Haslam 1972), but seedlings grown in 2% salinity died. Seedlings grown at 1% salinity died unless access to water was high (Haslam 1971). The seedlings that did survive, when grown under saline conditions, exhibited shorter stature (Haslam 1971).

Sedges and rushes

Cumbungi: *Typha orientalis* and *Typha domingensis*

The term cumbungi refers to species of *Typha*, which include the native *Typha orientalis* C.Presl (broadleaf cumbungi) and *T. domingensis* Pers. (narrow-leaved cumbungi) (Finlayson *et al.* 1985) as well as the introduced *T. latifolia* L., which has a limited distribution. Cumbungi is an emergent plant (Figure 2.10) and, although not a true sedge, rush or grass, may occur under similar water regimes as sedges, rushes and grasses. The inclusion of cumbungi with sedges and rushes is consistent with previous studies (Roberts and Marston 2000; Walker and Hopkins 1990).

Large stands of cumbungi are located in the Murray-Darling Basin within terminal wetlands such as the Macquarie Marshes (Roberts 2001). Rapid above-ground growth typically occurs in spring and early summer, while below-ground growth increases after mid-summer. New shoots emerge in autumn and winter (Roberts and Marston 2000). Due to higher temperatures, longer day lengths and higher daily irradiances in winter in inland Australia, cumbungi may exhibit a continuous growing season (Roberts and Ganf 1986).

Survival and maintenance

Cumbungi may form distinct communities in locations with relatively stable water levels (Blanch *et al.* 1999b; Brix *et al.* 1992; Walker *et al.* 1994). Cumbungi is more physiologically

Figure 2.10: Cumbungi, *Typha orientalis* and *T. domingensis*. Photograph: Tim Ralph (Macquarie University).

suited to stable water levels than to fluctuating water levels (Deegan *et al.* 2007; Matsui and Tsuchiya 2006; White *et al.* 2007). The rhizomes and roots of cumbungi are buried in anaerobic sediments but they can transport oxygen from shoots to roots for growth and respiration, by internal pressurisation and convective gas flow (Brix *et al.* 1992).

Cumbungi prefers water regimes from permanently wet (Blanch *et al.* 1999b) to seasonally or periodically dry and may survive dry conditions for three to four months (Roberts and Marston 2000). However, extended dry conditions increase the exposure of cumbungi to the desiccating effects of salinity. Cumbungi is considered moderately salt-tolerant: growth is reportedly reduced at sodium chloride concentrations of 50 mM, while individuals become severely damaged at 100 mM concentrations (Hocking 1981). Cumbungi may occur under a range of water regimes but differences occur between sites due to site-specific variation in abiotic factors and to interspecific interactions (Froend and McComb 1994).

Cumbungi has a high water requirement (Roberts and Marston 2000) and may survive in water depths of approximately 2 m (Sainty and Jacobs 1981). When water depth exceeds 2 m, adaptations to survive in aquatic environments fail and the roots and shoots become oxygen-starved, lose vigour and die (Roberts and Marston 2000). However, inundation is essential for survival and death will occur in prolonged drought periods. Ramets may survive dry conditions for three to four months following rapid growth in summer and rhizomes may remain viable for a few years when protected from desiccation (Roberts and Marston 2000). It is therefore apparent that above-ground productivity and reproduction may be limited at extremes of water depth or from prolonged drought or prolonged inundation. Biomass and inflorescence densities are generally greatest at intermediate water depths within a water regime gradient (Froend and McComb 1994). In spite of this, growth may vary in response to changes in water regime (Froend and McComb 1994), which may cause inundation gradients to shift temporarily.

Even though cumbungi prefers stable water levels, where conditions are seasonally or periodically dry the timing of inundation may be significant. The broadleaf cumbungi exhibits optimal growth in water temperatures of 25–28°C (Cary and Weerts 1984), which coincides with summer. Roberts and Marston (2000) reported from field observations that inundation in spring and summer is favoured as it correlates with the growing season for cumbungi. Periodic or seasonal inundation and fluctuations in water depth appear to provide ideal conditions for the growth of other emergent macrophytes, such as the common reed (Roberts and Marston 2000).

Reproduction and regeneration

Cumbungi is able to reproduce by vegetative expansion and sexually from seed (Nicol and Ganf 2000). Vegetative expansion is relatively slow and requires an existing cumbungi stand and conditions suitable for growth (Miao *et al.* 2001). Reproduction from seed facilitates rapid expansion, enabling cumbungi to become established at sites some distance from existing stands. Due to its ability to regenerate vegetatively and from seed as well as its broad water requirements for reproduction, cumbungi has a significant temporal and spatial window of opportunity for recruitment (Froend and McComb 1994; Nicol and Ganf 2000). However, conditions must be suitable for seed germination, survival, growth and propagation (Miao *et al.* 2001).

Vegetative growth appears to be related to depth of inundation. Froend and McComb (1994) indicated that commencement of growth and production of inflorescences occurred at the same time at wetlands in south-western Australia, but growth ceased earlier in sites with decreasing water depth. It is inferred that water stress, particularly during summer, may limit broadleaf cumbungi's ability to expand. This is partly supported by Nicol and Ganf (2000), who found that vegetative expansion was absent under rapid drawdown conditions at the waterline, static conditions at inundation depths of 30 cm and static and slow drawdown conditions at 80 cm depth.

Flowering of broadleaf cumbungi occurs during the warmer summer months from November to March, with seed production generally occurring during late summer from January to April (Froend and McComb 1994; Roberts 2001). Narrow-leaved cumbungi produces approximately 250 000 seeds per inflorescence (Nicol and Ganf 2000), each with a mass of 24–35 μg, while broadleaf cumbungi seeds have a mass of 35–49 μg (Roberts and Marston 2000). Light seed mass enables dispersal over large distances by wind (Froend and McComb 1994). Persistence of seeds within seed banks in excess of one year may be limited by the rapid loss of viability, particularly as seeds are small-sized and have thin coats (Miao *et al.* 2001).

Germination of cumbungi from seed requires light, moisture and temperatures in excess of 10°C (Roberts and Marston 2000). The opportunity for broadleaf cumbungi to reproduce from seed is substantial, as germination may occur during prolonged drawdown conditions when vegetative growth is limited or while cumbungi is submerged in shallow water of approximately 5 cm (Froend and McComb 1994; Roberts 2001). Germination of narrow-leaved cumbungi is greater under saturated soil conditions than under flooded conditions (Miao *et al.* 2001). Nicol and Ganf (2000) found that narrow-leaved cumbungi had broad niche requirements and that seeds could germinate in all experimental hydrologic regimes ranging from rapid drawdown conditions to static conditions at various inundation depths up to 80 cm. Germination was minimal at depths of 80 cm under slow drawdown or static conditions and moderate under rapid drawdown conditions. Moderate germination was evident under all drawdown conditions at depths of 30 cm and was minimal under static conditions. Approximately 50% of seeds germinated under static and slow drawdown conditions at 0 cm inundation depth, yet germination was minimal under rapid drawdown conditions. Accordingly, germination is optimal under static or slow drawdown conditions at 0 cm elevation.

Once seeds have germinated, seedlings can continue growth under drawdown conditions. Nicol and Ganf (2000) showed that optimal growth occurs at the waterline in slow drawdown or static conditions and that vegetative growth from newly formed rhizomes became evident after approximately six weeks. Under optimal conditions that are nutrient-rich and warm, cumbungi may reach 1 m height in a few months (Roberts 2001; Roberts and Marston 2000). However, continued survival of seedlings depends on the maintenance of hydrologic regimes and warmer conditions (Miao *et al.* 2001).

Rushes: *Juncus* species

Numerous rush species occur within floodplain wetlands of the Murray-Darling Basin, some of the more prevalent being the gold or yellow rush (*Juncus flavidus*), giant rush (*J. ingens*), tussock rush (*J. aridicola)* and billabong rush (*J. usitatus*; Figure 2.11). Sainty and Jacobs (1981) indicated that each species occurs in seasonally wet or damp locations. The tussock rush may also inhabit permanently or periodically inundated locations in arid and semiarid areas (Sainty and Jacobs 1981). All species are regarded as native short rhizomatous perennials growing to heights of 1–1.5 m, while the giant rush grows to heights of 1.5–5 m.

Survival and maintenance

There is little scientific literature about the water requirements of rushes within the Murray-Darling Basin. Some information can be inferred from the distribution of species along flood gradients. Blanch *et al.* (1999b) found that the tussock rush preferred stable water levels near the river edge and that its distribution was consistent with a constant damp water regime. Tussock rush appears to tolerate water fluctuations to depths of up to 60 cm. Based on species ordination, it appears that the tussock rush is less tolerant of large flood depths (>200 cm). Roberts and

Figure 2.11: Billabong rush, *Juncus usitatus,* and tussock rush, *J. aridicola*. Photographs: Tim Ralph (Macquarie University).

Ludwig (1991) indicated that the tussock rush commonly occurs in association with river red gums, spike-rushes (*Eleocharis acuta*), spiny flat sedge (*Cyperus gymnocaulos*) and couch (*Cnydon dactylon*) in backwater and billabong settings on the Murray River. These settings are consistent with low current and wave conditions and are likely to have relatively stable water levels. Sainty and Jacobs (1981) reported that tussock rush growth depends on standing water.

The billabong rush is more prevalent throughout eastern New South Wales. Its most westerly populations are in the Macquarie Marshes area and along the Murray River (Harden 1993). Little is known of the billabong rush's optimal flood depths, but it commonly occurs close to the water's edge or in shallow water (Cunningham *et al.* 1992). Due to the narrow range of water depth in which the billabong rush occurs, it is likely to prefer relatively stable water levels; its tolerance of fluctuating water levels is unknown. Billabong rush is regarded as drought- and salt-tolerant and is able to compete with other species under saline conditions (CSIRO 2006b).

The giant rush forms monospecific stands in shallow water along the Murray River corridor. It does not grow in deep water, but tolerates flooding to depths of 1.5 m (Roberts and Marston 2000). There is little experimental information about the ecohydrology of the giant rush but it is hypothesised that the giant rush requires flooding on its roots and rhizomes and that, to enable respiration, water levels should not exceed two-thirds of the plant height (Young *et al.* 2003). Extensive stands of giant rush are located at Barmah Forest which, under natural conditions, are inundated every year for an average period of 8.7 months, with dry periods occurring approximately four times in every five years for an average duration of 3.9 months (Leitch 1989). However, reduced flood conditions in 1989 sustained these communities, despite dry periods lasting more than 10 months. It has been suggested that the giant rush is distributed along a flood gradient at locations with a mean flood duration of four months in most years (Roberts and Marston 2000) or a flood duration of four to nine months (Young *et al.* 2003). The optimal flood timing is between May and November (Young *et al.* 2003).

Other relatively common *Juncus* species within the Murray-Darling Basin include the gold rush (*J. flavidus*) and the pale rush (*J. pallidus*); however, there is relatively little information on their water requirements. The gold rush reportedly grows in seasonally or ephemerally wet locations (Harden 1993) and was observed in water depths ranging from dry to 24 cm (Cook *et al.* 2009). Roberts and Marston (2000) indicated that pale rush grows in shallow water to depths less than 1 m and where flooding occurs annually for four to eight months, or an optimal period of five months.

Reproduction and regeneration

Juncus rush species within the Murray-Darling Basin predominantly flower during spring and summer, with flowering likely to coincide with seasonal flooding (Harden 1993). This is particularly important for the tussock rush, which depends on standing water for flowering (Sainty and Jacobs 1981). Old flowers may remain on *Juncus* for some time (particularly on the billabong rush, Sainty and Jacobs 1981), but seeds are shed relatively quickly in summer and autumn. Seeds of the tussock rush, billabong rush and gold rush have been found within soil seed banks and were able to germinate under glasshouse conditions (James *et al.* 2007; King and Buckney 2001; McIntyre 1985).

Seeds of the tussock rush and billabong rush are relatively small (CSIRO 2006b; Sainty and Jacobs 1981). As in other *Juncus* species, it is likely that seeds are dispersed by water (CSIRO 2006b). Little information is available about the establishment of tussock rush, billabong rush and gold rush from seed and their growth as juveniles. However, McIntyre (1985) reported that the tussock rush and gold rush prefer germination in drained moist soil conditions. The giant rush is considered intolerant of extended flooding and regeneration is unlikely to occur in flooded areas (Chesterfield 1986). The preferred condition for germination is described as wet mud in spring and summer after the recession of late winter to spring floods (Young *et al.* 2003). While rushes are considered intolerant of extended flooding, it is probable that juveniles will die if the inter-flood dry-period is long (CSIRO 2006a).

Due to the rhizomatous nature of *Juncus* species, they may spread through vegetative growth (CSIRO 2006b), but vegetative growth may not result in spatially separated ramets (McIntyre *et al.* 1995). McIntyre *et al.* (1995) suggest that the billabong rush does not expand vegetatively.

Marsh club-rushes: *Bolboschoenus* species

Three native Bolboschoenus club-rush species occur within Australia and the Murray-Darling Basin: *Bolboschoenus medianus*, *B. caldwellii* and *B. fluviatilis*. *B. medianus* is relatively common throughout south-eastern Australia, while extensive stands of *B. fluviatilis* in the Murray-Darling Basin are largely limited to wetlands on the Gwydir River. *B. caldwelli* is a narrow-leaved perennial sedge occurring in freshwater wetlands throughout Australia and New Zealand. It has a triangular stem and may grow to a height of up to 1.2 m (Figure 2.12).

Survival and maintenance

The Bolboschoenus genus is regarded as a flood-reliant species, typically located at sites that are flooded annually for some time or are within a short distance of inundation. Blanch *et al.* (1999b) reported that *B. caldwellii* typically occurs at infrequently flooded sites located just above the waterline and flooded to depths of 60 cm, while *B. medianus* is more widespread and

Figure 2.12: Marsh club-rush, *Bolboschoenus fluviatilis*. Photograph: Jeff Kelleway (DECCW).

tolerant of flooding and exposure. *B. medianus* occurs in a range of water regimes (Blanch *et al.* 1999b), but its biomass is reportedly greatest at flood depths of +20 cm to –20 cm (Blanch *et al.* 1999a). This is supported by the experiments of Siebentritt and Ganf (2000), who reported optimal growth rates for *B. caldwellii* at water depths of +20 cm to 20 cm below the waterline, while growth of *B. medianus* was relatively strong at depths between +20 cm and –60 cm.

Both *B. caldwellii* and *B. medianus* can respond to increasing water depths by raising culm heights, presumably in an attempt to maintain their ability to absorb carbon dioxide and oxygen from the atmosphere. However, an increase in height does not appear to match rising water levels, with a greater proportion of the plant being submerged than emergent (Blanch *et al.* 1999a; Siebentritt and Ganf 2000).

Experimental studies indicate that, at water depths of 60 cm, resources are allocated to increasing tuber size; it is hypothesised that this is an adaptation to increase tuber survival if water depths continue to rise (Siebentritt and Ganf 2000). Death occurs when plants are entirely submerged (Blanch *et al.* 1999b; Siebentritt and Ganf 2000) as the lack of emergent tissue above the waterline is likely to limit plant access to atmospheric carbon dioxide, resulting in carbon starvation (Siebentritt and Ganf 2000). However, tubers may be viable for two or more years (Cizkova-Koncalova *et al.* 1992; Grace 1993) after death from inundation. Similarly, tubers may remain viable in the soil provided there is enough moisture to sustain them.

There is limited information about the ideal flood duration for optimal growth of *B. caldwellii. B. caldwellii* located on the Murray River in South Australia predominantly occurs in areas that are flooded for 87–140 days over a two-year period, or up to 163 days over a two-year period (Blanch *et al.* 1999b). This period seems to reflect the species' optimal location just above or near the waterline and may be partly controlled by competition with *B. medianus*, which occurs in the same location (Siebentritt and Ganf 2000). Field studies and observations indicate that, where the two species occupy the same site, *B. caldwellii* dominates in higher regions and has shorter flood durations than *B. medianus* (Blanch *et al.* 1999b; Siebentritt and Ganf 2000). Experimental studies indicate that, in the absence of *B. medianus*, *B. caldwellii* is able to sustain moderate yields at 60 cm water depths (Siebentritt and Ganf 2000). Presumably in the absence of *B. medianus*, *B. caldwellii* would be able to survive longer periods of inundation due to reduced competition. *B. medianus* predominantly occurs in areas that are flooded for 163–432 days over a two-year period, or up to 12 months flooding (Blanch *et al.* 1999b).

Maximum exposure periods of approximately 315 days were evident for both *B. caldwellii* and *B. medianus* on the Murray River in South Australia (Blanch *et al.* 1999b). There is some evidence to suggest increased shoot production when *B. caldwellii* is submerged (Hayball and Pearce 2004). In order to maintain flood depths and durations and to enhance shoot production, inundation should occur over the growing season from spring into summer.

Information about the water requirements of *B. fluviatilis* is inferred from the response of the species to environmental flows in the Gwydir Wetlands. McCosker (1999) indicated that rapid growth occurs with spring flooding for a period of at least two months. It is also evident that adult *B. fluviatilis* requires winter–spring flooding to promote seed set and completion of the regeneration cycle (McCosker 1999).

Reproduction and regeneration

Bolboschoenus species can reproduce both vegetatively and sexually from seed. There is some evidence to suggest that asexual reproduction in *B. medianus* and *B. caldwellii* is limited at extremes of inundation (Siebentritt and Ganf 2000). Increased water levels have been associated with a decrease in proportional allocation of biomass to tubers and with an increase in the proportion of above-ground biomass. Siebentritt and Ganf (2000) suggested that this may represent a shift from asexual reproduction to survival.

McCosker (1999) observed that 'daughter' *B. fluviatilis* plants arising from asexual reproduction did not reach sexual maturity or set seed when flooding in the year following reproduction did not meet their water requirements. Flooding in late winter at the Gwydir Wetlands promoted both sexual and asexual reproduction, and reproduction was observed to shift from sexual to asexual throughout a flooding season (McCosker 1999).

Bolboschoenus species commonly flower in spring and summer (Harden 1993) but may flower between August and March (WA Dept of Environment and Conservation 1993). McCosker (1999) indicated that seed set for *B. fluviatilis* occurs only under favourable conditions for growth, implying a flood requirement during late winter and spring. Fruit and seed production usually occurs not long afterwards, with seed commonly falling in late spring to summer (McCosker 1999). This period coincides with flood drawdown, creating moist conditions suitable for seedling establishment. Flooding in the year following establishment should be optimal to ensure the survival of *Bolboschoenus* to sexual maturity.

Sedges: *Cyperus* species

Numerous *Cyperus* species may be found within floodplain wetlands of the Murray-Darling Basin, including Downs nutgrass (*Cyperus bifax*), trim flat-sedge (*C. concinnus*), rice sedge (*C. difformis*), tall flat-sedge (*C. exaltatus*; Figure 2.13), spiny flat-sedge (*C. gymnocaulos*) and curly flat-sedge (*C. rigidellus*). Each species has its own height range: tall flat-sedge is the tallest, up to 1.8 m high (Harden 1993). All species are described as rhizomatous tufted perennials, and curly flat-sedge may also have a slender annual life-form. All species grow in seasonally wet or ephemerally wet situations such as floodplains and are commonly regarded as cotton and rice field weeds (Harden 1993).

Figure 2.13: Tall flat-sedge, *Cyperus exaltatus*. Photograph: Sharon Bowen (DECCW).

Survival and maintenance

With the exception of curly flat-sedge, tall flat-sedge and spiny flat-sedge, there is very little specific information about the optimal water regime for the growth and survival of *Cyperus* species within the Murray-Darling Basin. Studies indicate that optimal water regimes for *Cyperus* differ for each species (Blanch *et al.* 1999b, 2000) as each species has varying degrees of tolerance to flooding and exposure. Based on the distribution of species in ephemerally or seasonally wet locations, it is apparent that *Cyperus* prefer water levels fluctuating between flooded and exposed. Timing of flooding should coincide with the growth season during spring and summer.

Curly flat-sedge predominantly established in weir pools on the Lower Murray River and occurred along a flood gradient that exhibited a flood duration of 225–328 days over a two-year period and an optimal flooding depth up to 30 cm (Blanch *et al.* 2000). Inter-flood dry-periods can be inferred from measures of longest exposure: it is evident that curly flat-sedge has some tolerance to exposure, with a median value of 139 days exposed over a two-year period. However, it may survive in excess of 195 days over two years (Blanch *et al.* 2000).

Tall flat-sedge predominantly occurred on the Lower Murray River at locations with a flood depth of less than 60 cm (Blanch *et al.* 1999b), a median flood duration of 332 days over a two-year period and a range of 10–700 days over a two-year period (Blanch *et al.* 1999b, 2000). Tall flat-sedge exhibits a flood exposure duration of approximately 232–250 days over a two-year period, but may survive up to 660 days of exposure (Blanch *et al.* 1999b, 2000).

Spiny flat-sedge occurred in locations with a flood duration of 162–392 days over a two-year period or up to 691 days in a two-year period. It may survive exposure for periods of 233–290, or up to 730, days (Blanch *et al.* 1999b, 2000).

Reproduction and regeneration

Cyperus species within the Murray-Darling Basin generally flower during spring and summer, thereby coinciding with seasonal water availability (Harden 1993). Very little is known of the life-cycle of *Cyperus* species, with the exception of rice sedge. As rice sedge is regarded as one of the world's 10 most important weeds (Holm *et al.* 1977), some research into its life-cycle has been undertaken (Sanders 1994).

Seeds of rice sedge are relatively small, with a length of 0.6 mm (Navie *et al.* 1997) and a mean weight of 41 µg (Sanders 1994). Rice sedge produces abundant seed, with one study reporting up to 50 000 seeds per plant (Jacometti 1912, cited in Sanders 1994). Even under poor conditions, rice sedge has been observed producing seed (K.M. Wilson, cited in Sanders 1994). Seeds are long-lived and survive up to six years in field conditions (Sanders 1994). Due to their longevity, seeds are readily detected within seed banks. An Australian study of seed reserves in rice fields found that rice sedge seeds dominated the seed bank (McIntyre 1985).

Rice sedge has been described as 'a highly opportunistic species that will germinate and reproduce as soon as possible depending on prevailing conditions' (Sanders 1994, p. 1035). Germination studies indicate that rice sedge requires standing water for germination but, once established, seedlings may survive some time (up to 60 days) without inundation provided soil conditions remain moist (Cox 1984, cited in Sanders 1994). McIntyre (1985) supported this view, observing significantly higher rates of germination under flood conditions rather than damp soil conditions. There is some evidence that wide temperature fluctuations result in higher levels of germination and may end seed dormancy (Sanders 1994).

Growth from seedling to mature adult is rapid. Studies indicate that flowering may occur within six weeks of germination and ripe seeds may be available for dispersal after eight weeks (Sanders 1994). Rapid growth may be in response to the marked change in photosynthetic ability once a seedling has emerged from the water surface (Sanders 1994).

There is some indication that rice sedge may expand through vegetative growth over a period of only one month (Vaillant 1967, cited in Sanders 1994), however, little else is known of its ability to expand vegetatively. Due to the effort given to producing seed and developing a long-lived seed bank, it is likely that sexual reproduction is the primary form of reproduction and regeneration in rice sedge.

Spike-rushes: *Eleocharis* species

A number of *Eleocharis* species occur in floodplain wetlands of the Murray-Darling Basin: the common spike-rush (*E. acuta*; Figure 2.14), pale spike-rush (*E. pallens*), flat spike-rush (*E. plana*), small spike-rush (*E. pusilla*) and tall spike-rush (*E. sphacelata*). Most *Eleocharis* species within the Murray-Darling Basin are distributed in wet or submerged areas, such as permanent or ephemeral wetland and riparian habitats (Murphy *et al.* 2007). All species grow in moist conditions of varying degree, with flat spike-rush growing in seasonally wet locations and tall spike-rush growing in relatively still water of at least 5 m depth (Harden 1993). *Eleocharis* is considered a rhizomatous perennial sedge and grows to heights of 2–90 cm, with the exception of tall spike-rush, which grows up to 5 m (Harden 1993).

Survival and maintenance

The majority of *Eleocharis* species are limited to wet or submerged parts of permanent and ephemeral wetlands (Murphy *et al.* 2007) and spike-rush species appear to have varying degrees of tolerance and reliance on water for survival and maintenance. Information about water requirements for *Eleocharis* within the Murray-Darling Basin is largely limited to common spike-rush and tall spike-rush. No scientific literature was found about the water requirements of flat spike-rush or pale spike-rush.

Common spike-rush has been observed on the Lower Murray River in areas with infrequent flooding that remained for some time after floodwaters receded. This corresponded to a

Figure 2.14: Common spike-rush, *Eleocharis acuta*. Photograph: Jordan Iles (DECCW).

water regime characterised by flooding for a median of 163 days and a range of 88–243 days over a two-year period (Blanch *et al.* 1999b). Roberts and Ludwig (1991) found that common spike-rush was established in billabong and backwater settings where waters were relatively calm and remained for some time. This correlates well with the functional classification of common spike-rush by Casanova and Brock (2000) as 'amphibious: fluctuation-tolerators' as it does not show major morphological change in response to flooding (Nicol *et al.* 2003). Nicol *et al.* (2003) suggested that common spike-rush was an indicator species of a drier water regime, and Reid and Quinn (2004) found that abundance of common spike-rush decreased when low flow frequencies increased. Common spike-rush was found at sites in northern Victoria where shallow flooding occurred for an optimal duration of eight months, but within the range of three to 10 months (Ward 1996, cited in Roberts and Marston 2000).

Tall spike-rush reportedly prefers long periods of inundation on a semi-regular basis. It can maintain dominance over the aquatic weed Lippia (*Phyla canescens*) under wetter water regimes (Mawhinney 2004).

Little information is available about the optimal flood timing for *Eleocharis* species, but flowering appears to coincide with peak flood timing. Ward (1996, cited in Roberts and Marston 2000) indicated that optimal flood timing for common spike-rush is during spring and summer.

Tolerance of deep or shallow flooding is unique to each *Eleocharis* species. Common spike-rush was not observed on the Murray River at depths in excess of 2 m and it predominantly occurred at flood depths less than 60 cm (Blanch *et al.* 1999b). In northern Victoria, common spike-rush reportedly prefers flooding to depths less than 10 cm (Ward 1996, cited in Roberts and Marston 2000). A glasshouse study of common spike-rush found that immersion to depths of 15 cm reduced above-ground biomass, number of shoots and shoot length compared to plants maintained in damp soil conditions (Blanch and Brock 1994). This is supported by the distribution of common spike-rush at lakes in New Zealand, where it was observed at relatively shallow depths of less than 25 cm or entirely above the waterline (Tanner *et al.* 1986).

Tall spike-rush and small spike-rush appear to have a wider depth tolerance than common spike-rush, with many tall spike-rush observed at depths ranging from 0 m to 2.25 m (Tanner *et al.* 1986) and small spike-rush observed at depths of 0.5 m to 3.5 m (Wells *et al.* 1998). Tall spike-rush biomass increased at shallow water depths of 32 cm, compared to water levels of 6 cm below the substrate (Sorrell *et al.* 2002). However, both tall and small spike-rushes are regarded as a short-growing shallow-water species at lake settings in New Zealand (Wells *et al.* 1998). While small spike-rush has a wider water depth tolerance, it is likely that its optimal range is quite shallow (Bell and Clarke 2004). Sorrell and Tanner (2000) indicated that convective flow of oxygen decreases with water depth, thereby limiting the occurrence of tall spike-rush in greater water depths.

In terms of tolerance of dry conditions, common spike-rush appears able to survive relatively long periods of up to 290 days without inundation (Blanch *et al.* 1999b). It can be inferred from the distribution of tall spike-rush and its preference for a wetter water regime (Mawhinney 2004) that it has a high water requirement and lower tolerance of dry conditions. The presence of small spike-rush seeds at temporary freshwater wetlands in arid Australia and their absence from saline or permanently inundated wetlands indicate that small spike-rush has some tolerance of dry conditions (Porter *et al.* 2007).

Common spike-rush appears to have some salinity tolerance, but growth was adversely affected at salinities in excess of 1000 mg L^{-1}. At salinities of up to 5000 mg L^{-1} height extension continued, but at a slower rate. Heights of common spike-rush declined at salinities greater than 7000 mg L^{-1}, as the stems had died (James and Hart 1993). Seed bank studies of fresh and saline wetlands in arid Australia found that small spike-rush seeds were absent from saline wetlands (Porter *et al.* 2007).

Reproduction and regeneration

Eleocharis species tend to flower during spring and summer in the Murray-Darling Basin and flowering may occur in response to flooding. Seeds appear to mature by autumn. Based on seed densities in the soil, it is likely that *Eleocharis* seeds persist within soil seed banks (Bell and Clarke 2004). This is supported by recruitment studies from soil seed banks which have shown the emergence of small spike-rush (Brock 1998; Porter *et al.* 2007), common spike-rush (Brock 1998; Nicol *et al.* 2003) and tall spike-rush (Brock 1998). These seeds appear to remain viable within seed banks for a consistently long period and exhibit a half-life in excess of 50 years (Bell and Clarke 2004).

Spike-rush seeds appear to require fluctuating temperatures and light for optimal germination. In addition, some *Eleocharis* seeds may undergo a period of seed dormancy (Bell and Clarke 2004).

Water depths required for germination of seeds reflects the distribution of mature spike-rush along a water gradient. Germination of common and small spike-rush seeds was restricted to shallow water depths of 0–18 cm. Tall spike-rush seedlings emerged at a greater range of water depths, with greatest emergence at water depths of approximately 30 cm (Bell and Clarke 2004) and seedlings have been observed germinating at depths of 45 cm (M. Casanova, cited in Bell and Clarke 2004). Common and small spike-rushes prefer drawdown conditions for establishment (Bell and Clarke 2004), but establishment may be inhibited when drawdown is rapid (Nicol *et al.* 2003). Rapid drawdown may exceed the rate of root extension in common spike-rush, or water levels may fall rapidly below the depth of root development.

While all species of spike-rush may be found at greater water depths than those in the germination experiment of Bell and Clarke (2004), this is likely to result from vegetative expansion rather than from germination at great water depths. Recruitment of seedlings of common, small and tall spike-rushes primarily occurs during spring, with some recruitment during autumn (Bell and Clarke 2004). It is likely that water, temperature and light requirements are suitable for germination at these times.

Aquatic macrophytes

Ribbonweed: *Vallisneria* species

The taxonomy of ribbonweed species within the Murray-Darling Basin is problematic: specimens have been referred to as *Vallisneria gigantea, V. spiralis* or *V. americana* (Jacobs and Frank 1997). The taxonomy of *Vallisneria* is in need of revision, but for the purposes of this chapter all Australian references to the species will be accumulated and referred to as ribbonweed. Ribbonweed is described as a stoloniferous perennial with strap-shaped leaves extending to 3 m length (Harden 1993; Figure 2.15). Leaves may extend to 5 m length (Roberts and Marston 2000). Ribbonweed typically occurs in stationary or flowing fresh water up to 7 m depth, and flowers during warmer months (Harden 1993).

Survival and maintenance

As a submerged aquatic macrophyte, ribbonweed is dependent on water availability for its survival and maintenance and is regarded as exposure-intolerant (Blanch *et al.* 1999b). Although ribbonweed is regarded as a perennial (Aston 1973; Harden 1993; Sainty and Jacobs

Figure 2.15: Ribbonweed, *Vallisneria* spp. Photograph: S. Jacobs. ©Royal Botanic Gardens and Domain Trust, Sydney.

2003) its growth cycle is similar to that of an annual species (Briggs and Maher 1985) in that canopy development must be complete before individuals can reproduce and that there is a marked decline in growth once reproduction has ceased. Briggs and Maher (1985) found that canopy growth commenced in spring and peaked in the warmer months of summer and autumn, with maximum growth at temperatures of 25°C or more (Sainty and Jacobs 1981). However, when summer drying of waters is extensive, growth may cease. Growth appears to be temperature-dependent, with dieback of leaves over winter (Briggs and Maher 1985). Leaf stumps may overwinter at temperatures of 5°C or more (Sainty and Jacobs 1981).

Ribbonweed appears to have a wide tolerance of flood frequency and has been observed growing under permanently flooded stable water levels (Blanch *et al.* 1999b), at sites which exhibit drying cycles (Briggs and Maher 1985; Crosslé and Brock 2002) or with complete drying following one growing season. However, it is apparent that flooding should occur at least annually.

Ribbonweed populations can be maintained under a permanent flooding regime, and flood duration is an important component of the water regime where flooding is seasonal. Flood duration should be long enough for ribbonweed to complete its growth cycle. Maximum biomass has been reported with flood durations that cover the growing season from spring to autumn; growth may prematurely cease when flooding does not extend throughout summer (Briggs and Maher 1985).

Flood timing should coincide with initiation of growth, and observations of ribbonweed growth indicate that cover is greatest when flooded during summer rather than spring (Nielsen and Chick 1997). This correlates with the observations of Sainty and Jacobs (1981) that summer growth of ribbonweed commonly followed the spring growth of some pondweed (*Potamogeton*) species.

Ribbonweed can occur at a range of flood depths. Specimens have been observed in waters to depths of 7 m (Harden 1993) and in depths as shallow as damp conditions, provided they are not sustained (Crosslé and Brock 2002). The optimal depth range appears to be less than 1–2 m (Briggs and Maher 1985; Blanch *et al.* 1999b). Depth tolerance depends on light attenuation, water turbidity (Walker *et al.* 1994; Blanch *et al.* 1998) and reductions in dissolved oxygen associated with shading (Morris *et al.* 2004). Where turbidity is in the range of 100–600 NTUs, growth will only be sustained at depths of less than 1 m (Walker *et al.* 1994).

Ribbonweed does not have a preference or requirement for drying or drawdown conditions. However, where populations behave as annuals, maintenance may occur with a maximum drying period of a few months over winter (Briggs and Maher 1985). In tank trials, ribbonweed was not sustained under 16-week damp conditions, but was able to survive and in some cases reproduce when drying occurred for shorter periods (Crosslé and Brock 2002).

Reproduction and regeneration

Ribbonweed appears able to reproduce and regenerate both vegetatively and sexually from seed. However, as a stoloniferous plant, ribbonweed primarily regenerates through vegetative growth (Sainty and Jacobs 1981). The trigger for expansion is unknown, but it has been observed during the primary growing season (Roberts and Marston 2000) and is therefore likely to be temperature-dependent.

Requirements for sexual reproduction are unknown. Flowering and seed production appear to be rapid (Crosslé and Brock 2002) and seeds may be maintained within soil seed banks (Britton and Brock 1994). Germination is rapid, does not appear to be dependent on light (Aston 1973) and may occur in any season (Crosslé and Brock 2002). Germination from seed occurred in tank trials at water depths of 60 cm but not under damp conditions (Crosslé and Brock 2002).

Herbs and forbs

Isotomes: *Isotoma* species

A number of *Isotoma* species occur in the Murray-Darling Basin, the most prevalent being the rock isotome (*I. axillaris*), swamp isotome (*I. fluviatilis*; Figure 2.16) and *I. tridens*. All *Isotoma* species are regarded as perennial herbs. Rock isotome generally flowers between September and May and grows in sandy soil on slopes and around rock waterholes, while swamp isotome flowers in late spring to summer and grows in moist sand or mud near stream edges or in seepage areas, and *I. tridens* flowers between November and May and grows at the edge of lakes, swamp and streams with slow-moving water (Harden 1992).

Little is known of the water requirements of isotomes. Ward (1996, cited in Roberts and Marston 2000) indicated that *I. fluviatilis* is found in areas of northern Victoria where flooding is shallow (<10 cm) and has a relatively short duration of one to three months during spring and summer. *I. fluviatilis* is regarded as an amphibious tolerating species (Casanova and Brock 2000; Leck and Brock 2000).

Primrose: *Ludwigia* species

The water primrose (*Ludwigia peploides*; Figure 2.17) and willow primrose (*L. octovalvis*) are prostrate perennial herbs that grow on the margins of lakes and streams or in wet to seasonally

Figure 2.16: Swamp isotome, *Isotoma fluviatilis*. Photograph: D. Hardin. ©Royal Botanic Gardens and Domain Trust, Sydney.

Figure 2.17: Water primrose, *Ludwigia peploides*. Photograph: Steve Jacobs (DECCW).

Figure 2.18: Nardoo, *Marsilea drummondii,* and water couch, *Paspalum distichum,* complex. Photograph: Tim Ralph (Macquarie University).

wet locations (Harden 1991). The water primrose flowers in summer and autumn, while the willow primrose flowers between summer and winter.

In northern Victoria, the water primrose occurs where flooding lasts for extended periods of eight to 10 months throughout winter to spring (Ward 1996, cited in Roberts and Marston 2000).

Seeds are able to germinate under water and on wet soils, in suitable light and temperature conditions. Germination is optimal at 30°C and absent at temperatures of 10°C. Germination is responsive to flooding, occurring within a day of flooding (Yen and Myerscough 1989). Based on the optimal temperature requirement for germination, it is likely that optimal flood timing for regeneration is during late spring and summer.

Nardoo: *Marsilea* species

Five species of the aquatic fern nardoo occur within the Murray-Darling Basin: narrow-leaf nardoo (*Marsilea costulifera*), common nardoo (*M. drummondii*), *M. exarata*, *M. hirsute* and *M. mutica*. All species are described as perennial with creeping rhizomes and floating leaves. Common nardoo is the most prevalent *Marsilea* species throughout the Murray-Darling Basin (Harden 1990; Figure 2.18). While there is relatively little literature available on most nardoo species, their distribution overlaps the range of common nardoo, on which there is some ecological literature.

Survival and maintenance

The distribution of common nardoo in moist sites indicates that it prefers damp mud, but it has been observed in floodwaters up to 1 m depth (Roberts and Marston 2000). Common nardoo has also been observed persisting in drying muds (CSIRO 2004b). Sainty and Jacobs (2003) referred to common nardoo as a 'resurrection plant' as apparently-dead plants can rehydrate and grow when moisture becomes available. Under drought conditions, leaves of common nardoo shrivel and the sporocarps detach, frequently lodging in cracks in soils (CSIRO 2004b). Common nardoo can persist as a perennial if conditions remain moist, otherwise it behaves as an annual (Cunningham *et al.* 1981). In northern Victoria, common nardoo occurs where shallow flooding of less than 10 cm lasts for one to six months (Ward 1996, cited in Roberts and Marston 2000). It also exhibits some degree of salinity tolerance, with a reported salinity range of 0.2–0.3 ppt total dissolved salts (Yezdani 1970, cited in Aston 1973).

Reproduction and regeneration

Fruit production generally occurs during drawdown conditions on drying muds (CSIRO 2004b). Sporocarp production has been recorded between November and March (Aston 1973). Spores are contained within sporocarps which split upon immersion in water. Under dry conditions, sporocarps may lodge in cracks in soils and remain viable for 20–30 years (CSIRO 2004b). This enables common nardoo spores to regenerate in response to favourable conditions, despite the possible absence of established common nardoo. Regeneration may be rapid, with sporocarps forming within three months (Cunningham *et al.* 1981).

Marshwort: *Nymphoides* species

Wavy marshwort *(Nymphoides crenata)* is one of five *Nymphoides* species within the Murray-Darling Basin, but is regarded as having the greatest abundance and distribution within the Basin. Other species include entire marshwort (*N. geminata*), water snowflake (*N. indica*), marshwort (*N. montana*) and marbled marshwort (*N. spinulosperma*). Marshwort and entire marshwort are regarded as rare in Australia, and wavy marshwort is listed as threatened in Victoria (*Flora and Fauna Guarantee Act 1988* (Vic.)). Marbled marshwort is listed as endangered in New South Wales (*Threatened Species Act 1995* (NSW)) and threatened in Victoria. Wavy marshwort is described as a robust perennial with floating leaves, distributed in slow-moving water to depths of approximately 1.5 m (Figure 2.19). It typically prefers a mud

Figure 2.19: Wavy marshwort, *Nymphoides crenata*. Photograph: Steve Jacobs (DECCW).

substrate, but has the ability to persist on drying mud (Aston 1973; Sainty and Jacobs 1981, 2003). Flowering and fruiting occurs between September and May (Sainty and Jacobs 2003).

Little literature exists about the water requirements of wavy marshwort. Ward (1996, cited in Roberts and Marston 2000) observed wavy marshwort where winter to summer flooding occurred for nine to 10 months to depths of up to 1 m. Lloyd *et al.* (2003) referred to the decline of wavy marshwort due to a lack of summer drying. It is unconfirmed whether wavy marshwort requires drying, to enable reproduction.

Pratia species

Two *Pratia* species are regarded as relatively common within the Murray-Darling Basin: poison pratia (*Pratia concolor*; Figure 2.20) and whiteroot (*P. purpurascens*). Both are described as dioecious perennial herbs. Poison pratia usually occurs on heavy soil in moist depressions and flowers between January and April, while whiteroot grows in shady wet areas such as wet sclerophyll forest, woodland and grassland, and flowers between October and June (Harden 1992).

Blanch *et al.* (1999b) referred to poison pratia occurring in areas that are infrequently flooded, where flooding does not exceed 60 cm and occurs for short periods of 47–106 days over a two-year period or up to 162 days over a two-year period. Poison pratia was observed to survive exposure for periods of 290–730 days over a two-year period. Ward (1996, cited in

Figure 2.20: Poison pratia, *Pratia concolor*. Photograph: Jordan Iles (DECCW).

Roberts and Marston 2000) observed poison pratia in areas of northern Victoria where shallow flooding of less than 10 cm occurred in spring and summer for periods of one to three months. Little else is known of the water requirements of *Pratia* species.

Ranunculus species

Numerous *Ranunculus* species may be found in the Murray-Darling Basin, including a number of introduced species such as the sharp buttercup (*R. muricatus*) and celery buttercup (*R. sceleratus*). River buttercup (*R. inundatus*) is most prevalent in floodplain wetlands of the Murray-Darling Basin, but ferny buttercup (*R. pumilio*) and swamp buttercup (*R. undosus*; Figure 2.21) are also relatively common. River buttercup is described as a perennial herb growing to heights of 30 cm, which generally flowers in spring and summer and occurs on wet muds or in ponds and streams. Ferny buttercup is an annual herb growing to heights of 40 cm, flowering between late winter and early summer and situated in intermittently moist sites. Swamp buttercup is a perennial herb growing to heights of 45 cm, flowering in spring and summer and situated on mud in intermittently wet sites (Harden 1990).

Based on the occurrence of these species in moist or intermittently wet sites (Harden 1990), it is likely that all have a flooding requirement for survival. Information about the water requirements of *Ranunculus* species is limited to the river buttercup located in northern Victoria (Ward 1996, cited in Roberts and Marston 2000). River buttercup was found to occur where shallow flooding of less than 10 cm occurred during winter and spring for one to nine months, the optimal duration being three months. River buttercup is regarded as an amphibious tolerator within the wetting and drying model (Brock and Casanova 1997; Leck and Brock 2000).

Summary of water requirements

Floodplain wetland plant species within the Murray-Darling Basin exhibit unique survival and maintenance niches (Table 2.1) and reproduction and regeneration niches (Table 2.2) within a

Figure 2.21: Swamp buttercup, *Ranunculus undosus*. Photograph: Tim Ralph (Macquarie University).

floodwater gradient. The plants may be classified on the basis of functional strategies for coping with flood and drought stress or cycles of wetting and drying. The CSR model (Grime 2001; Menges and Waller 1983), environmental sieve model (van der Valk 1981) and wetting and drying model (Brock and Casanova 1997; Casanova and Brock 2000) have been applied to species profiled in this chapter (Tables 2.3 and 2.4). It should be emphasised that the water requirements provided in this study are a general guide based on information within scientific literature, including ecohydrology literature from study sites outside the Murray-Darling Basin. Actual water requirements for specific wetlands within the Murray-Darling Basin may differ from those presented here, due to site-specific factors that influence the ability of vegetation to respond to flooding.

The response of plants to flooding and drought has a significant influence on the distribution of species across a flood gradient. While plant species exhibit unique adaptations and responses to flooding, it is apparent that some species exhibit similar water requirements and therefore may occur as plant associations within a flood gradient. Plant associations based on water requirements are in Table 2.5. These plant associations result in distinct zonation patterns along a flood gradient (Figure 2.22), which may be broadly grouped into three categories:

- permanent wetlands, which exhibit permanent to near-permanent flooding and are characterised by aquatic plants such as *Vallisneria* species and *Nymphoides crenata*;
- semi-permanent wetlands, which exhibit flooding on an annual basis and are characterised by *Paspalum distichum*, *Phragmites australis*, *Ludwigia peploides* and species of *Typha*, *Juncus*, *Bolboschoenus*, *Cyperus*, *Isotoma*, *Marsilea* and *Ranunculus*;
- ephemeral wetlands, which exhibit irregular flooding and significant drying periods and are characterised by *Muehlenbeckia florulenta*, *Eucalyptus camaldulensis*, *E. largiflorens*, *E. coolabah* and other dryland species.

Understanding the water regime requirements of floodplain wetland plants allows us to estimate the response of species to water regime alterations. Preliminary predictions of the

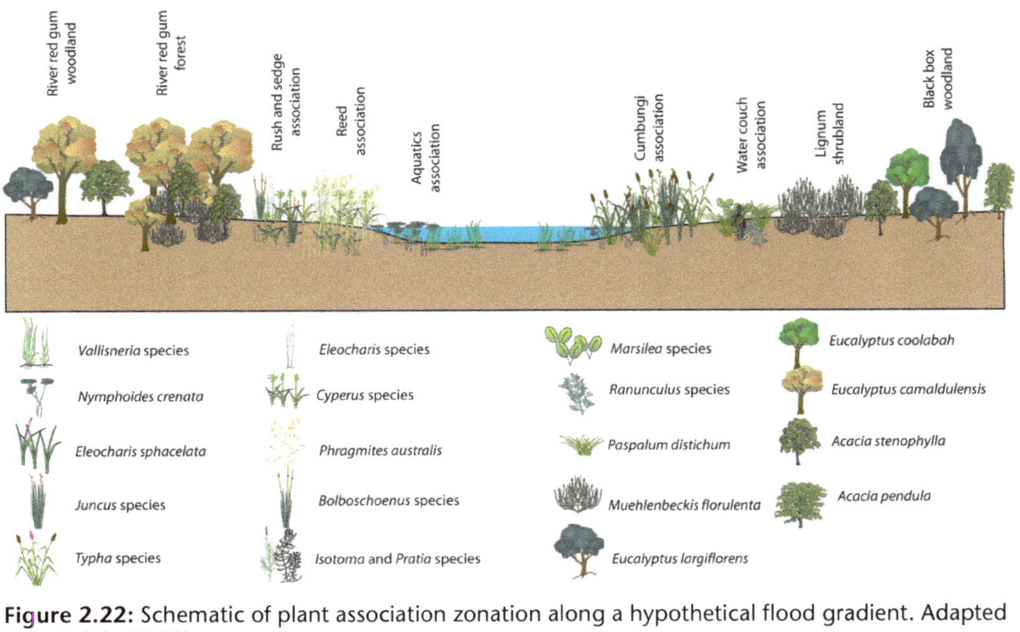

Figure 2.22: Schematic of plant association zonation along a hypothetical flood gradient. Adapted from Ralph (2008).

Table 2.1: Summary of the water requirements for the maintenance and survival of selected plant species in the Murray-Darling Basin

Plant group	Common name	Scientific name	Ideal flood frequency	Ideal flood duration	Maximum flood duration	Ideal flood timing	Maximum flood timing	Ideal flood depth	Maximum flood depth	Ideal inter-flood dry-period	Maximum inter-flood dry-period
Trees	River red gum	*Eucalyptus camaldulensis*	1–3 years[1,2]	2–8 months[1,2]	24 months[2]	Winter to spring	Winter to early summer	na	na	5–15 months	36–48 months[2]
	Black box	*Eucalyptus largiflorens*	1 in 2–5 years[3]	2–4 months	5 months	Any[4]	Any[4]	na	na	Variable[5]	Unknown[5]
	Coolibah	*Eucalyptus coolabah*	1 in 10–20 years[6]	2–5 weeks	Unknown[7]	Summer to autumn	Any	na	na	10–20 years	10–20 years
	Weeping myall	*Acacia pendula*	Unknown[8]	Unknown[8]	Unknown[8]	Unknown[8]	Unknown[8]	na	na	Unknown[8]	Unknown[8]
	River cooba	*Acacia stenophylla*	Unknown[9]	Unknown[9]	Unknown[9]	Unknown[9]	Unknown[9]	na	na	Unknown[9]	Unknown[9]
Shrubs	Lignum	*Muehlenbeckia florulenta*	3–10 years	1–6 months	12 months	Spring to early summer	–	0–60 cm	–	1–10 years	–
Grasses	Water couch	*Paspalum distichum*	Annual[10]	1–2 months or 299–440 days/2 years	163–513 days/2 years	Summer	Spring to summer	<60 cm	200 cm	236 days	290 days
	Common reed	*Phragmites australis*	1–2 years[11]	~6 months	12 months or permanent	Spring[12]	Any	Static: –20 to 50 cm Fluctuating: ±30 cm	200 cm	Few months	12 months[11]
Sedges and rushes	Broad-leaf cumbungi	*Typha orientalis*	Annual	9–12 months	12 months or permanent	Late winter to summer	Any	0–200 cm	200 cm	0–3 months	3–4 months[13]
	Narrow-leaved cumbungi	*Typha domingensis*	Annual	9–12 months	12 months or permanent	Late winter to summer	Any	0–200 cm	200 cm	0–3 months	3–4 months[13]
	Gold rush	*Juncus flavidus*	Annual	Unknown[14]	Unknown[14]	Late winter to summer	Any	0–24 cm	Unknown	Unknown[14]	Unknown[14]
	Giant rush	*Juncus ingens*	Annual	4–9 months	Unknown	May to November	Any	0–150 cm	150 cm	3–4 months	10 months
	Tussock rush	*Juncus aridicola*	Annual	12 months	12 months or permanent	Late winter to summer	Any	0–60 cm	200 cm	0 months	Unknown
	Billabong rush	*Juncus usitatus*	Annual	Unknown[15]	12 months or permanent	Late winter to summer	Any	Shallow	Unknown	0 months	Unknown

Plant group	Common name	Scientific name	Ideal flood frequency	Ideal flood duration	Maximum flood duration	Ideal flood timing	Maximum flood timing	Ideal flood depth	Maximum flood depth	Ideal inter-flood dry-period	Maximum inter-flood dry-period
	Pale rush	*Juncus pallidus*	Annual	5 months	4–8 months	Winter to spring	Any	Shallow	<100 cm	4–8 months	Unknown
	Marsh club-rush	*Bolboschoenus caldwellii*	Annual and fluctuating[16]	1–3 months or 87–140 days/2 years	3 months or 22–163 days/2 years	Spring to summer	Any	±20 cm	60 cm	4–6 months or 290–313 days/2 years	6 months or 313 days/2 years
	Marsh club-rush	*Bolboschoenus fluviatilis*	Annual and fluctuating[16]	2 months	Unknown	Winter to spring	Any	Unknown	Unknown	Unknown	Unknown
	Marsh club-rush	*Bolboschoenus medianus*	Annual and fluctuating[16]	2–8 months or 163–432 days/2 years	Permanent or 12–730 days/2 years	Late winter to summer	Any	–20 to 60 cm	Up to 200 cm	3–6 months or 226–315 days/2 years	6 months or 315 days/2 years
	Downs nutgrass	*Cyperus bifax*	Annual and fluctuating	Unknown	Unknown	Spring to summer	Any	Unknown	Unknown	4 months	11–12 months
	Trim flat-sedge	*Cyperus cincinnus*	Annual and fluctuating	Unknown	Unknown	Unknown	Any	Unknown	Unknown	Unknown	Unknown
	Rice sedge	*Cyperus diffomis*	Annual and fluctuating	Unknown	Unknown	Unknown	Any	Unknown	Unknown	Unknown	Unknown
	Tall flat-sedge	*Cyperus exaltus*	Annual and fluctuating	4–7 months or 273–399 days/2 years	~12 months or 10–700 days/2 years	Late winter to summer	Any	<60 cm	Up to 200 cm	1–11 months or 9–660 days/2 years	11 months or 660 days/2 years
	Spiny flat-sedge	*Cyperus gymnaucolos*	Annual and fluctuating	2–7 months or 162–392 days/2 years	~12 months or 24–691 days/2 years	Late winter to summer	Any	<60 cm	Up to 200 cm	1–12 months or 14–730 days/2 years	~12 months or 730 days/2 years
	Curly flat-sedge	*Cyperus rigidellus*	Annual and fluctuating	3–6 months or 225–328 days/2 years	Unknown	Late winter to summer	Any	<30 cm	Unknown	0–4 months or 3–195 days/2 years	4 months or 195 days/2 years
	Common spike-rush	*Eleocharis acuta*	Annual and fluctuating	<8 months or 88–243 days/2 years	3–10 months	Spring to summer	Any	Drawdown or <60 cm	<200 cm	4 months[1]	10 months
	Pale spike-rush	*Eleocharis pallens*	Annual and fluctuating	Unknown	Unknown	Unknown[17]	Any	Unknown	Unknown	Unknown	Unknown
	Flat spike-rush	*Eleocharis plana*	Annual and fluctuating	Unknown	Unknown	Unknown[17]	Any	Unknown	Unknown	Unknown	Unknown
	Small spike-rush	*Eleocharis pusilla*	Annual and fluctuating	Unknown	Unknown	Unknown[17]	Any	0.5–350 cm^3	<350 cm	<4 months	<4 months

Plant group	Common name	Scientific name	Ideal flood frequency	Ideal flood duration	Maximum flood duration	Ideal flood timing	Maximum flood timing	Ideal flood depth	Maximum flood depth	Ideal inter-flood dry-period	Maximum inter-flood dry-period
	Tall spike-rush	*Eleocharis sphacelata*	Annual and fluctuating	8–12 months	Unknown	Unknown[17]	Any	<225 cm^2	<225 cm	<4 months	<4 months
Aquatic macro-phytes	Ribbon-weed	*Vallisneria gigantea, V. spiralis, V. americana*[18]	Annual	9–12 months[19]	12 months	Spring to summer	Any	<200 cm	<700 cm	0 months	<4 months[19]
Herbs and forbs	Rock isotome	*Isotoma axillaris*	Annual	1–3 months	Unknown	Spring to summer	Any	<10 cm	Unknown	Unknown	Unknown
	Swamp isotome	*Isotoma fluviatilis*	Annual	Unknown	Unknown	Unknown	Any	Unknown	Unknown	Unknown	Unknown
	Isotome species	*Isotoma tridens*	Annual	Unknown	Unknown	Unknown	Any	Unknown	Unknown	Unknown	Unknown
	Water primrose	*Ludwigia peploides*	Annual	8–10 months	Unknown	Winter to spring	Any	Unknown	Unknown	Unknown	Unknown
	Willow primrose	*Ludwigia octovalvis*	Annual	Unknown	Unknown	Unknown	Any	Unknown	Unknown	Unknown	Unknown
	Common nardoo	*Marsilea drummondii*	Annual	1–6 months	Unknown	Late winter to summer	Any	Damp, <10 cm	Drying mud to <100 cm	0–6 months[20]	Unknown
	Narrow-leaf nardoo	*Marsilea costulifera*	Unknown	Unknown	Unknown	Unknown	Unknown	Unknown	Unknown	Unknown	Unknown
	Marsilea species	*Marsilea exarata*	Unknown	Unknown	Unknown	Unknown	Unknown	Unknown	Unknown	Unknown	Unknown
	Marsilea species	*Marsilea hirsute*	Unknown	Unknown	Unknown	Unknown	Unknown	Unknown	Unknown	Unknown	Unknown
	Marsilea species	*Marsilea mutica*	Unknown	Unknown	Unknown	Unknown	Unknown	Unknown	Unknown	Unknown	Unknown
	Wavy marshwort	*Nymphoides crenata*	Annual	9–12 months	Unknown	Winter to summer	Unknown	<100 cm	Unknown	Unknown[21]	Unknown[21]
	Entire marshwort	*Nymphoides geminata*	Unknown	Unknown	Unknown	Unknown	Unknown	Unknown	Unknown	Unknown	
	Water snowflake	*Nymphoides indica*	Unknown	Unknown	Unknown	Unknown	Unknown	Unknown	Unknown	Unknown	

Plant group	Common name	Scientific name	Ideal flood frequency	Ideal flood duration	Maximum flood duration	Ideal flood timing	Maximum flood timing	Ideal flood depth	Maximum flood depth	Ideal inter-flood dry-period	Maximum inter-flood dry-period
	Marshwort	*Nymphoides montana*	Unknown	Unknown	Unknown	Unknown	Unknown	Unknown	Unknown	Unknown	
	Marbled marshwort	*Nymphoides spinulosperma*	Unknown	Unknown	Unknown	Unknown	Unknown	Unknown	Unknown	Unknown	
	Poison pratia	*Pratia concolor*	Annual	1–3 months or 47–106 days/2 years	3 months or 162 days/2 years	Spring to summer	Any	<10 cm	<60 cm	4–12 months or 290–730 days/2 years	~12 months or 730 days/2 years
	Whiteroot	*Pratia purpurascens*	Unknown	Unknown	Unknown	Unknown	Unknown	Unknown	Unknown	Unknown	Unknown
	River buttercup	*Ranunculus inundatus*	Annual	3 months	1–9 months	Winter to spring	Unknown	<10 cm	Unknown	Unknown	Unknown
	Ferny buttercup	*Ranunculus pumilio*	Unknown	Unknown	Unknown	Unknown	Unknown	Unknown	Unknown	Unknown	Unknown
	Swamp buttercup	*Ranunculus undosus*	Unknown	Unknown	Unknown	Unknown	Unknown	Unknown	Unknown	Unknown	Unknown

[1]Frequency and duration reduced when watertable is shallow or trees have access to permanent water. Frequency and duration also reduced when setting is more characteristic of a woodland, compared to a forest. [2]Flood frequency and inter-flood dry-period may be higher for woodlands, compared to forests, while duration may be reduced. [3]Minimum flood frequency of 1 in 8–10 years. [4]No evidence that timing influences growth. [5]Dependent on groundwater availability, depth and salinity. Observations in excess of 30 years. [6]Increased frequency may ensure survival at saline sites. [7]Death when waterlogged. [8]Unknown, regarded as drought- and flood-tolerant. [9]Unknown, regarded as drought- and flood-tolerant. Water regime requirements suspected to lie between those for river red gum and black box. [10]Rhizomes may survive longer frequencies. [11]Survives without flooding when there is access to other water, e.g. shallow groundwater. [12]Inferred from seasonal flood regime at Macquarie Marshes. [13]Rhizomes may remain viable with dry-periods of 2–3 years. [14]Likely to exhibit some drought tolerance due to occurrence in ephemerally wet locations [15]Regarded as drought- and salt-tolerant. Tolerance of fluctuating water levels unknown. [16]Tubers may survive 2 years without flooding; minimum flood frequency of 1–2 years. [17]Optimal flood timing likely to be late winter to summer to coincide with growing and flowering season. [18]Vallisneria species are collectively referred to as ribbonweed due to taxonomic ambiguity. Taxonomy of Vallisneria requires revision. [19]May grow according to annual growth habit where drying occurs in the winter months only. [20]Plants with dead appearance can rehydrate when water becomes available. Otherwise readily regenerates from spores lodged in deep cracks in soils. 21May require some summer drying to promote seedling reproduction.

Table 2.2: Summary of the water requirements for the reproduction and regeneration of selected plant species in the Murray-Darling Basin

Plant group	Common name	Scientific name	Ideal germination timing	Maximum germination timing	Ideal flood duration	Maximum flood duration	Ideal flood timing	Maximum flood timing	Ideal flood depth	Maximum flood depth
Trees	River red gum	*Eucalyptus camaldulensis*	Spring to early summer	Late winter to mid-summer	2–5 weeks	14 weeks	Late spring to early summer	Late winter to mid-summer	5–50 cm	200 cm
	Black box	*Eucalyptus largiflorens*	North: May to October South: November to March	Any	30 days	30–60 days	November to March[1]	Any	1–30 cm	No complete immersion
	Coolibah	*Eucalyptus coolabah*	Late summer	Any	Unknown[1]	Unknown[1]	Late summer	Any	Unknown[1]	Unknown[1]
	Weeping myall	*Acacia pendula*	Unknown[1]	Unknown[1]	Unknown[1]	Unknown[1]	Unknown[1]	Unknown[1]	Unknown[1]	Unknown[1]
	River cooba	*Acacia stenophylla*	Spring to early summer	Any	Unknown[1]	Unknown[1]	Unknown[1]	Unknown[1]	Unknown[1]	Unknown[1]
Shrubs	Lignum	*Muehlenbeckia florulenta*	Spring to summer	Winter to summer	8–12 weeks	Unknown	Spring to summer	Winter to summer[2]	0–60 cm	Unknown
Grasses	Water couch	*Paspalum distichum*	Summer to autumn	Unknown	<3 months	Unknown	Spring to summer	Unknown	0 cm^3 or drawdown	Few cm[4]
	Common reed	*Phragmites australis*	Spring to summer	Any	Variable[5]	<7 weeks or variable[6]	Spring to summer	Unknown	Drawdown[5]	Variable[6] Substantial[7]
Sedges and rushes	Broad-leaf cumbungi	*Typha orientalis*	Summer to autumn	Any	Drawdown[8] 9–12 months[9,10]	10–12 weeks[8] 12 months[9,10]	Spring to summer	Unknown	Drawdown[8] 5–15 cm^9 0–10 cm[10]	Up to 80 cm Unknown[9] Variable[10]
	Narrow-leaved cumbungi	*Typha domingensis*	Summer to autumn	Any	Drawdown[8] 9–12 months[9,10]	10–12 weeks[8] 12 months[9,10]	Spring to summer	Unknown	Drawdown[8] 5–15 cm^9 0–10 cm[10]	Up to 80 cm Unknown[9] Variable[10]
	Gold rush	*Juncus flavidus*	Summer to autumn	Any	Drawdown[8]	Unknown[11]	Spring to summer	Spring to summer	Unknown[12]	Unknown[12]
	Giant rush	*Juncus ingens*	Summer to autumn	Any	Drawdown[8]	Unknown[11]	Spring to summer	Spring to summer	Unknown[12]	Unknown[12]
	Tussock rush	*Juncus aridicola*	Summer to autumn	Any	Drawdown[8]	Unknown[11]	Spring to summer	Spring to summer	Unknown[12]	Unknown[12]

Plant group	Common name	Scientific name	Ideal germination timing	Maximum germination timing	Ideal flood duration	Maximum flood duration	Ideal flood timing	Maximum flood timing	Ideal flood depth	Maximum flood depth
	Billabong rush	*Juncus usitatus*	Summer to autumn	Any	Drawdown[8]	Unknown[11]	Spring to summer	Spring to summer	Unknown[12]	Unknown[12]
	Pale rush	*Juncus pallidus*	Summer to autumn	Any	Drawdown[8]	Unknown[11]	Spring to summer	Spring to summer	Unknown[12]	Unknown[12]
	Marsh club-rush	*Bolboschoenus caldwellii*	Summer to autumn	Spring to autumn	Drawdown[13] 1–3 months or 87–140 days/2 years[14]	Unknown[13,15]	Spring to summer		Drawdown[13] ±20 cm[14]	Unknown[13] 60 cm[14,15]
	Marsh club-rush	*Bolboschoenus fluviatilis*	Summer to autumn	Spring to autumn	Drawdown[13] 2 months[14]	Unknown[13,15]	Late winter to spring		Drawdown[13] Unknown[14]	Unknown[13,14,15]
	Marsh club-rush	*Bolboschoenus medianus*	Summer to autumn	Spring to autumn	Drawdown[13] 2-8 months or 163–432 days/2 years[14]	Unknown [13, 15]	Spring to summer		Drawdown[13] -20–60 cm[14]	Unknown[13] Up to 200 cm [14, 15]
	Downs nutgrass	*Cyperus bifax*	Spring to summer	Unknown	Unknown[16,17]	Unknown[16,17]	Spring to summer	Unknown	Unknown[16,17]	Unknown[16,17]
	Trim flat-sedge	*Cyperus cincinnus*	Spring to summer	Unknown	Unknown[16,17]	Unknown[16,17]	Spring to summer	Unknown	Unknown[16,17]	Unknown[16,17]
	Rice sedge	*Cyperus diffomis*	Spring to summer	Any	Unknown[16,17]	Unknown[16,17]	Spring to summer	Any	1–3 cm[16] Unknown[17]	Unknown[16,17]
	Tall flat-sedge	*Cyperus exaltus*	Spring to summer	Unknown	Unknown[16] 4–7 months or 273–399 days/2 years[17]	Unknown[16] ~12 months or 10–700 days/2 years[17]	Spring to summer	Unknown	Unknown[16] <60 cm[17]	Unknown[16] Up to 200 cm[17]
	Spiny flat-sedge	*Cyperus gym-naucolos*	Spring to summer	Unknown	Unknown[16] 2–7 months or 162–392 days/2 years[17]	Unknown[16] ~12 months or 24–691 days/2 years[17]	Spring to summer	Unknown	Unknown[16] <60 cm[17]	Unknown[16] Up to 200 cm[17]
	Curly flat-sedge	*Cyperus rigidellus*	Spring to summer	Unknown	Unknown[16] 3–6 months or 225–328 days/2 years[17]	Unknown[16,17]	Spring to summer	Unknown	Unknown[16] <30 cm[17]	Unknown[16,17]
	Common spike-rush	*Eleocharis acuta*	Spring to summer	Spring to autumn	Unknown[16] <8 months or 88–243 days/2 years[17]	Unknown[16] <8 months or 88–243 days/2 years[17]	Spring to autumn	Unknown	Drawdown[16] Drawdown to <60 cm[17]	18 cm[16] <200 cm[17]

Plant group	Common name	Scientific name	Ideal germination timing	Maximum germination timing	Ideal flood duration	Maximum flood duration	Ideal flood timing	Maximum flood timing	Ideal flood depth	Maximum flood depth
	Pale spike-rush	*Eleocharis pallens*	Spring to summer	Spring to autumn	Unknown[16,17]	Unknown[16,17]	Spring to autumn	Unknown	Unknown[16,17]	Unknown[16,17]
	Flat spike-rush	*Eleocharis plana*	Spring to summer	Spring to Autumn	Unknown [16, 17]	Unknown [16, 17]	Spring to autumn	Unknown	Unknown[16,17]	Unknown[16,17]
	Small spike-rush	*Eleocharis pusilla*	Spring to summer	Spring to autumn	Unknown[16,17]	Unknown[16,17]	Spring to autumn	Unknown	Drawdown[16] 0.5–350 cm[17]	18 cm[16] <350 cm[17]
	Tall spike-rush	*Eleocharis sphacelata*	Spring and autumn	Spring to autumn	Unknown[16] 8–12 months[17]	Unknown[16] 8–12 months[17]	Spring to autumn	Unknown	30 cm[16] <225 cm[17]	45 cm[16] <225 cm[17]
Aquatic macro-phytes	Ribbon-weed	*Vallisneria gigantea, V. spiralis, V. americana*[18]	Any	Any	9–12 months	12 months	Spring to summer	Any	>0 cm[16] <200 cm[17]	Unknown[16] <700 cm[17]
Herbs and forbs	Rock isotome	*Isotoma axillaris*	Unknown	Unknown	Unknown	Unknown	Unknown	Unknown	Unknown	Unknown
	Swamp isotome	*Isotoma fluviatilis*	Unknown	Unknown	Unknown	Unknown	Unknown	Unknown	Unknown	Unknown
	Isotome species	*Isotoma tridens*	Unknown	Unknown	Unknown	Unknown	Unknown	Unknown	Unknown	Unknown
	Water primrose	*Ludwigia peploides*	Spring to summer	Unknown	8–10 months	Unknown	Spring to summer	Unknown	≥0 cm	Unknown
	Willow primrose	*Ludwigia octovalvis*	Unknown	Unknown	Unknown	Unknown	Unknown	Unknown	Unknown	Unknown
	Common nardoo	*Marsilea drummondii*	November to March	Unknown	1–6 months	Unknown	Spring to Summer	Unknown	<10 cm	Unknown
	Narrow-leaf nardoo	*Marsilea costulifera*	Unknown	Unknown	Unknown	Unknown	Unknown	Unknown	Unknown	Unknown
	Marsilea species	*Marsilea exarata*	Unknown	Unknown	Unknown	Unknown	Unknown	Unknown	Unknown	Unknown
	Marsilea species	*Marsilea hirsute*	Unknown	Unknown	Unknown	Unknown	Unknown	Unknown	Unknown	Unknown
	Marsilea species	*Marsilea mutica*	Unknown	Unknown	Unknown	Unknown	Unknown	Unknown	Unknown	Unknown
	Wavy marshwort	*Nymphoides crenata*	Unknown	Unknown	Unknown	Unknown	Unknown	Unknown	Unknown	Unknown

Plant group	Common name	Scientific name	Ideal germination timing	Maximum germination timing	Ideal flood duration	Maximum flood duration	Ideal flood timing	Maximum flood timing	Ideal flood depth	Maximum flood depth
	Entire marshwort	*Nymphoides geminata*	Unknown	Unknown	Unknown	Unknown	Unknown	Unknown	Unknown	Unknown
	Water snowflake	*Nymphoides indica*	Unknown	Unknown	Unknown	Unknown	Unknown	Unknown	Unknown	Unknown
	Marshwort	*Nymphoides montana*	Unknown	Unknown	Unknown	Unknown	Unknown	Unknown	Unknown	Unknown
	Marbled marshwort	*Nymphoides spinulosperma*	Unknown	Unknown	Unknown	Unknown	Unknown	Unknown	Unknown	Unknown
	Poison pratia	*Pratia concolor*	Unknown	Unknown	Unknown	Unknown	Unknown	Unknown	Unknown	Unknown
	Whiteroot	*Pratia purpurascens*	Unknown	Unknown	Unknown	Unknown	Unknown	Unknown	Unknown	Unknown
	River buttercup	*Ranunculus inundatus*	Unknown	Unknown	Unknown	Unknown	Unknown	Unknown	Unknown	Unknown
	Ferny buttercup	*Ranunculus pumilio*	Unknown	Unknown	Unknown	Unknown	Unknown	Unknown	Unknown	Unknown
	Swamp buttercup	*Ranunculus undosus*	Unknown	Unknown	Unknown	Unknown	Unknown	Unknown	Unknown	Unknown

[1]Regeneration is associated with flooding and may occur following substantial rainfall. [2]Seedling development of lignum may be limited when germination occurs during winter. [3]Germination of water couch is almost entirely inhibited with seed immersion. [4]Vegetative expansion of water couch may occur with a few centimetres of flooding. [5]Drawdown conditions for germination of common reed. [6]Tolerance of common reed seedlings dependent on age and height, 7 weeks duration based on week-old seedlings. [7]Substantial flooding enables transport of common reed seed away from parent plant for germination. [8]Drawdown conditions for germination. [9]Flooding for seedling growth and establishment as rhizomes are poorly developed. [10]Flooding for vegetative expansion. [11]Conditions for vegetative expansion are unknown. [12]Juveniles likely to die with extended inter-flood dry-period. [13]Drawdown conditions for germination and seedling establishment. [14]Flooding conditions for vegetative expansion as per adult requirements. [15]Expansion limited at extremes of duration and depth. [16]Conditions for seedling establishment. [17]Conditions for vegetative expansion.

Table 2.3: Functional classifications of selected plant species in the Murray-Darling Basin according to the CSR model (Grime 2001), environmental sieve model (van der Valk 1981) and wetting and drying model (Brock and Casanova 1997)

Plant group	Species	Brief description	Grime 2001	van der Valk 1981	Brock & Casanova 1997
Trees	*Eucalyptus camaldulensis*	Perennial. Tolerates flooding and drying. Does not have long-lived seeds that are maintained in seed banks. Requires moisture for the germination and establishment of seedlings, no standing water	ST-C	PD-I	ATe [1]
	Eucalyptus largiflorens	Perennial. Tolerates flooding and drying. Does not have long-lived seeds that are maintained in seed banks. Requires moisture for the germination and establishment of seedlings, no standing water	ST-C	PD-I	ATe [1]
	Eucalyptus coolabah	Perennial. Tolerates flooding and drying. Does not have long-lived seeds that are maintained in seed banks. Requires moisture for the germination and establishment of seedlings, no standing water	ST-C	PD-I	ATe [2]
	Acacia pendula	Perennial. Tolerates flooding and drying. Long-lived seeds that are maintained in seed banks. Requires moisture for the germination and establishment of seedlings, no standing water	C	PS-I	TDa
	Acacia stenophylla	Perennial. Tolerates flooding and drying. Does not have long-lived seeds that are maintained in seed banks. Requires moisture for the germination and establishment of seedlings, no standing water	C	PD-I	TDa
Shrubs	*Muehlenbeckia florulenta*	Perennial, emergent. Tolerates flooding and drying. Does not have long-lived seeds that are maintained in seed banks. Requires flooding to reproduce vegetatively. Requires moisture for the germination and establishment of seedlings, no standing water	ST	PD-I	ATe
Grasses	*Paspalum distichum*	Perennial, emergent. Tolerates flooding and drying. Long-lived seeds that are maintained in seed banks. Reproduces sexually and vegetatively, vegetative reproduction greater than sexual. Requires moisture to reproduce vegetatively. Requires moisture for the germination and establishment of seedlings, no standing water	ST	VS-I	ATe
	Phragmites australis	Perennial, emergent. Tolerates flooding and drying. Does not have long-lived seeds that are maintained in seed banks. Reproduces sexually and vegetatively, vegetative reproduction greater than sexual. Requires moisture to reproduce vegetatively. Requires moisture for the germination and establishment of seedlings, no standing water	ST	VD-I	ATe
Sedges and rushes	*Typha orientalis*	Perennial, emergent. Tolerates flooding and drying. Long-lived seeds that are maintained in seed banks. Reproduces sexually and vegetatively, sexual reproduction greater than vegetative. Requires moisture for vegetative expansion. Requires moisture for the germination and establishment of seedlings. No standing water	ST	VS-I	ATe
	Typha domingensis	Perennial, emergent. Tolerates flooding and drying. Long-lived seeds that are maintained in seed banks. Reproduces sexually and vegetatively, sexual reproduction greater than vegetative. Requires moisture for vegetative expansion. Requires moisture for the germination and establishment of seedlings. No standing water	ST	VS-I	ATe
	Juncus flavidus	Perennial, emergent. Tolerates flooding and drying. Long-lived seeds that are maintained in seed banks. Reproduces sexually and vegetatively, sexual reproduction greater than vegetative. Requires moisture for vegetative expansion. Requires moisture for the germination and establishment of seedlings. No standing water	ST	VS-I	ATe
	Juncus usitatus	Perennial, emergent. Tolerates flooding and drying. Long-lived seeds that are maintained in seed banks. Reproduces sexually and vegetatively, sexual reproduction greater than vegetative. Requires moisture for vegetative expansion. Requires moisture for the germination and establishment of seedlings. No standing water	ST	VS-I	ATe

Plant group	Species	Brief description	Grime 2001	van der Valk 1981	Brock & Casanova 1997
	Juncus aridicola	Perennial, emergent. Tolerates flooding and drying. Long-lived seeds that are maintained in seed banks. Reproduces sexually and vegetatively, sexual reproduction greater than vegetative. Requires moisture for vegetative expansion. Requires moisture for the germination and establishment of seedlings. No standing water	ST	VS-I	ATe
	Bolboschoenus caldwellii	Perennial, emergent. Tolerates flooding and drying. Does not have long-lived seeds that are maintained in seed banks. Reproduces sexually and vegetatively. Requires moisture for the germination and establishment of seedlings, no standing water	ST	VD-I	ATe
	Cyperus biffax	Perennial, emergent. Tolerates flooding and drying. Long-lived seeds that are maintained in seed banks. Unknown if sexual reproduction is greater than vegetative expansion. Requires moisture for germination and establishment. Perhaps some standing water for germination	ST	VS-II or PS-II	Ate
	Cyperus concinnus	Perennial, emergent. Tolerates flooding and drying. Long-lived seeds that are maintained in seed banks. Unknown if sexual reproduction is greater than vegetative expansion. Requires moisture for germination and establishment. Perhaps some standing water for germination	ST	VS-II or PS-II	ATe
	Cyperus difformis	Perennial, emergent. Tolerates flooding and drying. Long-lived seeds that are maintained in seed banks. Unknown if sexual reproduction is greater than vegetative expansion. Requires moisture for germination and establishment. Perhaps some standing water for germination	ST	VS-II or PS-II	ATe
	Cyperus rigidellus	Perennial, emergent. Tolerates flooding and drying. Long-lived seeds that are maintained in seed banks. Unknown if sexual reproduction is greater than vegetative expansion. Requires moisture for germination and establishment. Perhaps some standing water for germination	ST	VS-II or PS-II	ATe
	Eleocharis acuta	Perennial, emergent. Tolerates flooding and drying. Long-lived seeds that are maintained in seed banks. Reproduces sexually and vegetatively, unknown which is greater. Requires flooding for germination	ST	VS-II	ATe
	Eleocharis pallens	Perennial, emergent. Tolerates flooding and drying. Long-lived seeds that are maintained in seed banks. Reproduces sexually and vegetatively, unknown which is greater. Requires flooding for germination	ST	VS-II	ATe
	Eleocharis plana	Perennial, emergent. Tolerates flooding and drying. Long-lived seeds that are maintained in seed banks. Reproduces sexually and vegetatively, unknown which is greater. Requires flooding for germination	ST	VS-II	ATe
	Eleocharis pusilla	Perennial, emergent. Tolerates flooding and drying. Long-lived seeds that are maintained in seed banks. Reproduces sexually and vegetatively, unknown which is greater. Requires flooding for germination	ST	VS-II	ATe
	Eleocharis sphacelata	Perennial, emergent. Tolerates flooding and small amount of drying. Long-lived seeds that are maintained in seed banks. Reproduces sexually and vegetatively, unknown which is greater. Requires flooding for germination	ST	VS-II	ARp
Aquatic macrophytes	*Vallisneria species*	Perennial, submerged. Tolerates flooding and some drying. Long-lived seeds that are maintained in seed banks. Primarily regenerates vegetatively, may reproduce sexually. Requires flooding for germination			S

Plant group	Species	Brief description	Grime 2001	van der Valk 1981	Brock & Casanova 1997
Herbs and forbs	*Isotoma axillaris*	Perennial, low-growing. Tolerates flooding and drying. Unknown if seeds are maintained in seed banks. Unknown if plant reproduces vegetatively. Requires moisture for germination and establishment. No standing water	ST	PD-I	ATl
	Isotoma fluviatilis	Perennial, low-growing. Tolerates flooding and drying. Unknown if seeds are maintained in seed banks. Unknown if plant reproduces vegetatively. Requires moisture for germination and establishment. No standing water	ST	PD-I	ATl
	Ludwigia peploides	Perennial, low-growing. Tolerates flooding and drying. Unknown if seeds are maintained in seed banks. Unknown if plant reproduces vegetatively. Requires flooding for germination	ST	PD-II	ATl
	Marsilea costulifera	Perennial, floating leaves. May behave as an annual when conditions are dry. Tolerates flooding and drying. Long-lived seeds that are maintained in seed banks. Unknown if plant reproduces vegetatively. Requires moisture for germination and establishment, no standing water	ST-R	PS-I	ARf/ARp
	Marsilea drummondii	Perennial, floating leaves. May behave as an annual when conditions are dry. Tolerates flooding and drying. Long-lived seeds that are maintained in seed banks. Unknown if plant reproduces vegetatively. Requires moisture for germination and establishment, no standing water	ST-R	PS-I	ARf/ARp
	Nymphoides crenata	Perennial, floating leaves. Tolerates flooding and some drying. Unknown if plant reproduces vegetatively. May require flooding for germination	ST	PD-II	ARf
	Pratia concolor	Perennial, low-growing. Tolerates flooding and drying. Unknown if plant reproduces vegetatively. Requires moisture for germination and establishment, no standing water	ST	PD-I	ATl
	Pratia purpurascens	Perennial, low-growing. Tolerates flooding and drying. Unknown if plant reproduces vegetatively. Requires moisture for germination and establishment, no standing water	ST	PD-I	ATl
	Ranunculus inundatus	Perennial, low-growing. Tolerates flooding and drying. Unknown if plant reproduces vegetatively. Requires moisture for germination and establishment, no standing water	ST	PD-I	ATl
	Ranunculus pumilio	Perennial, low-growing. Tolerates flooding and drying. Unknown if plant reproduces vegetatively. Requires moisture for germination and establishment, no standing water	ST	PD-I	ATl
	Ranunculus undosus	Perennial, low-growing. Tolerates flooding and drying. Unknown if plant reproduces vegetatively. Requires moisture for germination and establishment, no standing water	ST	PD-I	ATl

[1]May be regarded as TDa in temperate locations where rainfall is more reliable. [2]May be regarded as TDa where there is a marked rainy season.

Table 2.4: Legend of functional classification schemes

	Functional group description	Source
ST	Stress-tolerant	Grime (2001)
C	Competitor	
R	Ruderal	
TDr	Terrestrial dry species	Brock and Casanova (1997)
TDa	Terrestrial damp species	
ATe	Amphibious fluctuation tolerators, emergent species	
ATl	Amphibious fluctuation tolerators, low-growing species	
ARp	Amphibious fluctuation responders, morphologically plastic species	
ARf	Amphibious fluctuation responders, species with floating leaves	
S	Submerged	
AD-I	Annual, dispersal-dependent, requires no standing water for establishment	Van der Valk (1981)
AD-II	Annual, dispersal-dependent, requires standing water for establishment	
PD-I	Perennial, dispersal-dependent, requires no standing water for establishment	
PD-II	Perennial, dispersal-dependent, requires standing water for establishment	
VD-I	Vegetatively reproducing perennial, dispersal-dependent, requires no standing water for establishment	
VD-II	Vegetatively reproducing perennial, dispersal-dependent, requires standing water for establishment	
AS-I	Annual, seed bank species, requires no standing water for establishment	
AS-II	Annual, seed bank species, requires standing water for establishment	
PS-I	Perennial, seed bank species, requires no standing water for establishment	
PS-II	Perennial, seed bank species, requires standing water for establishment	
VS-I	Vegetatively reproducing perennial, seed bank species, requires no standing water for establishment	
VS-II	Vegetatively reproducing perennial, seed bank species, requires standing water for establishment	

Table 2.5: Plant associations based on species' water requirements

Association name	Associated species	Approximate flood regime				Floodplain elevation	Environment
		Frequency	Duration	Timing	Depth		
Black box woodland	*Eucalyptus largiflorens* *Eucalyptus coolabah* *Acacia pendula* *Acacia stenophylla* Other dryland species	1 in 10 years	2 months	Summer to autumn		Upper	Outer limits of inundation. Marginal floodplain. Rarely inundated
River red gum woodland	*Eucalyptus camaldulensis* *Eucalyptus largiflorens* *Acacia stenophylla*	1 in 3–5 years	2–4 months	Spring to summer		Upper	Outer edges of floodplain, less regularly inundated
River red gum forest	*Eucalyptus camaldulensis* *Muehlenbeckia florulenta* *Acacia stenophylla*	1 in 3 years	2–6 months	Spring to summer		Upper	Depressions on floodplain, banks of watercourses, billabongs. Regular inundation
Lignum shrubland	*Muehlenbeckia florulenta* Dryland annuals Saltbush, roly-poly, burr	3–10 years	1–6 months	Spring to summer	60 cm	Mid	Elevated parts of floodplain receiving regular inundation to greater depths
Rush and sedge association	*Eleocharis acuta* *Bolboschoenus caldwellii* *Cyperus* species	Annual	2–4 months	Spring to summer	±20 cm	Mid–lower	Wetland sites with fluctuating water levels and regular flooding and drying
Water couch association	*Paspalum distichum* *Isotoma* species *Marsilea* species *Pratia* species	Annual	2–3 months	Spring to summer	shallow	Mid–lower	Floodplain sites with shallow regular flooding and drying
Reed association	*Phragmites australis* *Eleocharis* species *Cyperus* species	Annual	6 months	Spring to summer	±30 cm	Lower	Floodplain sites with shallow fluctuating flooding and drying
Cumbungi association	*Typha* species *Juncus* species *Eleocharis sphacelata* *Eleocharis pusilla* *Nymphoides crenata*	Annual	9–12 months	Spring to summer	0–200 cm	Lower	Edges of watercourses. Sites with permanent to regular flooding to some depth
Aquatic association	*Vallisneria* species *Nymphoides crenata*	Annual	12 months	Spring to summer	50–100 cm	Water-course	Below the waterline

response of floodplain wetland plants to climate change-induced alterations of the water regime at the Macquarie Marshes suggests that conditions will favour dryland species and amphibious species with an adaptive capacity, such as vegetative expansion and soil seed banks. Trees (e.g. river red gum and black box) and shrubs (e.g. lignum) will be particularly vulnerable to decreases in flood frequency and variability. Amphibious grasses, sedges and rushes, herbs and forbs will have to adapt to a reduced ecotone and increased competition for space and resources. Submerged aquatic macrophytes may have limited capacity to increase extent. The response of vegetation to water regime alterations will be discussed further in Chapter 7.

References

Aguiar FC, Ferreira MT, Albuquerque ANT, Oacute N and Bernez I (2005) Invasibility patterns of knotgrass (*Paspalum distichum*) in Portuguese riparian habitats 1. *Weed Technology* **19**, 509–516.

Akeroyd MD, Tyerman SD, Walker GR and Jolly ID (1998) Impact of flooding on the water use of semi-arid riparian eucalypts. *Journal of Hydrology* **206**, 104–117.

Alvarez-Cobelas M and Cirujano S (2007) Multilevel responses of emergent vegetation to environmental factors in a semiarid floodplain. *Aquatic Botany* **87**, 49–60.

Amsberry L, Baker MA, Ewanchuk PJ and Bertness MD (2000) Clonal integration and the expansion of *Phragmites australis*. *Ecological Applications* **10**, 1110–1118.

Armstrong J, Afreen-Zobayed F, Blyth S and Armstrong W (1999) *Phragmites australis*: effects of shoot submergence on seedling growth and survival and radial oxygen loss from roots. *Aquatic Botany* **64**, 275–289.

Aston HI (1973) *Aquatic Plants of Australia*. Melbourne University Press: Melbourne.

Bacon PE, Stone C, Binns DL, Leslie DJ and Edwards DW (1993) Relationships between water availability and *Eucalyptus camaldulensis* growth in a riparian forest. *Journal of Hydrology* **150**, 541–561.

Bell DM and Clarke PJ (2004) Seed-bank dynamics of *Eleocharis*: can spatial and temporal variability explain habitat segregation? *Australian Journal of Botany* **52**, 119–131.

Bennett M and Green J (1993) 'Preliminary assessment of Gwydir Wetlands water needs'. Department of Water Resources, Technical Services Division.

Bennett M and McCosker R (1994) Estimating environmental flow requirements of wetlands. In *Environmental Flows Seminar*. Canberra, ACT. Australian Water and Wastewater Association Inc.

Blanch SJ and Brock MA (1994) Effects of grazing and depth on two wetland plant species. *Australian Journal of Marine and Freshwater Research* **45**, 1387–1394.

Blanch SJ, Ganf GG and Walker KF (1998) Growth and recruitment in *Vallisneria americana* as related to average irradiance in the water column. *Aquatic Botany* **61**, 181–205.

Blanch SJ, Ganf GG and Walker KF (1999a) Growth and resource allocation in response to flooding in the emergent sedge *Bolboschoenus medianus*. *Aquatic Botany* **63**, 145–160.

Blanch SJ, Ganf GG and Walker KF (1999b) Tolerance of riverine plants to flooding and exposure indicated by water regime. *Regulated Rivers: Research and Management* **15**, 43–62.

Blanch SJ, Walker KF and Ganf GG (2000) Water regimes and littoral plants in four weir pools of the River Murray, Australia. *Regulated Rivers: Research and Management* **16**, 445–456.

Boland DJ, Brooker MIH and Turnbull JW (1986) 'Eucalyptus seed'. CSIRO Australia: Melbourne.

Boxshall B and Jenkyn T (2001) Eumong-River Cooba, *Acacia stenophylla*. Farm forestry species profile for North Central Victoria. Department of Primary Industries, Victoria.

Bramley H, Hutson J and Tyerman SD (2003) Floodwater infiltration through root channels on a sodic clay floodplain and the influence on a local tree species *Eucalyptus largiflorens*. *Plant and Soil* **253**, 275–286.

Bren LJ (1987) The duration of inundation in a flooding river red gum forest. *Australian Forestry Research* **17**, 191–202.

Bren LJ and Gibbs NL (1986) Relationships between flood frequency, vegetation and topography in a river red gum forest. *Australian Forestry Research* **16**, 357–370.

Briggs SV and Maher MT (1983) Litter fall and leaf decomposition in a river red gum (*Eucalyptus camaldulensis*) swamp. *Australian Journal of Botany* **31**, 307–316.

Briggs SV and Maher MT (1985) Limnological studies of waterfowl habitat in south-western New South Wales. II. Aquatic macrophyte productivity. *Australian Journal of Marine and Freshwater Research* **36**, 707–715.

Briggs S and Townsend G (1993) 'Restoration and management of Nearie Lake: final report'. NSW National Parks and Wildlife: Sydney.

Britton DL and Brock MA (1994) Seasonal germination from wetland seed banks. *Australian Journal of Marine and Freshwater Research* **45**, 1445–1457.

Brix H, Sorrell BK and Orr PT (1992) Internal pressurization and convective gas flow in some emergent freshwater macrophytes. *Limnology and Oceanography* **37**, 1420–1433.

Brock MA (1998) Are temporary wetlands resilient? Evidence from seed banks of Australian and South African wetlands. In *Wetlands for the Future: Contributions from INTECOL's 5th International Wetlands Conference.* (Eds AJ McComb and JA Davis) pp. 193–206. Gleneagles Publishing: Adelaide.

Brock MA and Casanova MT (1997) Plant life at the edge of wetlands: ecological responses to wetting and drying patterns. In *Frontiers in Ecology: Building the Links.* (Eds NK Klomp and I Lunt) pp. 181–192. Elsevier Science: Oxford, UK.

Brooker MIH, Connors JR, Slee AV and Suffy S (2002) *EUCLID: Eucalypts of southern Australia.* CSIRO Publishing: Melbourne.

Capon SJ (2003) Flow variability and vegetation dynamics in a large arid floodplain: Cooper Creek, Australia. BSc (Hons) thesis. Faculty of Environmental Sciences, Griffith University.

Capon SJ (2005) Flood variability and spatial variation in plant community composition and structure on a large arid floodplain. *Journal of Arid Environments* **60**, 283–302.

Capon SJ and Brock MA (2006) Flooding, soil seed bank dynamics and vegetation resilience of a hydrologically variable desert floodplain. *Freshwater Biology* **51**, 206–223.

Cary PR and Weerts PGJ (1984) Growth and nutrient composition of *Typha orientalis* as affected by water temperature and nitrogen and phosphorus supply. *Aquatic Botany* **19**, 105–118.

Casanova MT and Brock MA (2000) How do depth, duration and frequency of flooding influence the establishment of wetland plant communities? *Plant Ecology* **147**, 237–250.

Chesterfield EA (1986) Changes in the vegetation of the river red gum forest at Barmah, Victoria. *Australian Forestry* **49**, 4–15.

Chisholm L and Stone C (2003) 'Ground calibration of river red gum (*Eucalyptus camaldulensis*) health associated with airborne video imagery'. Wollongong.

Chong C and Walker KF (2005) Does lignum rely on a soil seed bank? Germination and reproductive phenology of *Muehlenbeckia florulenta* (Polygonaceae). *Australian Journal of Botany* **53**, 407–415.

Cizkova-Koncalova H, Kvet J and Thompson K (1992) Carbon starvation: a key to reed decline in eutrophic lakes. *Aquatic Botany* **43**, 105–113.

Clayton WD (1968) The correct name of the common reed. *Taxon* **17**, 168–169.

Clevering OA and Lissner J (1999) Taxonomy, chromosome numbers, clonal diversity and population dynamics of *Phragmites australis*. *Aquatic Botany* **64**, 185–208.

Cook D, Bayes E, Jolly K and Backstrom A (2009) 'Ecological response of four wetlands to the application of environmental water: final report on monitoring from May to December 2008'. Prepared by Australian Ecosystems for Goulburn Broken Catchment Management Authority: Melbourne.

Coops H and Velde G (1995) Seed dispersal, germination and seedling growth of six helophyte species in relation to water-level zonation. *Freshwater Biology* **34**, 13–20.

Coops H, van den Brink FWB and van der Velde G (1996) Growth and morphological responses of four helophyte species in an experimental water-depth gradient. *Aquatic Botany* **54**, 11–24.

Cowan RS (1996) Acacia group 2. In *Flora of Victoria*. (Eds NG Walsh and TJ Entwisle). Inkata Press: Melbourne.

Cox B (1984) Weeds. In *Growing Rice in NSW*. (Ed. A Currey). NSW Dept of Agriculture/Rice Research Committee: Griffith, NSW.

Craig AE, Walker KF and Boulton AJ (1991) Effects of edaphic factors and flood frequency on the abundance of lignum (*Muehlenbeckia florulenta* Meissner) (Polygonaceae) on the River Murray floodplain, South Australia. *Australian Journal of Botany* **39**, 431–443.

Crosslé K and Brock MA (2002) How do water regime and clipping influence wetland plant establishment from seed banks and subsequent reproduction? *Aquatic Botany* **74**, 43–56.

CSIRO (2004a) Taxon attribute profiles: *Acacia stenophylla* A.Cunn. ex Benth. CSIRO: Canberra.

CSIRO (2004b) Taxon attribute profiles: *Marsilea drummondii* a.Braun. CSIRO: Canberra.

CSIRO (2006a) Taxon attribute profiles: *Juncus ingens* N.A.Wakef. Giant Rush. CSIRO Water for a Healthy Country: Canberra.

CSIRO (2006b) Taxon attribute profiles: *Juncus usitatis* L.A.S.Johnson. CSIRO Water for a Healthy Country: Canberra.

Cunningham GM, Mulham WE, Milthorpe PE and Leigh JH (1981) *Plants of Western New South Wales*. Conservation Service of New South Wales: Sydney.

Cunningham GM, Mulham WE, Milthorpe PL and Leigh JH (1992) *Plants of Western New South Wales*. Inkata Press: Sydney.

Dalton K (1990) 'Managing our river red gums'. Soil Conservation Service of New South Wales: Sydney.

Deegan BM, White SD and Ganf GG (2007) The influence of water level fluctuations on the growth of four emergent macrophyte species. *Aquatic Botany* **86**, 309–315.

Dexter BD (1978) Silviculture of the river red gum forests of the Central Murray flood plain. *Proceedings of the Royal Society of Victoria* **90**, 175–191.

Doran JC and Boland DJ (1984) Effects of temperature on germination of *Eucalyptus microtheca*. *Australian Forestry Research* **14**, 49–55.

Faucon P (2005) *Acacia pendula*. Desert-Tropicals, Arizona. http://www.desert-tropicals.com/Plants/Fabaceae/Acacia_pendula.html.

Finlayson M, Forrester RI, Mitchell DS and Chick AJ (1985) Identification of native *Typha* species in Australia. *Australian Journal of Botany* **33**, 101–107.

Frankenberg J (1997) 'Guidelines for growing Phragmites for erosion control'. Cooperative Research Centre for Freshwater Ecology, Murray-Darling Freshwater Research Centre, Albury, NSW.

Froend RH and McComb AJ (1994) Distribution, productivity and reproductive phenology of emergent macrophytes in relation to water regimes at wetlands of south-western Australia. *Australian Journal of Marine and Freshwater Research* **45**, 1491–1508.

George AK (2004) Eucalypt regeneration on the Lower Murray floodplain, South Australia. PhD thesis. University of Adelaide.

George AK, Walker KF and Lewis MM (2005a) Functioning floodplains or tree museums? In *Proceedings of the 8th International Riversymposium*. Brisbane, Australia. 6–9 September 2005. International Water Forum, Brisbane.

George AK, Walker KF and Lewis MM (2005b) Population status of eucalypt trees on the River Murray floodplain, South Australia. *River Research and Applications* **21**, 271–282.

Grace JB (1993) The adaptive significance of clonal reproduction in angiosperms: an aquatic perspective. *Aquatic Botany* **44**, 159–180.

Greenwood ME and MacFarlane GR (2006) Effects of salinity and temperature on the germination of *Phragmites australis*, *Juncus kraussii*, and *Juncus acutus*: implications for estuarine restoration initiatives. *Wetlands* **26**, 854–861.

Grime JP (2001) *Plant Strategies, Vegetation Processes and Ecosystem Properties*. John Wiley and Sons: Chichester, UK.

Grose RJ and Zimmer WJ (1957) Preliminary laboratory studies on light requirements for the germination of some eucalypt seeds. *Australian Forestry* **21**, 76–80.

Grose RJ and Zimmer WJ (1958) Some laboratory germination responses of the seeds of river red gum, *Eucalyptus camaldulensis* Dehn. Syn. *Eucalyptus rostrata* Schlecht. *Australian Journal of Botany* **6**, 129–153.

Harden GJ (Ed.) (1990) *Flora of New South Wales*. Vol. 1. New South Wales University Press: Sydney.

Harden GJ (Ed.) (1991) *Flora of New South Wales*. Vol. 2. UNSW Press: Sydney.

Harden GJ (Ed.) (1992) *Flora of New South Wales*. Vol. 3. UNSW Press: Sydney.

Harden GJ (Ed.) (1993) *Flora of New South Wales*. Vol. 4. UNSW Press: Sydney.

Haslam SM (1969a) The development and emergence of buds in *Phragmites communis* Trin. *Annals of Botany* **33**, 289–301.

Haslam SM (1969b) The development of shoots in *Phragmites communis* Trin. *Annals of Botany* **33**, 695–709.

Haslam SM (1970) The performance of *Phragmites communis* Trin. in relation to water supply. *Annals of Botany* **34**, 867–877.

Haslam SM (1971) The development and establishment of young plants of *Phragmites communis* Trin. *Annals of Botany* **35**, 1059–1072.

Haslam SM (1972) *Phragmites communis*: biological flora of the British Isles. *Journal of Ecology* **1972**, 585–610.

Haslam SM (1975) The performance of *Phragmites communis* Trin. in relation to temperature. *Annals of Botany* **39**, 881–888.

Hayball N and Pearce M (2004) Influences of simulated grazing and water-depth on the growth of juvenile *Bolboschoenus caldwellii, Phragmites australis* and *Schoenoplectus validus* plants. *Aquatic Botany* **78**, 233–242.

Heinrich P (1990) 'The eco-physiology of riparian river red gum (*Eucalyptus camaldulensis*)'. Australian Water Resources Advisory Council.

Hocking PJ (1981) Response of *Typha domingensis* to salinity and high levels of manganese in the rooting medium. *Australian Journal of Marine and Freshwater Research* **32**, 907–919.

Hocking PJ (1989a) Seasonal dynamics of production, and nutrient accumulation and cycling by *Phragmites australis* (Cav.) Trin. ex Steudel in a nutrient-enriched swamp in inland Australia. I. Whole plants. *Marine and Freshwater Research* **40**, 421–444.

Hocking PJ (1989b) Seasonal dynamics of production, and nutrient accumulation and cycling by *Phragmites australis* (Cav.) Trin. ex Steudel in a nutrient-enriched swamp in inland Australia. II. Individual shoots. *Marine and Freshwater Research* **40**, 445–464.

Holland KL, Tyerman SD, Mensforth LJ and Walker GR (2006) Tree water sources over shallow, saline groundwater in the lower River Murray, south-eastern Australia: implications for groundwater recharge mechanisms. *Australian Journal of Botany* **54**, 193–205.

Holm LG, Plucknett DL, Pancho JV and Herberger JP (1977) *The World's Worst Weeds.* University Press of Hawaii: Hawaii.

Hsiao AI and Huang WZ (1989) Effects of flooding on rooting and sprouting of isolated stem segments and on plant growth of *Paspalum distichum* L. *Weed Research* **29**, 335–344.

Huang WZ and Hsiao AI (1987) Factors affecting seed dormancy and germination of *Paspalum distichum. Weed Research* **27**, 405–415.

Huang WZ, Hsiao AI and Jordan L (1987) Effects of temperature, light and certain growth regulating substances on sprouting, rooting and growth of single-node rhizome and shoot segments of *Paspalum distichum* L. *Weed Research* **27**, 57–67.

Ishii J and Kadono Y (2002) Factors influencing seed production of *Phragmites australis. Aquatic Botany* **72**, 129–141.

Jacobs MR (1955) 'Growth habits of the eucalypts'. Forestry and Timber Bureau: Canberra.

Jacobs SWL and Frank KA (1997) Notes on *Vallisneria* (Hydrocharitaceae) in Australia, with descriptions of two new species. *Telopea* **7**, 111–118.

Jacometti G (1912) Le erbe che infestano le risaie italiano. Atti del Congresso Risicolo Internazionale. *Vercelli* **4**, 57–91.

James KR and Hart BT (1993) Effect of salinity on four freshwater macrophytes. *Marine and Freshwater Research* **44**, 769–777.

James CS, Capon SJ, White MG, Rayburg SC and Thoms MC (2007) Spatial variability of the soil seed bank in a heterogenous ephemeral wetland system in semi-arid Australia. *Plant Ecology* **190**, 205–217.

Jensen AE (2006) Using phenology of eucalypts to determine environmental watering regimes for the River Murray floodplain, South Australia. In *Australian Rivers: Making a Difference. Proceedings of the 5th Australian Conference on Stream Management.* (Eds AL Wilson, RL Dehaan, RJ Watts, KJ Page, KH Bowmer and A Curtis) pp. 175–180. Charles Sturt University: Albury.

Jensen AE (2008) The role of seed banks and soil moisture in recruitment of semi-arid floodplain plants: the River Murray, Australia. PhD thesis. School of Earth and Environmental Sciences, University of Adelaide.

Jensen AE, Walker KF and Paton DC (2006) The secret life of tangled lignum, *Muehlenbeckia florulenta* (Polygonaceae): little-known plant of the floodplains. In *Wetlands of the Murrumbidgee Catchment.* (Eds I Taylor, P Murray and S Taylor) pp. 79–85. Murrumbidgee Catchment Management Authority: Leeton.

Johns C *et al.* (2009) 'Native trees of the River Murray floodplain: literature review and experimental designs to examine effects of flow enhancement and floodwater retention'. Murray-Darling Freshwater Research Centre: Wodonga.

Jolly ID (1996) The effects of river management on the hydrology and hydroecology of arid and semi-arid floodplains. In *Floodplain Processes.* (Eds MG Anderson, DE Walling and PD Bates) pp. 577–609. John Wiley and Sons: New York.

Jolly ID and Walker GR (1996) Is the field water use of *Eucalyptus largiflorens* F.Muell. affected by short-term flooding? *Austral Ecology* **21**, 173–183.

Jolly ID, Walker GR and Thorburn PJ (1993) Salt accumulation in semi-arid floodplain soils with implications for forest health. *Journal of Hydrology* **150**, 589–614.

Junk WJ, Bayley PB and Sparks RE (1989) The flood pulse concept in river-floodplain systems. *Canadian Special Publication of Fisheries and Aquatic Sciences* **106**, 110–127.

Kidson R, Martin W, Witts T and Raison G (2000a) 'Monitoring of river red gum (*Eucalyptus camaldulensis*) in the Macquarie Marshes'. Riverine Environment Unit, Central West Region, NSW Department of Land and Water Conservation: Dubbo, NSW.

Kidson R, Witts T and Martin W (2000b) 'Historical vegetation mapping in the Macquarie Marshes 1949–1991'. NSW Department of Land and Water Conservation: Dubbo, NSW.

King SA and Buckney RT (2001) Exotic plants in the soil-stored seed bank of urban bushland. *Australian Journal of Botany* **49**, 717–720.

Koppitz H and Kühl H (2000) The importance of genetic diversity of *Phragmites australis* in the development of reed stands. *Wetlands Ecology and Management* **8**, 403–414.

Koppitz H, Kühl H, Hesse K and Kohl J-G (1997) Some aspects of the importance of genetic diversity in *Phragmites australis* (Cav.) Trin. ex Steudel for the development of reed stands. *Botanica Acta* **110**, 217–223.

Kozlowski TT (1984) *Flooding and Plant Growth*. Academic Press: Orlando.

Leck MA and Brock MA (2000) Ecological and evolutionary trends in wetlands: evidence from seeds and seed banks in New South Wales, Australia and New Jersey, USA. *Plant Species Biology* **15**, 97–112.

Leitch C (1989) 'Towards a strategy for managing the flooding of Barmah Forest'. Dept of Conservation, Forests and Lands, Benalla, Vic.

Lichtenthaler HK (1996) Vegetation stress: an introduction to the stress concept in plants. *Journal of Plant Physiology* **148**, 4–14.

Lloyd N, Quinn G, Thoms MC, Arthington AH, Gawne B, Humphries P and Walker KF (2003) 'Does flow modification cause geomorphological and ecological reponse in rivers? A literature review from an Australian perspective'. CRC for Freshwater Ecology, Canberra, ACT.

Manuel JS, Mercado BL and Lubigan RT (1979) Approaches to the control of *Paspalum distichum* L. in lowland rice. *Philippine Agriculturalist* **62**, 255–261.

Marcar N, Crawford D, Leppert P, Jovanovic T, Floyd R and Farrow R (1995) 'Trees for saltland'. CSIRO: Canberra.

Marcar NE, Crawford DF, Hossain AKMA and Nicholson AT (2003) Survival and growth of the tree species and provenances in response to salinity on a discharge site. *Australian Journal of Experimental Agriculture* **43**, 1293–1302.

Matsui T and Tsuchiya T (2006) Root aerobic respiration and growth characteristics of three Typha species in response to hypoxia. *Ecological Research* **21**, 470–475.

Mauchamp A, Blanch S and Grillas P (2001) Effects of submergence on the growth of *Phragmites australis* seedlings. *Aquatic Botany* **69**, 147–164.

Mawhinney WA (2004) Restoring biodiversity in the Gwydir Wetlands through environmental flows. *Water Science and Technology* **48**, 73–81.

McCosker RO (1999) 'Gwydir Wetlands ecological response to flooding: 1998'. Landmax: Kangaroo Point, Qld.

McCosker RO (2001) 'Gwydir Wetlands ecological response to flooding: 2000–2001'. Landmax: Kangaroo Point, Qld.

McIntyre S (1985) Seed reserves in temperate Australian rice fields following pasture rotation and continuous cropping. *Journal of Applied Ecology* **22**, 875–884.

McIntyre S, Lavorel S and Tremont RM (1995) Plant life-history attributes: their relationship to disturbance response in herbaceous vegetation. *Journal of Ecology* **85**, 31–44.

McKee J and Richards AJ (1996) Variation in seed production and germinability in common reed (*Phragmites australis*) in Britain and France with respect to climate. *New Phytologist* **133**, 233–243.

MDBC (2003) 'Preliminary investigations into observed river red gum decline along the River Murray below Euston'. Murray-Darling Basin Commission: Canberra.

Meeson N, Robertson AI and Jansen A (2002) The effects of flooding and livestock on post-dispersal seed predation in river red gum habitats. *Journal of Applied Ecology* **39**, 247–258.

Menges ES and Waller DM (1983) Plant strategies in relation to elevation and light in floodplain herbs. *American Naturalist* **122**, 454–473.

Mensforth LJ, Thorburn PJ, Tyerman SD and Walker GR (1994) Sources of water used by riparian *Eucalyptus camaldulensis* overlying highly saline groundwater. *Oecologia* **100**, 21–28.

Metcalfe L, Sivertsen DP, Tindall D and Ryan KM (2003) Natural vegetation of the New South Wales wheat-belt (Cobar-Nyngan-Gilgandra, Nymagee-Narromine-Dubbo 1:250 000 vegetation sheets). *Cunninghamia* **8**, 253–284.

Miao SL, McCormick PV, Newman S and Rajagopalan S (2001) Interactive effects of seed availability, water depth, and phosphorus enrichment on cattail colonization in an Everglades wetland. *Wetlands Ecology and Management* **9**, 39–47.

Middleton BA (1990) Effect of water depth and clipping frequency on the growth and survival of four wetland plant species. *Aquatic Botany* **37**, 189–196.

Middleton B (1999) Succession and herbivory in monsoonal wetlands. *Wetlands Ecology and Management* **6**, 189–202.

Middleton BA, van der Valk AG, Mason DH, Williams RL and Davis CB (1991) Vegetation dynamics and seed banks of a monsoonal wetlands overgrown with *Paspalum distichum* L. in northern India. *Aquatic Botany* **40**, 239–259.

Miller AC, Watling JR, Overton IC and Sinclair R (2003) Does water status of *Eucalyptus largiflorens* (Myrtaceae) affect infection by the mistletoe *Amyema miquelii* (Loranthaceae)? *Functional Plant Biology* **30**, 1239–1247.

Morris K, Harrison KA, Bailey PC and Boon PI (2004) Domain shifts in the aquatic vegetation of shallow urban lakes: the relative roles of low light and anoxia in the catastrophic loss of the submerged angiosperm *Vallisneria americana*. *Marine and Freshwater Research* **55**, 749–758.

Murphy LR, Barroca J, Franceschi VR, Lee R, Roalson EH, Edwards GE and Ku MSB (2007) Diversity and plasticity of C_4 photosynthesis in *Eleocharis* (Cyperaceae). *Functional Plant Biology* **34**, 571–580.

Navie SC, Rogers RW and Rogers RW (1997) The relationship between attributes of plants represented in the germinable seed bank and stocking pressure in a semi-arid subtropical rangeland. *Australian Journal of Botany* **45**, 1055–1071.

Nicol JM and Ganf GG (2000) Water regimes, seedling recruitment and establishment in three wetland plant species. *Marine and Freshwater Research* **51**, 305–309.

Nicol JM, Ganf GG and Pelton GA (2003) Seed banks of a southern Australian wetland: the influence of water regime on the final floristic composition. *Plant Ecology* **168**, 191–205.

Nielsen DL and Chick AJ (1997) Flood-mediated changes in aquatic macrophyte community structure. *Marine and Freshwater Research* **48**, 153–157.

NT Dept of NREA (2006) 'Fact sheet: native plants of central Australia'. Department of Natural Resources Environment and Arts: Darwin, NT.

Okuma M and Chikura S (1984) Ecology and control of a subspecies of *Paspalum distichum* L. 'Chikugo-suzumenohie', growing in creeks in the paddy area on the lower reaches of the Chikugo River in Kyushu. IV. Possibility of reproduction by seeds. *Weed Research Japan* **29**, 45–50.

Overton I and Jolly I (2004) 'Integrated studies of floodplain vegetation health, saline groundwater and flooding on the Chowilla floodplain, South Australia'. CSIRO Land and Water: Canberra.

Overton IC, Jolly ID, Slavich PG, Lewis MM and Walker GR (2006) Modelling vegetation health from the interaction of saline groundwater and flooding on the Chowilla floodplain, South Australia. *Australian Journal of Botany* **54**, 207–220.

Pagter M, Bragato C and Brix H (2005) Tolerance and physiological responses of *Phragmites australis* to water deficit. *Aquatic Botany* **81**, 285–299.

Paul KI, Polglase PJ, O'Connell AM, Carlyle JC, Smethurst PJ, Khanna PK and Worledge D (2003) Soil water under forests (SWUF): a model of water flow and soil water content under a range of forest types. *Forest Ecology and Management* **182**, 195–211.

Pettit NE and Froend RH (2001) Availability of seed for recruitment of riparian vegetation: a comparison of a tropical and temperate river ecosystems in Australia. *Australian Journal of Botany* **49**, 515–528.

Porter J, Kingsford R and Brock M (2007) Seed banks in arid wetlands with contrasting flooding, salinity and turbidity regimes. *Plant Ecology* **188**, 215–234.

Pryor LD and Briggs JD (1981) 'Australian endangered species: eucalypts'. NSW National Parks and Wildlife Service: Canberra.

Pudney S (1998) Seed dispersal in *Eucalyptus camaldulensis*: implications for natural regeneration. Hons thesis. University of Adelaide.

Ralph TJ (2008) Channel breakdown and floodplain wetland morphodynamics in the Macquarie Marshes, south-eastern Australia. PhD thesis. Department of Physical Geography, Macquarie University.

Reid MA and Quinn GP (2004) Hydrologic regime and macrophyte assemblages in temporary floodplain wetlands: implications for detecting responses to environmental water allocations. *Wetlands* **24**, 586–599.

Roberts J (1993) Regeneration and growth of coolibah, *Eucalyptus coolabah* subsp. *arida*, a riparian tree, in the Cooper Creek region of South Australia. *Australian Journal of Ecology* **18**, 345–350.

Roberts J (2001) Large plants. In *Rivers as Ecological Systems: The Murray-Darling Basin*. (Ed. WJ Young) pp. 187–222. Murray-Darling Basin Commission: Canberra.

Roberts J and Ganf GG (1986) Annual production of *Typha orientalis* Presl. in inland Australia. *Australian Journal of Marine and Freshwater Research* **37**, 659–668.

Roberts J and Ludwig JA (1991) Riparian vegetation along current-exposure gradients in floodplain wetlands of the River Murray, Australia. *Journal of Ecology* **79**, 117–127.

Roberts J and Marston F (2000) 'Water regime of wetland and floodplain plants in the Murray-Darling Basin'. CSIRO Land and Water: Canberra.

Roberts J, Young B and Marston F (2000) 'Estimating the water requirements for plants of floodplain wetlands: a guide'. Occasional Paper 04/00. Land and Water Resources Research and Development Corporation: Canberra.

Robertson AI, Bacon P and Heagney G (2001) The responses of floodplain primary production to flood frequency and timing. *Journal of Applied Ecology* **38**, 126–136.

Sainty GR and Jacobs SWL (1981) 'Waterplants of New South Wales'. NSW Water Resources Commission: Sydney.

Sainty GR and Jacobs SWL (2003) 'Waterplants in Australia'. Sainty and Associates: Sydney.

Saltmarsh A, Mauchamp A and Rambal S (2006) Contrasted effects of water limitation on leaf functions and growth of two emergent co-occurring plant species, *Cladium mariscus* and *Phragmites australis*. *Aquatic Botany* **84**, 191–198.

Sanders BA (1994) The life cycle and ecology of *Cyperus difformis* (rice weed) in temperate Australia: a review. *Australian Journal of Experimental Agriculture* **34**, 1031–1038.

Sharley T and Huggan C (1995) 'Chowilla resource management plan: final report'. Murray-Darling Basin Commission: Canberra.

Shepheard M (1992) 'Flooding frequency and *Eucalyptus largiflorens* (black box) wetland health in the Wah Wah District'. NSW Dept of Water Resources: Sydney.

Siebentritt MA and Ganf GG (2000) Influence of abiotic and biotic factors on two co-occurring species of *Bolboschoenus. Marine and Freshwater Research* **51**, 73–80.

Slavich PG, Walker GR, Jolly ID, Hatton TJ and Dawes WR (1999) Dynamics of *Eucalyptus largiflorens* growth and water use in response to modified watertable and flooding regimes on a saline floodplain. *Agricultural Water Management* **39**, 245–264.

Soetaert K, Hoffmann M, Meire P, Starink M, Oevelen DV, Regenmortel SV and Cox T (2004) Modeling growth and carbon allocation in two reed beds (*Phragmites australis*) in the Scheldt estuary. *Aquatic Botany* **79**, 211–234.

Sorrell BK and Tanner CC (2000) Convective gas flow and internal aeration in *Eleocharis sphacelata* in relation to water depth. *Journal of Ecology* **88**, 778–789.

Sorrell BK, Tanner CC and Sukias JPS (2002) Effects of water depth and substrate on growth and morphology of *Eleocharis sphacelata*: implications for culm support and internal gas transport. *Aquatic Botany* **73**, 93–106.

Stone C and Bacon PE (1994) Relationships among moisture stree, insect herbivory, foliar cineole content and the growth of river red gum *Eucalyptus camaldulensis. Journal of Applied Ecology* **31**, 604–612.

Stone C and Bacon PE (1995a) Influence of insect herbivory on the decline of black box (*Eucalyptus largiflorens*). *Australian Journal of Botany* **43**, 555–564.

Stone C and Bacon PE (1995b) Leaf dynamics and insect herbivory in a *Eucalyptus camaldulensis* forest under moisture stress. *Austral Ecology* **20**, 473–481.

Tanner CC, Clayton JS and Harper LM (1986) Observations on aquatic macrophytes in 26 northern New Zealand lakes. *New Zealand Journal of Botany* **24**, 539–551.

Thorburn PJ and Walker GR (1993) The source of water transpired by *Eucalyptus camaldulensis*: soil, groundwater, or streams? In *Stable Isotopes and Plant Carbon–Water Relations.* (Eds JR Ehleringer, AE Hall and GD Farquhar) pp. 511–527. Academic Press: New York.

Thorburn PJ and Walker GR (1994) Variations in stream water uptake by *Eucalyptus camaldulensis* with differing access to stream water. *Oecologia* **100**, 293–301.

Thorburn PJ, Hatton TJ and Walker GR (1993) Combining measurements of transpiration and stable isotopes of water to determine groundwater discharge from forests. *Journal of Hydrology* **150**, 563–587.

Thorburn PJ, Mensforth LJ and Walker GR (1994) Reliance of creek-side river red gums on creek water. *Australian Journal of Marine and Freshwater Research* **45**, 1439–1443.

Treloar GK (1959) Some factors affecting seedling survival of *Eucalyptus largiflorens* F.Muell. *Australian Forestry* **23**, 46–48.

Turnbull JW and Doran JC (1987) Seed development and germination in the Myrtaceae. In *Germination of Australian Native Plant Seed.* (Ed. P Langkamp) pp. 186–193. Inkata Press: Melbourne.

Vaillant A (1967) Chemical control of annual weeds in rice. *World Crops* **19**, 38–44.

Valk AG, Pederson RL and Davis CB (1992) Restoration and creation of freshwater wetlands using seed banks. *Wetlands Ecology and Management* **1**, 191–197.

Van der Sommen F (1980) Ecology of floodplain plant communities. *Riverlander (Murray Valley League)* **March**, 22–23.

van der Valk AG (1981) Succession in wetlands: a Gleasonian approach. *Ecology* **62**, 688–696.

Walker J and Hopkins MS (1990) Vegetation. In *Australian Soil and Land Survey: Field Handbook*. (Eds RC McDonald, RF Isbell, JG Speight, J Walker and MS Hopkins) pp. 55–86. Inkata Press: Melbourne.

Walker KF, Boulton AJ, Thoms MC and Sheldon F (1994) Effects of water-level changes induced by weirs on the distribution of littoral plants along the River Murray, South Australia. *Australian Journal of Marine and Freshwater Research* **45**, 1421–1438.

Ward K (1996) 'Flood requirements of vegetation for the rehabilitation of Lake Mokoan'. Draft report. Vic. Dept of Natural Resources and Environment: Melbourne.

Watt S, García-Berthou E and Vilar L (2007) The influence of water level and salinity on plant assemblages of a seasonally flooded Mediterranean wetland. *Plant Ecology* **189**, 71–85.

Weisner SEB, Graneli W and Ekstam B (1993) Influence of submergence on growth of seedlings of *Scirpus lacustris* and *Phragmites australis*. *Freshwater Biology* **29**, 371–375.

Wells RDS, Clayton JS and de Winton MD (1998) Submerged vegetation of lakes Te Anau, Manapouri, Monowai, Hauoko, and Poteriteri, New Zealand. *New Zealand Journal of Marine and Freshwater Research* **32**, 621–638.

WA Dept of Environment and Conservation (1993) '*Bolboschoenus caldwellii* (V.J.Cook) Sojak'. Dept of Environment and Conservation: Perth.

White DA, Turner NC and Galbraith JH (2000) Leaf water relations and stomatal behaviour of four allopatric *Eucalyptus* species planted in Mediterranean south-west Australia. *Tree Physiology* **20**, 1157–1165.

White SD, Deegan BM and Ganf GG (2007) The influence of water level fluctuations on the potential for convective flow in the emergent macrophytes *Typha domingensis* and *Phragmites australis*. *Aquatic Botany* **86**, 369–376.

Windmill Outback Nursery (2003) *Acacia pendula*. Windmill Outback Nursery: Louisa, Virginia, USA. http://www.australiaplants.com/Acacia_pendula.htm.

Yen S and Myerscough PJ (1989) Co-existence of three species of amphibious plants in relation to spatial and temporal variation: investigation of plant responses. *Australian Journal of Ecology* **14**, 305–318.

Yezdani GH (1970) A study of the Quaternary vegetation history in the volcanic lakes region of western Victoria. PhD thesis. Dept of Geography, Monash University.

Young WJ, Scott AC, Cuddy SM and Rennie BA (2003) 'Murray flow assessment tool: a technical description'. CSIRO Land and Water: Canberra.

Chapter 3

Waterbirds

Kerrylee Rogers

Introduction

Within floodplain wetlands, flooding is recognised as the 'principal driving force responsible for the existence, productivity and interactions of the major biota' (Junk *et al.* 1989, p. 110). In addition, it has been proposed that biota in variable habitats develop strategies to adapt to variable flow (Walker *et al.* 1995). This chapter explores some of the strategies that waterbirds have developed, such as opportunism and flexibility, in response to variable flows. The adaptive capacity of waterbirds to flow variability, which is driven by Australia's variable climatic patterns, is particularly evident when comparing Australian waterbirds to those in the northern hemisphere. Australian waterbirds tend not to have clearly predictable patterns of movement, feeding ecology, reproduction, moult and habitat use (Kingsford and Norman 2002); rather, they exhibit many aspects of opportunism. In particular, many waterbirds have great dispersal capabilities, thereby providing the opportunity to search for ideal wetland habitats within the fluctuating mosaic of wetlands in arid and semiarid Australia (Roshier *et al.* 2001a, 2001b). In fact, the response of waterbirds to flooding and their innate ability to track floodwaters has long fascinated bird watchers in Australia (see, e.g., the account of the silver gull by Hobbs 1961).

Reproductive performance is the most prominent response of waterbirds to the flood pulse. As the flood pulse stimulates productivity throughout a wetland, prey items of waterbirds become abundant. This enables waterbirds to store fat for sustenance throughout their breeding seasons and stimulates gonadal development and egg formation (Miller 1980; Norman and Hurley 1984). Flooding acts as a stimulus for breeding in most waterbirds: only two species in the Murray-Darling Basin, the musk duck (*Biziura lobata*) and blue-billed duck (*Oxyura australis*), are identified as purely seasonal breeders (Briggs 1990).

As top-order consumers, waterbirds also have an indirect link to the flood pulse, which is mediated by the response of habitats and food items to aspects of the flood pulse. Due to the trophic link between waterbirds and their ecosystems, poor reproductive performance may signify long-term environmental change related to reduced ecosystem productivity at lower trophic levels (Kushlan 1993). It is therefore essential when considering the response of waterbirds to flooding to also consider the influence of flooding on prey items and habitats used for breeding, foraging and roosting.

Despite the importance of water availability to the reproductive performance of waterbirds and the productivity of wetland ecosystems, waterbird-related literature fails to quantify the water requirements for the survival and maintenance (see Glossary) of waterbird populations. A monumental effort has been made to summarise the habitat, distribution, movement, food, social organisation and behaviour, breeding, appearance and morphology literature about birds, including waterbirds, in the *Handbook of Australian, New Zealand and Antarctic Birds* (HANZAB) (Higgins 1999; Higgins and Davies 1996; Higgins and Peter 2002; Higgins *et al.* 2006, 2001; Marchant and Higgins 1990, 1993). Considerable effort has been given to exploring the stimulus for waterbird breeding (e.g. Crome 1988; Frith 1967; Fullagar *et al.* 1988; Schodde 1982) with accounts now generally agreeing on a condition model dominated by food availability whereby flooding typically provides conditions favourable for successful breeding (Maher 1991). The relationship between flood duration and waterbird breeding has been considered (Kingsford and Auld 2005; Kingsford and Johnsons 1998); however, with some exceptions (Briggs *et al.* 1991; Crome 1988; Halse and Jaensch 1989; Kingsford and Auld 2005; Leslie 2001), very few studies have considered the effect of other aspects of the flood regime such as the rate of fall of water, water depth, inter-flood dry-period and flood frequency (see Glossary), particularly with reference to waterbird breeding success. This chapter aims to quantify scientific knowledge relating to the water requirements of waterbirds, with specific reference to aspects of the water regime. Information is also provided on the habitats utilised by waterbirds for breeding, foraging and roosting. The water requirements for the maintenance of these habitats may be developed by referring to the vegetation species profiles provided in Chapter 2.

Australia supports in excess of 860 bird species (Christidis and Boles 2008), and a recent survey of waterbirds in the Murray-Darling Basin identified 98 species (Straw, cited in Scott 1997). This chapter discusses only 48 species, for a number of reasons, the most important being that they are regarded as waterbirds. Waterbirds are an ecological grouping based on an essential flooding requirement, rather than a strict taxonomic grouping. Waterfowl (Anatidae) such as ducks, swans and geese are obvious inclusions, as are many *Ardea* species, such as herons, egrets, ibis and spoonbills. Further inclusions become difficult when considering some birds, such as the swamp harrier (*Circus approximans*), which opportunistically inhabit wetlands but are not restricted to those habitats and do not necessarily require flooding to feed and reproduce. This chapter employs the definition of waterbirds provided by Maher (1991) – waterbirds are those species that depend on free-standing water for feeding by swimming, wading or diving, or for the provision of nesting sites. Species were included if they were considered typical species that would breed within the Murray-Darling Basin, not predominantly considered coastal dwellers or regarded as infrequent visitors or vagrants.

A specific methodology was used when preparing the waterbird species profiles. First, waterbirds were classified according to the functional feeding groups provided by Roshier *et al.* (2002). These functional groups include the fish-eaters, deep-water foragers, dabbling ducks, grazing waterfowl, shoreline foragers, large waders and small waders (see Glossary). Second, key aspects of waterbird ecology were considered important and incorporated into the species profiles. The profiles include a brief description of the habitat and dietary requirements of waterbirds, but we recommend that readers refer to the comprehensive accounts provided in the *Handbook of Australian, New Zealand and Antarctic Birds* (HANZAB) volumes for waterbirds (Higgins and Davies 1996; Marchant and Higgins 1990, 1993). Reference is made to the breeding stimulus, breeding season, breeding lag time before the commencement of breeding and breeding duration (see Glossary). Stimulus for breeding refers to whether breeding commences in response to flooding, season or rainfall, and may largely be driven by the conditions required for maximising food availability. Similarly, breeding season or the timing of breeding is largely driven by food availability. Some waterbirds require a lag time before the commencement of breeding so that they can build up fat reserves to increase reproductive success, establish breeding pairs and/or collect material to construct nests. The duration of this lag time may depend on the seasonality of flooding, with shorter lag times after spring flooding than after autumn flooding (Briggs and Thornton 1999). Most waterbird species generally require flooding for the entire breeding duration, which incorporates the time required to build nests, lay and incubate eggs and fledge young to independence.

Aspects of the water regime that were considered important and incorporated into the species profiles include:

- *flood duration* (see Glossary) – for waterbirds this incorporates the required lag time before the commencement of breeding, and breeding duration;
- *rate of fall of water* – may be essential for breeding success of some waterbirds that require continuous flooding throughout the breeding duration. In this chapter, water recession has been scaled against flood duration, whereby waterbirds requiring flooding for six to 12 months have been prescribed a slow rate of fall, while waterbirds that exhibit breeding success with flooding less than three months are prescribed a fast rate of fall;
- *depth of flooding* – may be essential for facilitating fish movements and the dietary requirements of some species;
- *flood frequency* – refers to the frequency at which flooding should occur to maintain habitats, provide breeding opportunities and ultimately aid the survival of waterbird populations, which depends on a balance between recruitment and mortality (Kingsford and Norman 2002). In this chapter, flood frequency primarily refers to the frequency of flooding required to maintain foraging, roosting and breeding habitats. Large flood frequency refers to the frequency required to ensure large breeding events and to boost adult population numbers so that the recruitment of adults at least equals the mortality of adults. In the absence of population viability analyses of waterbird species, the required frequency of large floods is derived from values of life expectancy in the wild or longevity in captivity. It is estimated that a minimum of two large floods should occur in Murray-Darling Basin wetlands within the life expectancy of wild waterbirds, or four large breeding events within the captive longevity of waterbirds. It is anticipated that this should account for waterbird population losses due to poor condition or predation;
- *inter-flood dry-period* – refers to the period between flooding. While this aspect is partly encompassed within flood frequency, there is some evidence that breeding success may be enhanced when wetlands undergo complete drying prior to the breeding season (Briggs and Maher 1985; Crome 1986, 1988; Maher and Carpenter 1984).

Finally, other factors aside from aspects of the flood regime, such as competition, predation, hunting and habitat loss and degradation, may influence the breeding success of waterbird species. In this chapter emphasis is given to flood and drought perturbations, due to the overarching importance of water availability and the flood regime on the maintenance of breeding populations of waterbirds.

Fish-eaters

Great crested grebe: *Podiceps cristatus*

The great crested grebe is a medium to large diving waterbird that sits low in the water when swimming and diving. It has a straight bill, a long neck, a distinctive black double crest on its head and a matching chestnut frill fringed in black on its cheeks (Figure 3.1). The upper parts of its plumage are dark brown and the underparts are satin white. Both sexes have a similar

Figure 3.1: Great crested grebe, *Podiceps cristatus*. Photograph: Silva Vaughan-Jones (Hunter Bird Observers Club).

appearance; however, males are generally slightly larger. They have a length of 48–61 cm and a weight of approximately 1.1 kg (Marchant and Higgins 1990). They are not highly gregarious birds. They exhibit a preference for large open freshwater bodies, but may be observed in a range of inland and coastal habitats. The great crested grebe is widely distributed throughout Europe, Africa, Asia and Australia, where their distribution is largely limited to the east coast, the Murray-Darling Basin and south-west Western Australia (Marchant and Higgins 1990).

Habitat and diet

The great crested grebe has been observed in rivers, lakes, swamps, lagoons, reservoirs, salt fields, estuaries and other coastal locations; however, they exhibit a preference for large deep open freshwater sites (Corrick 1981, 1982; Marchant and Higgins 1990; Wheeler 1957). They are rarely seen on land or in flight. Foraging and roosting commonly occurs on open water, but may be observed among surface vegetation or at the edge of well-flooded vegetation cover (Marchant and Higgins 1990). Nesting commonly occurs on or near vegetated margins of large open waters. Nests consist of a platform or mound of dead plant material that is usually attached to a fixed item such as reeds, branches or stumps, but nests have also been observed floating freely (Marchant and Higgins 1990).

The diet of the great crested grebe mostly comprises fish, but may include insects, plants, molluscs and crustaceans (Marchant and Higgins 1990; O'Donnell 1982; Vestjens 1977b). Fish species consumed by great crested grebes at Lake Cowal was dominated by flathead gudgeon (*Philypnodon grandiceps*) (Vestjens 1977b).

Breeding and flood requirements for breeding

Flooding is identified as the primary breeding stimulus for great crested grebes (Briggs 1990); however, this may reflect the sedentary nature of this species and its preference for large deep open freshwater habitats (Marchant and Higgins 1990), rather than flooding to stimulate gonadal development or food supply. Some studies identify relationships between breeding and aspects of the flood regime such as inter-flood drying (Crome 1988) and flood depth (Halse and Jaensch 1989). Other studies indicate that the abundance of great crested grebes correlates with season or water depth (Halse *et al.* 1993; Harper 1990). It is therefore suggested that the stimulus for breeding is likely to be more seasonal in deep permanent waterbodies, or related to an increase in water depth in semi-permanent or fluctuating waterbodies as a result of flooding. Great crested grebes breed in simple pairs, but may form loose colonies or small groups (Marchant and Higgins 1990). Breeding of great crested grebes primarily occurs between November and February (Marchant and Higgins 1990), but may occur as early as August (Buller 1942) and as late as March (Napier 1973). Nests are positioned 2–10 m from the shore and in water depths of 1–2 m (Marchant and Higgins 1990).

Identifying a lag time for flooding is difficult for this species as great crested grebes prefer permanent waterbodies. Under these conditions breeding is likely to be stimulated by season, therefore a lag time to enable increases of food supply and fattening would not be necessary. However, breeding is most successful when there are two months of deep water and a one-month lag (Halse and Jaensch 1989). This lag may relate to the time needed to reach a water depth threshold or for fish populations to establish in sufficient numbers in semi-permanent environments or sites with fluctuating water levels. It is suggested that a short lag time of one to three months may be required when initial water depths are too low to stimulate breeding. In deep more permanent waterbodies, a lag time may not be required to induce breeding.

The clutch size generally ranges between five and seven eggs and replacement eggs may be laid after loss (Bright and Taysom 1932; Marchant and Higgins 1990). Eggs are laid at intervals of 48 hours and are incubated by both sexes for 25–31 days. Chicks have been observed swimming two days after hatching, and diving about one week after hatching. The period until sexual maturity is unknown, although breeding duration is estimated at one to two months. Ideal flood duration is estimated at two to five months (Marchant and Higgins 1990).

Water depth appears to be an important factor for stimulating breeding in great crested grebes. Water depths of approximately 1.8 m have been associated with great crested grebe breeding (Crome 1988; Halse and Jaensch 1989), although a depth of 1–2 m was provided by Marchant and Higgins (1990). Despite that, deep water is required for breeding, and the water needs to remain in a wetland for a minimum two to three months for successful breeding before recession commences. Therefore the rate of fall is cautiously described as moderate to fast. However, as the period until chicks are able to move between waterbodies remains unknown, flooding for longer periods is likely to sustain chicks until they are able to travel to permanent waterbodies.

The frequency at which conditions should stimulate breeding in great crested grebes to maintain population numbers is unknown and difficult to infer, given that there may not be a primary flood stimulus for breeding. Due to their preference for deep open water sites for breeding, breeding at semi-permanent sites may occur only opportunistically. Therefore, prescribing a flood frequency to maintain their population is not necessary as breeding is likely to be maintained at permanent water sites in the absence of flooding.

Crome (1988) indicated that great crested grebe breed most successfully when flooding follows drying. It is suggested that inter-flood drying of a few months may promote wetland productivity and breeding success in semi-permanent wetlands. In the absence of the required flooding and drying regime, breeding will most likely be maintained at permanent wetland sites.

Hoary-headed grebe: *Poliocephalus poliocephalus*

The hoary-headed grebe is a small grebe with a short stout bill, broad body and broad downy upturned rear (Figure 3.2). The plumage is darkish grey and white, with a distinctive narrow

Figure 3.2: Hoary-headed grebe, *Poliocephalus poliocephalus*. Photograph: Chris Herbert (Hunter Bird Observers Club).

black streak down the nape of the neck. During the breeding season the hoary-headed grebe develops white streaking over the entire head. It has an approximate length of 29–31 cm and a wingspan of 46 cm. The male is slightly larger, with a weight of 0.26 kg compared to the female weight of 0.22 kg. The hoary-headed grebe is regarded as a gregarious bird and may be observed in large flocks on large open waterbodies. While it may be observed in brackish and estuarine settings, it exhibits a preference for freshwater wetlands. Distribution is limited to Australia and New Zealand, and it is generally absent from the arid central parts of Australia (Marchant and Higgins 1990).

Habitat and diet

The hoary-headed grebe may be found in wetlands ranging from inland to estuarine and in temperate, tropical, arid and semiarid parts of Australia (Marchant and Higgins 1990). In inland settings it has been associated with extensive lakes, rivers, swamps, ponds, waterholes, dams and other man-made settings (Fjeldsa 1983, 1985; Marchant and Higgins 1990). It reportedly exhibits a preference for large sheets of open water with water depths of 0.5–3 m, with submerged vegetation. The hoary-headed grebe avoids waters dominated by submerged macrophytes. Roosting occurs near grassy banks, low sedges, scattered lignum, flooded trees or shrubs, or where submerged vegetation is near the water surface and food is obtained by diving among vegetation. Water clarity can be an important factor. Breeding occurs preferentially at permanent water or 'climax' flood stages of semi-permanent wetlands (Crome 1988). Colonies breed in the shallows of wetlands, among scattered sedges, reed, emergent vegetation, lignum and dense submerged vegetation. Nests consist of a small floating platform of water weeds or other plant material, with a depression in the centre (Fjeldsa 1988; Marchant and Higgins 1990).

The diet of the hoary-headed grebe consists of fish and arthropods such as insects, fish, crustaceans, frogs, spiders and earthworms, which are obtained by diving. Consumed crustaceans identified from stomach content analyses included *Daphnia, Mytilocypris henricae* and *Paratya australiensis*, while consumed fish included *Gambusia affinis* (Fjeldsa 1988).

Breeding and flood requirements for breeding

Breeding activity in the hoary-headed grebe is reportedly stimulated by flooding (Briggs 1990), a view supported by Crome (1988) who observed hoary-headed grebe nesting when flooding followed drying of a wetland. However, breeding of hoary-headed grebes was also correlated with photoperiod with a one-month lag (Halse and Jaensch 1989). In the absence of more detailed breeding information, it is suggested that breeding of hoary-headed grebes is stimulated by both flooding and season. The hoary-headed grebe breeds in simple pairs in colonies or loose colonies of up to 300 pairs (Marchant and Higgins 1990). Breeding primarily occurs between October and January, but may occur as early as August and as late as March (Halse and Jaensch 1989; Marchant and Higgins 1990).

Despite the strong correlation established between breeding of hoary-headed grebes and photoperiod, a relationship was also established between breeding and rainfall with a three-month lag (Halse and Jaensch 1989). This estimate is consistent with the lag time required for other grebe species, particularly for grebe breeding at semi-permanent wetlands.

Clutch sizes are usually four or five eggs, but may range from three to six eggs. There is no evidence indicating that second broods may be reared within a season, but replacement broods have been observed when loss occurs. Eggs are likely to be laid daily and incubated by both sexes for a period of 20–25 days (Marchant and Higgins 1990). Total breeding duration is estimated at one to two months. The age at fledging and parental independence is unknown; however, based on the numbers of immature birds observed at breeding sites, it is suggested that sexual maturity may take some time (Marchant and Higgins 1990). Based on the proposed breeding duration and lag time, it is therefore estimated that hoary-headed grebes require a minimum flood duration of three to four months, but breeding success may increase when flooding occurs for six to eight months. Floodwaters should recede at a moderate to slow rate to ensure breeding is successful.

As evident by their strong diving ability and a diet dominated by fish and crustaceans, it is likely that hoary-headed grebes prefer deep waters. This is supported by observations by Halse *et al.* (1993), who found positive correlations between hoary-headed grebe numbers and water depths at the start of breeding (September). This may be an artefact of their requirement for moderately long flood durations, although the hoary-headed grebe morphology is strongly suited to diving in deep waters (Storer 1987).

Similar to the great crested grebe, the frequency at which suitable conditions for hoary-headed grebe breeding should occur is unknown and difficult to infer, given that flooding may not be the primary stimulus for breeding. Hoary-headed grebes exhibit a preference for permanent water sources or 'climax' communities for breeding (Crome 1988; Marchant and Higgins 1990). As their abundance appears to correlate with deep permanent waters (Halse *et al.* 1993), it is likely that breeding in semi-permanent wetlands occurs opportunistically. In semi-permanent wetlands, drying for a few months may promote wetland productivity and enhance breeding success. In the absence of ideal flooding and drying conditions in semi-permanent and ephemeral wetlands, breeding may occur at sites which are permanently flooded.

Australasian grebe: *Tachybaptus novaehollandiae*

The Australasian grebe is the smallest grebe in the Australasian region. It has a short stout bill and neck, and a dumpy blunt-ended body (Figure 3.3). Males are generally heavier than females. The Australasian grebe has an approximate length of 23–25 cm, a wingspan of 39 cm

Figure 3.3: Australasian grebe, *Tachybaptus novaehollandiae*. Photograph: Chris Herbert (Hunter Bird Observers Club).

and a weight of 0.1–0.23 kg (Marchant and Higgins 1990). The non-breeding plumage is dark grey-brown on its upper parts and silver-grey below. There is a distinctive white oval patch of bare skin near the bill, a glossy black head and a rich chestnut stripe from the eye to the base of the neck during the breeding season. The Australasian grebe may be observed solitary or in pairs or small groups within shallow freshwater habitats. It is distributed throughout the Australasian region and has a wide Australian distribution, except in the drier arid regions. It has been observed in most parts of the Murray-Darling Basin (Marchant and Higgins 1990).

Habitat and diet

The Australasian grebe may be observed in many wetland types in a range of settings, but it exhibits a strong preference for freshwater wetlands. In inland New South Wales, it has commonly been observed on semi-permanent swamps and dams (Fjeldsa 1988; Vestjens 1977b). In Victoria it has been observed on shallow marshes of less than 0.5 m depth and dominated by herbs, in deep open marshes without emergent vegetation or marshes dominated by rushes, and in permanent natural lakes or reservoirs over 1 m deep (Corrick 1982; Corrick and Norman 1980). Foraging commonly occurs near steep banks or at the edges of reed beds, submerged vegetation, logs or floating vegetation (Fjeldsa 1985, 1988; Marchant and Higgins 1990). Roosting occurs on the water in pairs, families or small groups, near vegetation, below steep banks or near logs in the water. During the breeding season, roosting may occur on or beside the nest. Breeding occurs in the water among emergent, overhanging or submerged vegetation or fallen trees. Nests consist of a small platform constructed from water weeds, sedges and other plant material; eggs are positioned in a small depression in the centre of the nest (Marchant and Higgins 1990).

The diet of the Australasian grebe consists of fish, snails and aquatic arthropods, such as crustaceans, spiders, insects, amphipods and ostracods. Consumed items may include frogs, freshwater mussels, Poaceae seeds and other plant material (Fjeldsa 1988; Hobbs 1958; Lyle 1979; Marchant and Higgins 1990; Vestjens 1977b).

Breeding and flood requirements for breeding

Breeding in Australasian grebes is reportedly stimulated by flooding and season (Briggs 1990), with some studies supporting the season hypothesis (Gosper 1981; Halse and Jaensch 1989) and others supporting flooding (Crome 1988). Like other grebes, seasonal stimulus of breeding may be more important in permanent wetlands. The Australasian grebe is territorial and breeds in simple pairs. The peak breeding season for inland New South Wales reportedly is in November, but eggs have been observed as early as August and as late as April (Marchant and Higgins 1990).

The lag time for breeding is difficult to estimate. Peak breeding of Australasian grebe in south-western Australia coincided with maximum flood depth plus one or two months lag and there was a weak relationship between rainfall plus a three-month lag, indicating a substantial lag time (Halse and Jaensch 1989). Therefore it is estimated that the Australasian grebe requires a minimum lag time of one to two months from the flood peak or three months from the commencement of flooding. Breeding success will increase with a lag time of five to seven months.

Australasian grebe clutch size is likely to be four or five eggs but may range between one and nine eggs, though larger sizes may represent clutches from two nesting females in the one nest. Eggs are laid irregularly at intervals of up to 48 hours and are incubated by both parents for about 23 days. Young are independent of parents by eight weeks old, but the age until first flight is unknown (Marchant and Higgins 1990). The total breeding duration is estimated at three months. The minimum flood duration for breeding success should be six months; however, breeding success will be enhanced with longer flood durations of seven months or more and a slow recession of floodwaters.

The Australasian grebe has a diet heavily reliant on fish and other aquatic animals and morphologically is well-designed for diving in deep waters, as it has been observed in water depths of less than 0.5 m to over 1 m (Jaensch *et al.* 1988). This suggests that the Australasian grebe prefers deeper wetlands, a hypothesis supported by field surveys (Halse *et al.* 1993). It is therefore postulated that the Australasian grebe prefers deeper water depths that are more permanent.

Like other grebes, the frequency at which flooding should occur to maintain population numbers is difficult to establish, but may not be an essential characteristic of successful breeding. The use of semi-permanent wetlands may be opportunistic, as the Australasian grebe appears to prefer deep permanent wetlands (Halse *et al.* 1993). A flood frequency to maintain populations is not prescribed, as it is likely that, in the absence of flooding at semi-permanent sites, breeding will be sustained at deep permanent wetlands. Similarly, prescribing an inter-flood dry-period may not be necessary as breeding may occur at permanent wetlands in the absence of flooding. However, in semi-permanent and ephemeral wetlands, the Australasian grebe was found to breed best when flooding followed drying (Crome 1988). It is therefore suggested that drying in these wetlands may promote wetland productivity and the breeding success of the Australasian grebe.

Australian pelican: *Pelecanus conspicillatus*

The Australian pelican is a large, mainly white bird with short legs and tail, black and white wings, and a huge bill with a fleshy pouch (Figure 3.4). The sexes are similar in appearance, but females are generally slightly smaller. The Australian pelican has a length of 1.6–1.8 m, a wingspan of 2.3–2.5 m and a weight of 4.0–6.8 kg (Marchant and Higgins 1990). The Australian

Figure 3.4: Australian pelican, *Pelecanus conspicillatus*. Photograph: Kerrylee Rogers (DECCW).

pelican is regarded as gregarious and it may be seen in small to large flocks on rivers, estuaries, billabongs and lagoons. Its distribution is primarily limited to the Australasian region and it is generally absent from desert areas when water is not present (Marchant and Higgins 1990).

Habitat and diet

The Australian pelican may inhabit a range of settings including inland wetlands, estuarine and marine settings, and ephemeral wetlands. It generally prefers large sheets of open water that are free of dense vegetation, such as large lakes, reservoirs, billabongs, rivers, floodplains, deep open water, channels in swamps and lakes, shallow freshwater swamps and large dams (Corrick and Norman 1980; Fjeldsa 1985; Gosper 1981; Sharland 1957; Vestjens 1977b). Foraging typically occurs in open water areas where prey items are scooped into the bill, or in receding floodwaters or seasonal wetlands where prey is concentrated and easily caught (Crawford 1972; Hobbs 1961). Roosting and loafing occurs on stretches of open shoreline with little or no vegetation or on logs and rocks near water (Marchant and Higgins 1990). Breeding occurs on low secluded islands, islets or shores on bare ground or among patchy vegetation such as grass or lignum. Breeding reportedly requires an undisturbed site with abundant food for at least three months prior to commencement. Nests may be established just above the maximum water level or near the water's edge. Nests are constructed in shallow depressions that are lined or surrounded with plant stems, sticks, seaweed, water weed, ribbongrass, weeds and rubbish (Marchant and Higgins 1990; Vestjens 1977a).

The diet of the Australian pelican primarily consists of fish, but may include other insects, small crustaceans and even small ducks. It is known to cannibalise small live young pelicans

(Campbell and Sonter 1985; Hobbs 1961; Marchant and Higgins 1990; Smith and Munro 2008; Stone 1913; Vestjens 1977a).

Breeding and flood requirements for breeding

The Australian pelican breeds colonially in simple pairs, often in association with other colonial nesting waterbirds such as ibis, spoonbill, heron and cormorants. The timing of breeding is greatly dependent on flooding, rainfall and water levels at the breeding site (Briggs 1990; Marchant and Higgins 1990). Due to the relationship between flooding and the absence of quantitative studies of pelican breeding, it is difficult to identify a core breeding season. However, as breeding success is closely linked with rising and falling water levels and with food availability (MacGillvray 1922), it is suggested that environmental conditions, rather than season, are the primary determinants of breeding. Since flooding generally occurs within the Murray-Darling Basin in late winter to early summer, it is suggested that breeding is likely to occur in spring to summer to coincide with flooding.

The lag time for Pelecaniformes is estimated at two to three months (Scott 1997); Vestjens (1977a) and Nelson (2005) indicated that the Australian pelican requires three undisturbed months and an assured food supply for successful breeding. Its abundance has been found to correlate with rainfall occurring approximately three months prior to breeding (Woodall 1985).

Clutch size for the Australian pelican generally ranges from one to four eggs, with the majority laying one or two eggs. The first egg is laid two to three days after nest construction and subsequent eggs are laid two to four days later (Vestjens 1977a). Replacement laying may occur after loss (Marchant and Higgins 1990), but it has been suggested that second clutches are not laid within a season (MacGillvray 1922). The eggs are incubated by both parents for 32–35 days and young are cared for continuously for about 25 days before they are left alone in the nest. The first flight is attempted at about three months of age. The time required before young are completely independent or sexually mature is unknown (Marchant and Higgins 1990). The total breeding duration is estimated at four to five months. It is therefore estimated that flooding should be sustained for a period of seven to eight months.

Depth of flooding may not be important for pelicans, particularly as feeding behaviour does not require access to bottom sediments or edges of wetlands. However, fluctuating water levels have been implicated in reduced breeding success and chick mortality (Vestjens 1977a). Early drops in water level may expose nests to destruction or predation, and a rise in water levels after nest establishment will flood nests and eggs and drown the young (Vestjens 1977a). It is therefore suggested that pelicans require a consistent rate of water rise rather than fluctuating water levels, and that waters should recede at a relatively slow pace.

Information regarding the flood frequency needed to sustain breeding pelican populations is not available. In its absence, flood frequency is estimated from the longevity of wild Australian pelicans. The Australian pelican may live for 60 years in captivity (Nelson 2005), but longevity in the wild is much shorter. Banding-returns of pelicans indicated that they may live in excess of 16.8 years in the wild (Australian Bird and Bat Banding Scheme 1995b), but this is likely to be a maximum life expectancy. Anecdotal evidence indicates that the Australian pelican has a life expectancy of 10–25 years in the wild. Therefore, it is estimated that the Australian pelican requires a minimum large flood frequency of one flood every five years, with small maintenance floods every other year.

It is unlikely that inter-flood drying is an essential characteristic of Australian pelican breeding at inland wetlands, particularly as it reportedly exhibits a preference for deep permanent wetland settings (Halse *et al.* 1993). While drying may promote productivity of semi-permanent or temporary wetlands, provided that there are adequate fish stocks, drying may

Figure 3.5: Darter, *Anhinga melanogaster*. Photograph: Jennifer Spencer (DECCW).

not be necessary. Alternatively, breeding may be undertaken at more permanent wetlands when fish stocks are low.

Darter: *Anhinga melanogaster*

The darter is a large slender waterbird with a length of approximately 90 cm, a wingspan of 120 cm and a weight of 0.9–2.6 kg (Figure 3.5). Its bill is described as 'fine stiletto-like', its head is slender and it has an elongated glossy black body. Females may be distinguished from males as the white stripe from the gape along the side of the head has a black border and white to pale buff underparts. Darters are mainly solitary waterbirds, but nesting may occur in small colonies. The darter may intermingle with other colonial-nesting waterbird species such as cormorants, herons and ibises, and it may congregate where food is abundant. It has a widespread distribution throughout Africa, South-east Asia and Australia. It may be observed in all parts of the Murray-Darling Basin (Marchant and Higgins 1990).

Habitat and diet

The darter inhabits large areas of water such as lakes, estuaries and large rivers, but may also be observed in deeper parts of wetlands, shallow vegetated edges of large lakes, semi-permanent and seasonal freshwater swamps, and coastal and estuarine settings (Corrick and Norman 1980; Fjeldsa 1985; Gosper 1981; Jaensch *et al.* 1988; Marchant and Higgins 1990; Vestjens 1975a). It forages in areas where emergent and aquatic vegetation is sparse, probably to enable unobstructed diving and swimming (Marchant and Higgins 1990). Nesting occurs in trees or tall bushes growing in or over water; if water levels fall during breeding, the darter may move to deeper waters (Jaensch *et al.* 1988; Vestjens 1975a). Nesting occurs in the forks of branches on live or dead trees away from the main trunk and above water with a depth of at least 0.3 m (Vestjens 1975a). The darter may have a greater preference for dead river red gum (*Eucalyptus camaldulensis*) than for live river red gum (Briggs *et al.* 1997). Nests are constructed from a base of green twigs, built up with dry sticks and lined with green leaves. Old nests may be reused during subsequent breeding. The darter also requires fringing or projecting tree trunks, branches, stumps or other objects on which to roost, perch and dry its wings (Fjeldsa 1985; Marchant and Higgins 1990; Vestjens 1975a).

The diet of the darter primarily consists of fish and insects, occasionally aquatic animals such as tortoises, and plant matter (Dostine and Morton 1989b; McKeown 1943; Serventy 1938; Vestjens 1975a).

Breeding and flood requirements for breeding

Darter breeding appears to be stimulated by flooding (Briggs 1990) and therefore occurs erratically in response to water availability (Marchant and Higgins 1990). It is evident that the darter's nesting frequency correlates with flooding in excess of four months rather than with the degree of hydrological control within a wetland, which suggests that flooding is the stimulus for breeding (Briggs *et al.* 1997). Breeding generally tends to occur from November through to April, though this may represent the seasonal nature of flooding throughout the Murray-Darling Basin. Breeding occurs in solitary or small loose colonies with other colonial nesting waterbirds such as spoonbills, ibises and cormorants (Marchant and Higgins 1990; Vestjens 1975a).

Pelecaniformes, such as the darter, reportedly require no drainage of water or a long lag of two to three months (Crome 1988; Scott 1997). The mean duration between flooding under nests and peak number of darter broods was five months following winter/spring flooding and eight months following autumn flooding (Briggs and Thornton 1999). As the diet consists primarily of fish, it is likely that the lag time depends on the time required for fish populations to establish and to sustain the darter throughout breeding. Lag times may therefore depend on the season in which flooding occurs as well as whether complete drying of a wetland is required for maximum productivity. To achieve successful darter breeding events in semi-permanent wetlands, the lag should be long enough that wetlands become dominated by fish and other insects. It is suggested that, where complete drying occurs, a lag of at least a few months is essential (Crome 1988). When flooding occurs outside the breeding season and after complete drying, a longer lag period may be essential for the wetland to reach peak productivity.

Darter clutch size ranges between two and six eggs, with an average of four eggs (Vestjens 1975a). The first egg is laid two to three days after pair formation and subsequent eggs are laid at intervals of one to three days. Both parents incubate a clutch for an average period of 28 days. Chicks are fed by both parents for about four weeks, after which they clamber out of the nest, returning to feed and rest after short periods. After 50 days chicks may fly short distances; fledging generally occurs at about 60 days (Vestjens 1975a). No information is available about the period from fledging to maturity (Marchant and Higgins 1990) and breeding duration is estimated at approximately three months. In wetlands that undergo drying, and exhibit flooding during periods of peak productivity, flood duration should be a minimum of five to six months to sustain successful darter breeding events. Bimodal breeding (Halse and Jaensch 1989; Vestjens 1975a) may be evident when flooding is sustained for longer periods.

The depth of flooding around nest sites is reportedly at least 0.3 m with an average of 3.5 m (Vestjens 1975a). The rate of fall of water should be slow enough for flooding to be maintained around nest sites for the duration of breeding. While the darter has been known to relocate nests over deep water if the water level drops (Vestjens 1975a), this behaviour is limited to the lag period rather than breeding period.

There is little information about the flood frequency required to sustain darter populations. The darter has been kept in captivity for approximately 16 years (delHoyo *et al.* 1992) but its longevity in the wild is likely to be less. As the darter can disperse over long distances (>2000 km) during the non-breeding season then contracts to core breeding areas during the breeding season (Marchant and Higgins 1990), it is essential that flooding occurs within Murray-Darling Basin wetlands at a frequency that enables the darter to breed and maintain population numbers. It is estimated that darters require one large flood every four years, plus small floods every one to two years, for population maintenance. However, in the absence of flooding, the darter may be able to breed at permanent wetlands (Crome 1988) provided that resource needs are met.

There is conflicting evidence about whether the darter requires an inter-flood dry-period. Crome (1988) suggested a preference for highly productive climax wetlands with abundant fish and other insects to sustain darters for breeding, while others indicated that dry-periods are essential for maximising productivity (Briggs *et al.* 1997; Scott 1997). Since the darter and some of the fish it consumes are adapted to the semi-permanent conditions of wetlands within the Murray-Darling Basin, it is suggested that breeding success in semi-permanent settings may increase with drying for a few months.

Great cormorant: *Phalacrocorax carbo*

The great cormorant is a large mainly black cormorant with a stout bill that has a prominent terminal hook (Figure 3.6). It has a long neck, large wings, short stout legs and large black webbed feet. It has a length of 80–85 cm, a wingspan of 130–150 cm and a weight of 2–2.5 kg. Both sexes exhibit a similar appearance; a black crest on the nape and a white thigh patch develops when

Figure 3.6: Great cormorant, *Phalacrocorax carbo*. Photograph: Chris Herbert (Hunter Bird Observers Club).

breeding (Marchant and Higgins 1990). The great cormorant is regarded as a solitary bird or occurs within small groups, although it forms large colonies for flying and breeding. It has a widespread distribution throughout most of Australia except the arid interior. Breeding colonies of less than 10 to several thousand occur within the Murray-Darling Basin (Marchant and Higgins 1990).

Habitat and diet

The great cormorant can inhabit inland wetlands and coastal locations. At inland locations it has been observed on lakes, reservoirs, swamps, rivers, pools and billabongs (Marchant and Higgins 1990). It apparently favours large sheets of water, deep permanent water, major rivers or open water in deep marshes (Corrick and Norman 1980; Fjeldsa 1985; Hobbs 1961). Foraging occurs among a range of habitats and appears unrelated to salinity, turbidity or the type or occurrence of shoreline vegetation (Fjeldsa 1985). Roosting occurs on trees and bushes near water at inland locations and the great cormorant may be observed perching with its wings extended to dry. Nest sites may include lakes, swamps, rivers, billabongs and temporary floodwaters (Corrick 1982; Jaensch *et al.* 1988; Vestjens 1977b). Nests are constructed within trees, shrubs and reeds, such as river red gum, *Melaleuca* species, lignum and common reed, that are typically located in or near water. The great cormorant is most commonly observed nesting in trees at a height of 1–3 m above the water level, or up to 7 m above the water level. The nest consists of a platform constructed from dead sticks, twigs and other plant material and debris (Marchant and Higgins 1990).

The diet of the great cormorant largely consists of fish, as well as crustaceans and insects (Marchant and Higgins 1990). Analyses of guts from the great cormorant from inland locations indicate a range of fish species may be consumed, such as goldfish (*Carassius auratus*), redfin perch (*Perca fluviatilis*), short-finned eel (*Anguilla australis*), bony bream (*Nematlosa erebi*), tench (*Tinca tinca*), golden perch (*Macquaria ambigua*) and southern pygmy perch (*Nannoperca australis*) (Bright 1934; Hobbs 1961; McKeown 1943; Serventy 1938; Vestjens 1977b). Other items consumed include tadpoles, crustaceans, insects, spiders, frogs, turtles and ducklings (Brooker *et al.* 1979; Edwards 1920; Marchant and Higgins 1990; Vestjens 1977b).

Breeding and flood requirements for breeding

The great cormorant reportedly breeds in response to season and flooding (Briggs 1990), but breeding may occur at any time when food is plentiful (Marchant and Higgins 1990). It has preferentially bred at sites where river red gums were flooded in excess of four months (Briggs *et al.* 1997). The great cormorant is regarded as a colonial-nester, occurring in sites with other colonial-nesting waterbird species such as other species of cormorants, herons, spoonbills and ibises. Two main breeding seasons are reported for the Murray-Darling Basin: from September to January and from March to August (Marchant and Higgins 1990).

There is very limited information about the relationship between breeding success and flooding, including specific information on lag times. Scott (1997) indicated that Pelecaniformes require flooding with a lag of two to three months. Inland cormorant species reportedly exhibited greater breeding success with a minimum flood lag time of three months and a range of up to nine months (Kingsford and Auld 2005). It is cautiously suggested that the great cormorant requires a minimum flood lag of three months and an ideal flood lag of up to nine months. This extended lag period is supported by Briggs *et al.* (1997), who found that the great cormorant showed a preference for nesting in wetlands with prolonged inundation, even to the extent that flooding resulted in the death of river red gums. There were strong relationships between the number of breeding records and area of river red gum flooded in excess of four months, maximum area of dead gums and maximum area of open water in that study.

Great cormorant clutch size is generally three to five eggs, but up to six eggs may be produced per clutch. Laying occurs at intervals of 48 hours and two broods may be laid per year. No information is available about whether replacement laying occurs when a clutch is abandoned or damaged. Both parents incubate the clutch for a period of 27–31 days. Chicks have been observed leaving the nest at about four weeks old, with fledging occurring at about seven weeks old. Chicks are self-sufficient about four weeks after fledging. No information is available about the period from fledging to maturity, but some great cormorants may breed in their second year (Marchant and Higgins 1990). Breeding duration is estimated at approximately four months. A minimum flood duration of seven months is required for successful breeding, and an ideal flood duration in excess of 12 months may enhance breeding success and promote the rearing of multiple broods (Marchant and Higgins 1990).

Water depth is not an essential characteristic of great cormorant breeding, as nesting occurs in trees and prey items are caught by diving to depths of up to 30 m (Gremillet *et al.* 1998). However, waters should be deep enough to enable diving for prey items and the rate of fall of water should be slow enough to maintain flooding around the nest for the duration of breeding or a minimum of seven months.

The frequency of large floods needed to maintain breeding populations is estimated from the longevity of the great cormorant and its dispersal ability. Large cormorants, such as the great cormorant, may have a life-span of 20–30 years (Nelson 2005); however, longevity records indicate that the great cormorant has survived in the wild for approximately 18 years (delHoyo *et al.* 1992). As the great cormorant is dispersive following successful inland breeding (Marchant and Higgins 1990), it is suggested that it requires one large flood every five years and small floods every one to two years to maintain frequented habitats. However, due to a preference for large deep permanent waters, semi-permanent and ephemeral waters are likely to be used only opportunistically.

As a specialist fish-eater, the great cormorant is likely to have similar drying requirements as the darter. Dry-periods may be essential for increasing the productivity of consumed fish (Briggs *et al.* 1997; Leslie 2001; Scott 1997) and at least some of the fish species consumed have flood requirements for spawning and survival (see Chapter 4). Drying for a few months may promote the recruitment of fish consumed by the great cormorant. This requires further study, as the preference for large permanent waterbodies suggests drying may not be essential.

Little pied cormorant: *Phalacrocorax melanoleucos*

The little pied cormorant is a small squat black and white cormorant with a long tail. It has a length of 55–65 cm, a wingspan of 85–90 cm and a weight of 410–880 g (Marchant and Higgins 1990; Figure 3.7). It is regarded as a solitary bird, except during the breeding season when it may form colonies with other colonial-nesting birds such as other cormorants, herons, darters, spoonbills and ibises (Vestjens 1977b). It may also form small flocks when roosting and foraging. It is typically observed in marine and inland waters and has a widespread distribution throughout Australia, excluding arid central Australia, but may be observed in ephemeral settings in desert areas (Marchant and Higgins 1990).

Habitat and diet

In inland locations, the little pied cormorant may frequent lakes, reservoirs, swamps, small and large waterways, billabongs, pools, channels, dams and ornamental ponds (Marchant and Higgins 1990). It is less commonly observed in wet meadows, and densely vegetated swamps with reeds, sedges or trees (Corrick and Norman 1980; Fjeldsa 1985; Gosper 1981). Foraging typically occurs in open water and on large lakes but is limited to the shallow margins when native

Figure 3.7: Little pied cormorant, *Phalacrocorax melanoleucos*. Photograph: Chris Herbert (Hunter Bird Observers Club).

crustaceans are abundant (Miller 1979; Vestjens 1977b). Roosting and nesting occurs in trees and bushes near the water. The little pied cormorant may also be observed perched on trees, stumps, rocks, earth banks and artificial structures. Nesting primarily occurs in freshwater wetlands such as lakes, swamps, billabongs and rivers (Corrick and Norman 1980; Jaensch *et al.* 1988; Miller 1980) with nests constructed within forks, stems and branches of trees such as species of *Melaleuca, Casuarina* and *Eucalyptus*, in bushes such as lignum or on snags in swamps, lakes and along rivers in trees or bushes located in or near the water. The average height of nests above the water level is reportedly 2.8 m. Nests consist of a platform of sticks and bark, lined with leaves (Marchant and Higgins 1990). Breeding may be more successful in swamps with high levels of organic matter, complex flora and diverse invertebrate populations (Crome 1988).

The diet of the little pied cormorant primarily consists of crayfish and other crustaceans (Marchant and Higgins 1990). It may also include shrimp, yabbies, fish, insects, molluscs, amphibians and plant material (Dickinson 1951; Miller 1979; Serventy 1938; Vestjens 1977b).

Breeding and flood requirements for breeding

Breeding is stimulated by flooding and season (Briggs 1990), and may occur whenever conditions are suitable. The importance of flood and season stimuli was evident through strong relationships between flow, flood extent and nest numbers at the Macquarie Marshes (Kingsford and Auld 2005), and correlations between breeding, rainfall lag and the weight of European carp (*Cyprinus carpio*) ingested by cormorants at Lake Cargelligo and Lake Brewster in inland New South Wales (Miller 1980). Breeding occurs with other colonial-nesting waterbirds, such as egrets and herons (Vestjens 1977b). Little information on breeding season is available for this species, but it is reportedly more common in September to March at Lake Cowal, New South Wales (Vestjens 1977b).

Pelecaniformes, such as the little pied cormorant, reportedly require flooding with a lag time of two to three months (Scott 1997). Flow lags of three to nine months were correlated with the number of cormorant nests at the Macquarie Marshes (Kingsford and Auld 2005). The mean duration between flooding under nests and peak number of little pied cormorant broods was three months following winter/spring flooding and six months following autumn flooding (Briggs and Thornton 1999). Correlations between breeding records, area of live river red gum flooded in excess of four months and maximum area of river red gum at Barmah Forest indicate that the little pied cormorant has some lag time requirements. However, as it generally does not nest in dead river red gums it may be inferred that flood duration should not be long enough to kill the trees, as preferred by the great cormorant at wetlands on the Murrumbidgee floodplain (Briggs *et al.* 1997). It is therefore suggested that the little pied cormorant requires a minimum lag of three months, and up to nine months.

Clutch sizes for the little pied cormorant range from three to five eggs (Marchant and Higgins 1990, Vestjens 1977a). No information is available about the laying interval, period of incubation and period to fledging and maturity. It is also unknown whether juveniles can breed in their first year (Marchant and Higgins 1990). Assuming that breeding is similar to that of the little black cormorant, it is estimated that the little pied cormorant has a breeding duration of approximately three to four months, requires a minimum flood duration of six to seven months and an ideal flood duration of approximately 12 months. Nests may be abandoned in response to the sudden onset of drought conditions (Miller 1980). Based on the preferential selection of wetlands with living river red gums, it is inferred that the little pied cormorant does not prefer extended flood durations of more than approximately 12 months (Briggs *et al.* 1997).

While nests are built over water, specific water depths are not essential for little pied cormorant breeding as nesting occurs within trees and prey items are caught by diving. However, there is evidence to suggest that the little pied cormorant prefers to forage for food in shallow water habitats and is capable of diving only for short periods (Miller 1979). The rate of drawdown of water may have significant implications for breeding success. Crustaceans, particularly yabbies, bury into sediments and become inaccessible when waters recede and this has been associated with the abandonment of nests at Lake Cargelligo and Lake Brewster in inland New South Wales (Miller 1979, 1980). Waters should recede relatively slowly after a minimum flood duration of six to seven months.

There is no information on the flood frequency required to maintain little pied cormorant populations and longevity. However, low breeding success at Barmah Forest has been associated with smaller flood volumes since regulation in the mid 1970s (Leslie 2001). Based on the average life expectancy (eight to 10 years) for wild colonial waterbirds in Australia (Scott 1997), it is suggested that large floods should occur every four years with smaller habitat maintenance floods every one to two years.

The diet of the little pied cormorant primarily consists of crustaceans, such as yabbies and shrimp. As yabbies bury into the ground when floodwaters recede, it may be inferred that the little pied cormorant would have greater breeding success when flooding follows a period of drying. However, in permanent water and in the absence of flooding, the little pied cormorant will rely on more readily available food sources such as fish (Miller 1979). There is evidence to suggest that reproductive output of yabbies and their populations may increase in response to flooding (see Chapter 4). It is therefore proposed that an inter-flood dry-period should occur for a few months between breeding seasons so as to increase food availability.

Little black cormorant: *Phalacrocorax sulcirostris*

The little black cormorant is described as a small slender dark-coloured cormorant with a thin bill, dark facial skin and long tail (Figure 3.8). Its plumage is dull brown-black. It has a length of about 55–65 cm, a wingspan of 95–105 cm and a weight of 0.52–1.21 kg. The little black cormorant is a colonially nesting species and nests in association with other cormorants and waterbirds. It may be observed in large flocks foraging in inland waters and marine inlets. It has a limited distribution in Australasia but a widespread distribution throughout Australia. It does occupy ephemeral waters in deserts, but does not occur in arid central Australia (Marchant and Higgins 1990).

Habitat and diet

The little black cormorant has a widespread distribution in wetlands and favours inland settings, such as large lakes, rivers, billabongs and pools, dams and channels and swamps

Figure 3.8: Little black cormorant, *Phalacrocorax sulcirostris*. Photograph: Chris Herbert (Hunter Bird Observers Club).

(Marchant and Higgins 1990). Foraging mainly occurs in waters deeper than 1 m, particularly in lakes with abundant European carp and submerged beds of ribbonweed (Corrick and Norman 1980; Dostine and Morton 1988; Fjeldsa 1985; Miller 1979). It may be observed less frequently within vegetated water such as around the edge of lakes, seasonal freshwater swamps, or swamps vegetated with sedges, reeds or trees such as spike-rush, common reed, club-rush and *Melaleuca* species (Corrick and Norman 1980; Vestjens 1977b). Roosting occurs in trees or on stumps, posts and other artificial structures (Marchant and Higgins 1990). Breeding occurs in vegetated swamps, preferably in trees surrounded by water and at heights of up to 30 m above the water level (Miller 1980; Vestjens 1977b). Nests consist of a constructed platform of sticks, leaves, dry reeds, lignum and water weed and may be lined with paperbark, leaves and feathers (Marchant and Higgins 1990).

The diet of the little black cormorant primarily consists of fish including goldfish, European carp, gambusia (*Gambusia holbrooki*), Australian smelt (*Retropina semonii*), freshwater catfish (*Tandanus tandanus*), olive perchlet (*Ambassis agassizii*), redfin perch and gudgeon (*Hypseloeotris* spp.), as well as freshwater crayfish and other crustaceans, and insects (Marchant and Higgins 1990; Miller 1979; Vestjens 1977b).

Breeding and flood requirements for breeding

Season and flooding are the primary breeding stimuli in the little black cormorant (Briggs 1990). This is based on the response of its gonadal cycles to temperature, water level and food (Miller 1980). The role of flooding in breeding is evident by the reported absence of breeding at inland New South Wales sites in response to drought (Miller 1980). Relationships have also been observed between water flow, flood extent and number of nests in the Macquarie Marshes (Kingsford and Auld 2005). The little black cormorant breeds with other colonially nesting waterbirds such as other cormorants, herons, spoonbills, ibises and darters (Vestjens 1977b). Breeding generally occurs later in the season (summer to autumn) in northern parts of Australia, and primarily occurs in spring to autumn in southern locations. The main breeding season in New South Wales is between spring and autumn, but may extend as long as September to June (Vestjens 1977b) or occur any time when conditions are favourable (Marchant and Higgins 1990).

Flood lags of two to three months are common for Pelecaniformes (Scott 1997), while lags of between three and nine months have correlated with the number of cormorant nests at the Macquarie Marshes (Kingsford and Auld 2005). The mean duration between flooding under nests and peak number of little black cormorant broods was four months following winter/spring flooding and six months following autumn flooding (Briggs and Thornton 1999). Correlations between breeding records and area of live river red gum flooded in excess of four months indicate that the little black cormorant does have some lag time requirements (Briggs *et al.* 1997). However, as it generally does not nest in dead river red gums it may be inferred that flood duration should not be long enough to kill the trees (Briggs *et al.* 1997). It is therefore inferred that the little black cormorant requires a minimum lag of three months or an ideal lag of up to nine months.

The average clutch size has not been quantified, but appears to be three to six eggs. No information on the laying interval, period of incubation, period from hatching to maturity and age of first breeding is available (Marchant and Higgins 1990). The period until fledging is reportedly more than seven weeks (Nelson 2005). In the absence of more reliable breeding information, it is estimated that the little black cormorant has a breeding duration similar to that of the little pied cormorant, estimated at three to four months. Flood duration is likely to be a significant stimulus for breeding, as the little black cormorant has not been observed nesting during a drought in inland New South Wales (Miller 1980). However, based on the preferential selection of wetlands with living river red gum trees on the Murrumbidgee floodplain, there is evidence to suggest that the little black cormorant does not prefer extended flood durations of more than approximately 12 months (Briggs *et al.* 1997). Therefore, it requires a minimum flood duration of six to seven months for successful breeding and an ideal flood duration of approximately 12 months. The rate of water recession should be slow enough to maintain flooding around nests for a minimum period of six to seven months.

The little black cormorant does not exhibit a water depth preference, as nesting occurs in trees and prey items are caught by diving. However, diving does not usually occur to great depths, and is generally limited to waters with an average depth of 1–2 m (Nelson 2005).

Information on the flood frequency and longevity of the little black cormorant is poor; however, low breeding success at Barmah Forest has been associated with smaller flood volumes since regulation in the mid 1970s (Leslie 2001). Based on the average life expectancy of eight to 10 years for colonial waterbirds in the wild in Australia (Scott 1997), it is suggested that large floods should occur every four years with habitat maintenance floods every one to two years.

Crome (1988) suggested that the little black cormorant is suited to swamps that have been filled for some time and have a climax community dominated by ribbonweed and fish species. Hence, it may breed most successfully at sites with little or no inter-flood dry-period. However, some fish species consumed by the little black cormorant have some reliance on flooding for recruitment (King *et al.* 2003; see Chapter 4). Therefore, it is suggested that inter-flood dry-periods may be required in semi-permanent settings to promote fish recruitment, while drying may not be required in climax communities where there is high resource availability.

Pied cormorant: *Phalacrocorax varius*

The pied cormorant is a large black and white cormorant with a long neck and a long stout bill with prominent terminal hook (Figure 3.9). It has short broad wings, a short tail and legs and large black webbed feet. It has a length of approximately 65–85 cm, a wingspan of 110–130 cm

Figure 3.9: Pied cormorant, *Phalacrocorax varius*. Photograph: Chris Herbert (Hunter Bird Observers Club).

and a weight of 1.3–2.2 kg. During the breeding season, plumage on the forehead, crown and nape are black and contrast sharply with its white face and neck. Non-breeding adults exhibit faded dark-brown plumage (Marchant and Higgins 1990). The pied cormorant may be regarded as solitary or gregarious and typically feeds singularly, but roosts and breeds colonially. It is more commonly observed in marine locations but does occur on large inland waters. It is endemic to Australia and its distribution is limited to the eastern states (Queensland, New South Wales, Victoria and Tasmania) and the coastlines of other states. Breeding typically occurs in coastal locations (Marchant and Higgins 1990). The pied cormorant is regarded as near-threatened in Victoria (*Flora and Fauna Conservation Act 1988*).

Habitat and diet

The pied cormorant is primarily a marine waterbird; however, it also inhabits inland wetlands including lakes, swamps, rivers, billabongs, pools and sewage ponds (Fjeldsa 1985; Marchant and Higgins 1990; Vestjens 1977b). At inland locations it is associated with large areas of water such as permanent freshwater lakes, reservoirs and open water in deep freshwater marshes (Corrick 1982; Corrick and Norman 1980). It may be unaffected by salinity and turbidity levels and the occurrence of shoreline vegetation (Fjeldsa 1985). It is reportedly less common in shallow vegetated waters, semi-permanent swamps and swampy edges of lakes (Vestjens 1977b). Roosting occurs in trees near water. Breeding more commonly occurs in coastal locations, as wetlands are generally more permanent and less affected by drainage than inland wetlands. At inland locations, nests are located in trees in or near water and are constructed at heights of 7–17 m (Favaloro 1952). Nests are constructed from twigs, seaweed and sometimes rubbish (Marchant and Higgins 1990).

The diet of the pied cormorant primarily consists of fish and some crustaceans (Marchant and Higgins 1990; Serventy 1938).

Breeding and flood requirements for breeding

Season and flooding are the primary stimuli for breeding in the pied cormorant (Briggs 1990). As breeding primarily occurs within the coastal zone and is less affected by fluctuating water levels, it is likely to be more strongly influenced by season. Flooding may be more important at inland locations, where water levels may fluctuate. Breeding may occur at any time and is related to resource availability and water conditions, but the main breeding season is between spring and autumn (Marchant and Higgins 1990). The pied cormorant breeds colonially and often in association with other cormorant species, Australian pelicans and spoonbills.

The clutch size for the pied cormorant is typically two to four eggs although five eggs is not uncommon (Norman 1974). Laying of eggs occurs at intervals of 48 hours and eggs are incubated for a period of 25–33 days. Chicks may walk from the nest at about 34 days old, and return to the nest for about four weeks. Fledging occurs at about 47–60 days, but young remain dependent on parents until about 80 days old. The period from fledging to maturity is unknown. It is also unknown whether young can breed in their first year (Marchant and Higgins 1990).

Pied cormorants exhibit a preference for breeding sites with more stable water levels, such as those within the coastal zone. Breeding at inland locations is associated with trees standing in lakes and swamps (Marchant and Higgins 1990; Vestjens 1977b), including dead or living river red gums (Leslie 2001). The pied cormorant is likely to breed at inland locations opportunistically (Leslie 2001) in response to extended flooding conditions. No further information on the flood conditions required for successful breeding is available, except that the pied cormorant requires floods of extended duration and does not require an inter-flood dry-period (Crome 1988).

Great egret: *Ardea alba*

The great egret is described as a graceful white egret with a flat crown, a black beak which changes to yellow during courtship and breeding, and gape extending behind the eye (Figure 3.10). The great egret has a body length of 83–103 cm, a wingspan of approximately 150 cm and a weight of 700–1200 g. Males and females have a similar appearance and develop filamentous plumes during breeding. The great egret has been observed singly, in small groups or as large colonies on floodplains during the wet season. It has a widespread distribution throughout Australia and is regarded as scarce or absent from the drier parts of Western Australia, Northern Territory and South Australia. The majority of breeding occurs within the New South Wales and Victorian parts of the Murray-Darling Basin (Marchant and Higgins 1990). The international importance of great egret habitats is recognised through its inclusion in the Japan–Australia and China–Australia Migratory Bird Agreements (JAMBA and CAMBA). Great egrets are regarded as vulnerable in Victoria (*Flora and Fauna Conservation Act 1988*).

Habitat and diet

The great egret frequents inland, estuarine and littoral wetlands, and moist grasslands. It has been observed at a variety of inland sites but has a preference for permanent waterbodies on floodplains, such as billabongs, watercourses or pools (Marchant and Higgins 1990). It may also frequent the shallows of deep permanent lakes that are open or vegetated with shrubs or tall trees, semi-permanent swamps with tall emergent vegetation such as cumbungi and spike-rush, herb-dominated swamps, freshwater meadows, seepage springs, flooded grassland, deep

Figure 3.10: Great egret, *Ardea alba*. Photograph: Chris Herbert (Hunter Bird Observers Club).

swamps vegetated with grasses, sedges or trees such as *Eragrostis* species, common reed, club-rush and *Melaleuca* species, and farm dams and channels (Andrew 1963; Corrick 1982; Corrick and Norman 1980; Fjeldsa 1985; Gosper 1981; Marchant and Higgins 1990; Vestjens 1977b). Foraging occurs in open shallow water at depths of approximately 0.3 m, but fish may be taken from the surface of deeper waters. Roosting occurs in trees in or near wetlands, while breeding occurs in wetlands with fringing or flooded trees or other tall vegetation to build nests in (Marchant and Higgins 1990). Nests are constructed in forks in the tops or upper parts of trees standing in water, commonly at heights of 7–15 m. Nests are made from a shallow platform of sticks woven together. Nest-building is usually complete six to 15 days prior to courtship, but can extend into the nestling period (Marchant and Higgins 1990).

Adult gut content analyses indicate that the diet of the great egret largely consists of fish, including eels and tadpoles, and may include freshwater snails, shrimps, freshwater crayfish, insects, snakes, frogs and small birds (Cleland 1912; Marchant and Higgins 1990; Recher *et al.* 1983; Vestjens 1977b).

Breeding and flooding requirements for breeding

Flooding and season stimulate breeding in the great egret (Briggs 1990), with significant correlations reported with rainfall (Halse and Jaensch 1989) and area of live river red gum flooded in excess of four months (Briggs *et al.* 1997). There are significant differences between the success of breeding in wet and dry years (Maddock and Baxter 1991). It is likely that breeding commences sooner when flooding occurs during the main breeding season of November to early May. However, breeding may occur in response to flooding outside the breeding season, with breeding events occurring as early as September and as late as April. The great egret breeds in pairs, colonially and in association with other colonial-nesting waterbirds such as herons, ibises, spoonbills and cormorants (Marchant and Higgins 1990).

The lag time for the great egret is variable, with lags of three to nine months reported. However, most literature suggests that breeding is most successful after a lag time of approximately nine months, making the great egret one of the slowest-breeding waterbirds (Briggs and Thornton 1999; Kingsford and Auld 2005; Scott 1997). Differences in the minimum lag time required may reflect the season in which flooding occurs, with shorter lag periods perhaps coinciding with the main breeding season. It is suggested that a minimum lag time of three months is required when flooding coincides with the breeding season, and peak breeding occurs with a lag time of nine months.

The great egret has a clutch size of two to six eggs. No information is available about the period of egg-laying and incubation. Chicks start to scramble from the nest at 25–37 days old. The period from hatching to first flight is unknown; flights from the nesting tree occur for about 10 days before the juveniles finally depart the nest. The period from hatching to final departure is 55–88 days (Marchant and Higgins 1990). Breeding duration is estimated at approximately three to four months. The minimum flood duration required to generate successful great egret breeding is six to seven months, but peak breeding may occur after flooding for 12 months.

No maximum or minimum flood depth is required for breeding. However, it is likely that floods of great depth have a greater duration and will therefore promote breeding success. This is supported by Leslie (2001), who proposed that unsuccessful breeding of great egrets in Barmah Forest in 1996 was related to insufficient flood duration, which caused decreased nest security and food availability. Therefore, the rate of water recession should be slow enough to maintain water around nests for the minimum flood duration of six to seven months.

There is little information about the frequency of flooding required to maintain great egret populations. The observed longevity of the great egret in Australia has not been greater than seven years and one month (McKilligan 2005). Although it is a dispersive bird, there may be some differentiation between breeding colonies in the northern and southern parts of the Murray-Darling Basin (Marchant and Higgins 1990). It is estimated that large floods should occur approximately every three years in both the northern and southern parts of the Murray-Darling Basin, and smaller floods should occur every one to two years to maintain habitats.

The great egret has had greater breeding success when flooding follows a drying period. Crome (1988) suggested that the production of large populations of detritivous chironomids and other prey, stimulated after drying and refilling of swamps, may sustain successful breeding in the great egret. Maddock and Baxter (1991) indicated that greater breeding success occurred when flooding in a dry period was followed by flooding in a wet period. Dry periods should occur for a few months between breeding events to maintain the seasonal nature of breeding.

Intermediate egret: *Ardea intermedia*

The intermediate egret is a slender white egret with long neck and dark legs. It grows to a length of 56–70 cm and a weight of 0.4 kg (Figure 3.11). The sexes exhibit a similar appearance. The colours of the bare parts change during the breeding cycle; the bill changes from deep pink to bright red with a yellow stripe during courtship. The intermediate egret has been observed singly or in groups of hundreds. It has a wide distribution range between northern and eastern Australia and throughout the Murray-Darling Basin (Marchant and Higgins 1990). The intermediate egret is regarded as regionally extinct in Victoria (*Flora and Fauna Conservation Act 1988*).

Figure 3.11: Intermediate egret, *Ardea intermedia*. Photograph: Chris Herbert (Hunter Bird Observers Club).

Habitat and diet

The intermediate egret can be seen frequenting inland wetlands, wet grasslands and, occasionally, sheltered coastal habitats. Habitats include billabongs, pools, wet meadows, moist pasture, swamps and watercourses on floodplains, preferably among emergent vegetation of grasses, herbs, sedges, reeds or rushes such as water primrose, common reed, spike-rush and cumbungi. It may occur in swamps or along the margins of shallow lakes with emergent shrubs or trees such as lignum, river red gum or *Melaleuca* species (Corrick and Norman 1980; Crawford 1972; Fjeldsa 1985; Gosper 1981; Hobbs 1961; Marchant and Higgins 1990; Vestjens 1977b). Foraging occurs in still water less than 80 mm deep and among dense aquatic and emergent vegetation. Foraging may also occur in deeper waters by diving or where matted vegetation provides a platform. Roosting occurs in trees fringing wetlands, and nests are built in trees within or near water (Marchant and Higgins 1990). Breeding occurs in the forks of trees, on horizontal branches and in the canopies of trees, such as river red gums, in or beside water. Nests are commonly established 1.5–15 m above the water and comprise a shallow platform of loosely woven dry sticks and fresh twigs with some leaves attached (Marchant and Higgins 1990).

The diet of the intermediate egret mainly consists of fish, but frogs, lizards, insects, larvae and arachnids may also be consumed (Marchant and Higgins 1990; Mathews 1910; Recher *et al.* 1983; Vestjens 1977b).

Breeding and flood requirements for breeding

Breeding in the intermediate egret is stimulated by flooding and season (Briggs 1990) but may be strongly influenced by food availability. Strong relationships have been reported between breeding success and rainfall (Baxter 1994; Maddock and Baxter 1991), with number of nests and flow rate and wetland area (Kingsford and Auld 2005), and with the area of river red gum flooded in excess of four months (Briggs *et al.* 1997). Flooding that coincides with the peak breeding season will enhance fish productivity, thereby generating climax conditions (Crome 1988) and possibly promoting greater breeding success. The breeding season in northern parts of Australia is reportedly between December and March, while breeding in south-eastern Australia occurs between November and April. Breeding of the intermediate egret occurs in pairs among dense colonies with other waterbirds, such as herons, ibises, spoonbills and cormorants. Breeding is dependent on the occurrence of heavy rains and floodwaters to replenish breeding habitats (Marchant and Higgins 1990).

Similar to the great egret, the time needed for the intermediate egret to fatten up, establish breeding pairs and construct nests is significant: a minimum lag time of three months has been reported, and peak breeding occurs after a nine-month flood lag (Kingsford and Auld 2005). Baxter (1994) found a correlation between breeding success and rainfall after a five-month lag.

The clutch size of the intermediate egret is generally three or four eggs, with a range of two to six eggs recorded. No data are available about the period of egg-laying. Both sexes incubate the eggs for 24–27 days. Chicks are continuously brooded by both parents for 12 days and partially brooded for another eight to 14 days. Young have been observed scrambling from the nest at 24 days old. Fledging occurs at around 40 days, with a range of 37–53 days. Maturity is reached another three weeks after fledging (at 62–75 days) (Marchant and Higgins 1990). Age of first breeding is unknown, but observations at Shortlands Wetlands, in Newcastle, New South Wales, indicated that although one-year-old egrets displayed full breeding plumage, they were not seen nesting (Maddock and Baxter 1991). Breeding duration is estimated at about three to four months. The intermediate egret has a minimum flood duration requirement of six to seven months, and peak breeding success occurs with flood durations of 12 months.

Nests are established in trees over water, but other information on the minimum water depth required for breeding is limited (Marchant and Higgins 1990). Due to the relationship between flood depth and duration, it is likely that floods of greater depth have a greater duration, thereby enhancing the breeding success of the intermediate egret. In fact, unsuccessful breeding of egrets in 1996 at Barmah Forest was associated with rapid drawdown of water, which caused reduced nest security and food availability (Leslie 2001). It is therefore estimated that the intermediate egret requires slow drawdown of water so that nest sites are flooded for the minimum flood duration of six to seven months.

There is no information on the flood frequency required to maintain intermediate egret populations or its longevity. Low breeding success of egrets at Barmah Forest has been associated with smaller flood volumes since regulation in the mid 1970s (Leslie 2001). Based on the average life expectancy of eight to 10 years for colonial waterbirds in the wild (Scott 1997), it is suggested that large floods should occur every four years with sustaining small floods every one to two years.

Like the great egret, the intermediate egret is likely to show greater breeding success when flooding follows a period of drying in the wetland (Crome 1988; Maddock and Baxter 1991). It is therefore suggested that dry periods should occur for a few months outside the main breeding season to maintain breeding success and the seasonal nature of breeding.

Little egret: *Egretta garzetta*

The little egret is a small to medium-sized, wholly white, colonially nesting waterbird with a sleek appearance and black bill and legs (Figure 3.12). It grows to a length of approximately 60 cm and has a wingspan of 90 cm and a weight of approximately 900 g. The sexes exhibit a similar appearance, and plumes on the nape, breasts and back develop during breeding (Marchant and Higgins 1990). The little egret is observed singly or in mixed flocks of egrets

Figure 3.12: Little egret, *Egretta garzetta*. Photograph: Chris Herbert (Hunter Bird Observers Club).

and ibises in a range of coastal and inland settings. Internationally and nationally it has a widespread distribution and occurs in northern Australia, the eastern states and southern Australia (Marchant and Higgins 1990). The little egret is regarded as endanged in Victoria (*Flora and Fauna Conservation Act 1988*).

Habitat and diet

The little egret has been observed in inland wetlands and estuarine and littoral habitats including watercourses, pools, billabongs, floodwaters with low vegetation of grasses or sedges, shallow open areas in freshwater swamps with short or tall emergent vegetation, open lakes, reservoirs, shrubby or wooded lakes or swamps, wet meadows, flooded depressions and seepage springs (Corrick and Norman 1980; Crawford 1972; Fjeldsa 1985; Gosper 1981; Marchant and Higgins 1990; Vestjens 1977b). It forages in open water, on shores or exposed flats or banks, particularly in areas with soft substrates, aquatic vegetation and little or no emergent vegetation. Roosting occurs in trees on banks or, occasionally, on artificial structures. Breeding occurs in fresh, brackish or saline wetlands with trees to build nests in. Nesting occurs in forks of trees, on lateral limbs, often above standing water at a height of 3–7 m. Nests are usually concentrated in one area of a colony rather than scattered throughout. Nesting material consists of loosely woven sticks (Marchant and Higgins 1990).

The little egret consumes fish, some frogs and insects. Gut analyses of adult little egrets at Lake Cowal, New South Wales, found they consumed crayfish, shrimps, insects and gambusia (Vestjens 1977b). Nestling little egrets consumed small fish (gambusia and carp gudgeons, *Hypseleotris* species) and aquatic insects at Shortland Wetlands, Newcastle, New South Wales (Maddock 1986).

Breeding and flood requirements for breeding

Flooding is a strong stimulus for breeding in the little egret (Briggs 1990). Large fluctuations in breeding success were observed between wet and dry seasons (Maddock and Baxter 1991), and correlations have been observed between flood extent and number of little egret nests (Kingsford and Auld 2005). Breeding occurs between October and March (Halse and Jaensch 1989; Marchant and Higgins 1990), in pairs or colonially with other herons, ibises, spoonbills and cormorants.

The lag time for the little egret is not well established. Kingsford and Auld (2005) found no relationship between water flow with a lag and the number of little egret nests. Baxter (1994) suggested that the lack of relationship between rainfall in the previous five months of breeding and little egret fledging success may be due to its greater exploitation of other foraging habitats. The little egret consumes small freshwater fish, aquatic insects and crustaceans (Maddock and Baxter 1991). As it consumes smaller prey items than other egret species it may breed sooner, thereby making the lag time less important for successful breeding. However, a lag time would still be required to establish breeding pairs and nests. It is therefore suggested that the little egret requires a lag time of less than three months.

The clutch size for the little egret has not been accurately quantified but is presumed to be three or four eggs. Incubation of eggs is undertaken by both sexes and is estimated to occur for a period of 20–25 days. New chicks are brooded by both parents and have been observed scrambling from nests and perching on nearby branches at 21–32 days old. The nestling period from hatching to first flight is 31–46 days, but young remain dependent on parents for food for some time after this. The period from hatching to complete independence is unknown (Marchant and Higgins 1990). Breeding duration is estimated at three to four months, while ideal flood duration for successful breeding should be in the order of four to six months.

While the little egret prefers to nest in live trees over floodwaters, the influence of specific water depths on breeding success is minimal. Due to the relationship between flood depth and duration, it is likely that floods of greater depth will have a greater duration, thereby enhancing breeding success. Unsuccessful breeding of egrets in 1996 at Barmah Forest has been associated with rapid drawdown of water, which caused reduced nest security and food availability (Leslie 2001). It is therefore estimated that the little egret requires moderate drawdown of water so that nest sites are flooded for the minimum flood duration of four to six months.

Estimating the ideal flood frequency for the little egret is difficult. The longevity of the little egret is in excess of 22 years, based on an individual from France (Staav and Fransson 2008); however, longevity in the wild is likely to be eight to 10 years (Oiseaux 2009a; Scott 1997). Therefore, the little egret may require large floods at least every four years and small sustaining floods every one to two years to maintain habitats.

As the diet of the little egret consists of prey that are readily available following flooding, and it may prefer to nest in live river red gums (Leslie 2001), an inter-flood dry-period is likely to be important for the little egret. It is therefore suggested that dry-periods should occur for a few months outside the main breeding season so as to maintain habitats and little egret breeding success.

Pacific heron: *Ardea pacifica*

The Pacific heron is a large slate-grey to black heron with mostly white head (Figure 3.13). The bill, lores, legs, feet and skin around the eye are black to dark grey. It has a length of 76–106 cm, a wingspan of 147–160 cm and a weight of approximately 650–860 g. The sexes have a similar

Figure 3.13: Pacific heron, *Ardea pacifica*. Photograph: Chris Herbert (Hunter Bird Observers Club).

appearance, but females are slightly smaller. During breeding their dark colourings become darker and maroon tapering scapular plumes develop (Marchant and Higgins 1990). The Pacific heron may be seen singly or in small groups in freshwater swamps, dams or moist grasslands and may even be observed in brackish and saline environments. It has a widespread distribution throughout Australia, except in eastern Western Australia and western South Australia (Marchant and Higgins 1990).

Habitat and diet

The Pacific heron occurs in inland wetlands, grasslands and estuarine habitats. While it is less commonly observed in the interior of Australia, in these locations it frequents permanent or ephemeral waters. It primarily prefers freshwater locations and is opportunistic in its use of waters, including floodwaters, artificial waterbodies and transient wetlands from rainfall or seepage (Marchant and Higgins 1990). Preferred habitats include shallow swamps, pools, watercourses and flooded lands that are sparsely vegetated with grasses, sedges and rushes, such as water couch, spike-rush and cumbungi. It may frequent shallow parts of wooded lakes, swamps and watercourses vegetated with shrubs and trees such as lignum, river red gum, oak and *Melaleuca* species (Corrick 1981, 1982; Corrick and Norman 1980; Fjeldsa 1985; Gosper 1981; Jaensch *et al.* 1988; Marchant and Higgins 1990; Recher and Holmes 1982; Vestjens 1977b). Foraging occurs in wet grasslands with shallow water less than 70 mm deep (Recher and Holmes 1982). Foraging may also occur in deeper locations by plunging from a perch such as a steep bank (Marchant and Higgins 1990). Roosting occurs in trees; during the breeding season birds roost in pairs (Sedgwick and Morrison 1948). Nesting occurs in freshwater wetlands with flooded or fringing trees in which nests are built (Hobbs 1961; Marchant and Higgins 1990). Nesting is usually on twisted horizontal or sloping forks of limbs away from the centre of the tree. Nests are established 1–40 m above the water, and consist of a bulky and loosely constructed platform with a central depression. The nest material includes sticks and twigs of 1–2 cm diameter and it may have a lining of finer twigs. Nests may be reused from previous breeding seasons if they survive (Marchant and Higgins 1990).

The diet of the Pacific heron primarily consists of small aquatic animals and the occasional fish. There are no detailed studies of gut contents; however, the Pacific heron has been observed consuming freshwater mussels; crustaceans, including shrimp and crayfish; spiders and insects, fish, frogs, lizards; and young ducklings and water rats (Hopkins 1947; Jackson 1918; Marchant and Higgins 1990; Recher *et al.* 1983; Recher and Holmes 1982; Vestjens 1977b). Consumed fish are usually less than 3 cm in length (Recher *et al.* 1983; Recher and Holmes 1982).

Breeding and flood requirements for breeding

Breeding in the Pacific heron is stimulated by flooding and season (Briggs 1990). Numerous studies have found correlations between rainfall or other flood indicators and breeding (Briggs *et al.* 1997; Crome 1988; Halse and Jaensch 1989). Breeding in response to discharge that peaked unseasonally in late June and resulted in markedly less bird breeding success at wetlands on the Murrumbidgee floodplain (Briggs *et al.* 1997) also highlights the role of season as a breeding stimulant. Nesting can occur over a long period between early August and early March, with most records indicating that breeding in northern Australia (>29°S latitude) generally occurs later in the season (December to March) and that breeding in southern Australia may occur earlier (Barnard 1913; Chaffer 1940; Elliott 1940; Frith and Davies 1961; Hobbs 1961; Marchant and Higgins 1990).

The mean duration between flooding under nests and peak number of Pacific heron was five months following winter/spring flooding and seven months following autumn flooding

(Briggs and Thornton 1999). Briggs *et al.* (1997) found relationships between Pacific heron nests and the area of dead river red gum flooded in excess of four months, maximum area of dead river red gum and maximum area of open water, with the strongest relationship between area of river red gum flooded in excess of four months. As the Pacific heron is known to use sites with living river red gum (Leslie 2001) and dead river red gum (Briggs *et al.* 1997), it is difficult to make any inferences of flood duration and lag time required for breeding. The Pacific heron exhibits a minimum lag preference of one to two months and the ideal lag period is five to seven months, depending on the seasonality of flooding.

Clutch size is two to six eggs (Lowe 1989). Eggs are laid at intervals of 48 hours and incubation is likely to be undertaken by both parents, although observations do not confirm this (Marchant 1988). Incubation is undertaken for about 28 days, after which parents brood chicks continuously for about one week and guard them for a few days until they can clamber out of the nest. Chicks clamber from the nest at about three weeks old and first flight occurs at about six to seven weeks old. The period to complete fledging, maturity and breeding is unknown (Marchant and Higgins 1990); however, the breeding duration is estimated at approximately three months. Therefore, the minimum flood duration is four to five months and the ideal flood duration is eight to 10 months.

As the Pacific heron nests in trees and generally forages in shallow water, water depth is not an important characteristic for breeding success. However, Halse *et al.* (1993) found significant associations between Pacific heron occurrence and deep water in September. This may be because deep waters subside slowly and thus are likely to remain for the duration of breeding. It is therefore suggested that the rate of recession of water should be moderate so that breeding sites are inundated for the minimum four to five months required for successful breeding.

Definitive information on the flood frequency required for successful breeding and the longevity of Pacific heron is limited. It is likely that the Pacific heron in the wild has a life-span of eight to 10 years (Scott 1997). It is therefore estimated that it requires large floods every four years to boost population numbers and small floods every one to two years to maintain population numbers.

Crome (1988) indicated that drying is as important as flooding to breeding success of the Pacific heron. Complete drying for one month in Murrumbidgil Swamp in New South Wales reportedly promoted the productivity of the small prey items consumed by the Pacific heron. It is therefore suggested that complete drying for at least one month will increase the breeding success of the species.

White-faced heron: *Egretta novaehollandiae*

The white-faced heron is a blue-grey heron with a white face and upper throat, and yellow legs (Figure 3.14). It grows to a height of 66 cm, has a wingspan of 106 cm and a weight of approximately 550 g. The sexes have a similar appearance, but females are slightly smaller, and the plumes on their back and breast become brighter and more numerous when breeding (Marchant and Higgins 1990). The white-faced heron is generally regarded as a solitary bird but may occur in flocks of up to 60 (Lowe 1983b). It is commonly seen in Australia except in eastern Western Australia, small parts of the Northern Territory and South Australia (Marchant and Higgins 1990).

Habitat and diet

The white-faced heron may occur in a variety of habitats including littoral and estuarine habitats, inland wetlands and grasslands. In inland wetlands, it has been observed in pools, billabongs, watercourses and floodwaters on floodplains. It may also frequent freshwater meadows,

Figure 3.14: White-faced heron, *Egretta novaehollandiae*. Photograph: Kerrylee Rogers (DECCW).

shallow swamps with low emergent vegetation such as spike-rush, cumbungi and common reed, wet or dry pasture, shallow parts of permanent lakes and reservoirs, and lakes, swamps or floodwaters vegetated with trees and shrubs such as lignum, river red gum, oak and *Muehlenbeckia* species (Corrick 1982; Corrick and Norman 1980; Fjeldsa 1985; Gosper 1981; Hobbs 1956b, 1961; Recher and Holmes 1982; Vestjens 1977b). Foraging commonly occurs in open areas in shallow water, along shores or exposed surfaces in wetlands or in moist and dry vegetation (Lowe 1983b; Moore 1984; Recher and Holmes 1982). The white-faced heron roosts in trees beside wetlands (Lowe 1983b; Moore 1984). Breeding generally occurs solitarily in fresh or saline habitats with fringing or flooded trees (Gosper 1981; Hobbs 1961; Lowe 1983b; Moore 1984), and may occur in trees away from water (Marchant and Higgins 1990). Nests are built in trees or shrubs, with river red gum and black box favoured at inland locations (Marchant and Higgins 1990). Breeding was reportedly more extensive at highly productive sites with complex flora and diverse fauna populations (Crome 1988). Nests are usually constructed in tree forks in the upper canopy at an average height of 10.1 m, or approximately 4.5 m below the tree top, or on horizontal branches well away from the tree trunk. Old nests of ravens and cormorants may be reused, or nest sites may be reused in subsequent years. Nests are a flimsy construction of coarse sticks lined with finer twigs and may include green sticks from live trees. Building of the nest may occur up to 12 days before laying of eggs (Marchant and Higgins 1990).

The white-faced heron has a varied diet of aquatic invertebrates and vertebrates (Marchant and Higgins 1990) and may consume crustaceans (including amphipods, isopods, shrimps, crabs and crayfish), fish, insects, spiders, frogs and gastropods (Lowe 1983b; Marchant and Higgins 1990; Vestjens 1977b).

Breeding and flood requirements for breeding

Breeding in the white-faced heron is primarily stimulated by flooding and season (Briggs 1990). It has been associated with area of live river red gum flooded in excess of four months and maximum area of river red gum (Briggs *et al.* 1997), flooding following drying of wetlands and breeding success (Crome 1988), rainfall with a one- to two-month lag (Halse and Jaensch 1989), and season (Halse and Jaensch 1989; Marchant and Higgins 1990). Similarly, relationships have been found between the abundance of white-faced heron, rainfall and discharge with a two-month lag (Gosper *et al.* 1983) and winter rainfall (Woodall 1985). The main breeding season for the white-faced heron is between August and October, with some breeding as early as June and as late as March when conditions are favourable (Marchant and Higgins 1990). This large range in the breeding season emphasises the influence of climate and wetland availability on successful breeding.

Like the Pacific heron, the lag time for the white-faced heron is unknown but may be inferred from observed relationships between flooding and breeding. Briggs *et al.* (1997) found relationships between white-faced heron breeding and area of live river red gum flooded in excess of four months, while Halse and Jaensch (1989) found strong relationships between commencement of egg-laying and rainfall with a one- or two-month lag. As the white-faced heron prefers sites with living river red gum (Briggs *et al.* 1997), this suggests that lag times should not be so long as to cause the death of trees. It is therefore suggested that the white-faced heron prefers a minimum lag time of one to four months.

The clutch size for the white-faced heron is generally three to four eggs. There are reports of multiple broods being reared per season, but this is not clear (Marchant and Higgins 1990). Replacement laying has been observed after loss of eggs at an early stage. The laying interval is unknown, but eggs are incubated by both sexes for about 22–30 days. The nestling period is also unknown, but chicks have been observed clambering onto branches beside nests 10 days after hatching and fledging was observed at approximately 43 days after hatching. After fledging, young return to the nest site to feed. This behaviour has been observed 18 days after fledging (Marchant and Higgins 1990). Breeding duration is estimated at approximately three months. Therefore, the white-faced heron is likely to require a minimum flood duration of four to seven months with a maximum duration of approximately seven months, but not markedly more.

The white-faced heron does not exhibit a water depth preference, but significant relationships have been found between the occurrence of white-faced heron and deep water in wetlands (Halse *et al.* 1993). The relationship between water depth, duration and recession of waters may be the main reason for this correlation. It is therefore suggested that the white-faced heron requires a moderate rate of drawdown so as to maintain flood conditions for the duration of the lag time and breeding period.

Information about the flood frequency, longevity or average life expectancy of the white-faced heron is limited. Based on the likely life-span of wild colonial waterbirds in Australia, of eight to 10 years (Scott 1997), it is estimated that the white-faced heron requires a large flood frequency of one in four years and small floods every one to two years to maintain population numbers.

Drying prior to flooding is an important characteristic of the flood regime that promotes successful white-faced heron breeding (Crome 1988). It is therefore likely that complete drying for at least one month will increase its breeding success.

Rufous night heron: *Nycticorax caledonicus*

The rufous night heron is a stocky, rufous-backed heron with black crown, thick neck and legs, short square tail and heavy bill (Figure 3.15). It has a length of 55–65 cm, a wingspan of 95–110 cm and a weight of approximately 800 g. Males and females exhibit a similar appearance and there are slight seasonal changes in appearance (Marchant and Higgins 1990). The rufous night heron has been observed singly or in large groups, though it is generally regarded

Figure 3.15: Rufous night heron, *Nycticorax caledonicus*. Photograph: Chris Herbert (Hunter Bird Observers Club).

as a solitary bird with nocturnal feeding habits. It has been observed in a range of coastal and inland habitats and has a wide distribution in northern, eastern and south-western Australia, being most numerous throughout the Murray-Darling Basin (Marchant and Higgins 1990). Rufous night herons are regarded as near-threatened in Victoria (*Flora and Fauna Conservation Act 1988*).

Habitat and diet

The rufous night heron primarily occurs in littoral and estuarine habitats, and inland wetlands and grasslands (Marchant and Higgins 1990). In inland regions, the rufous night heron prefers permanent waterbodies on floodplains, billabongs, watercourses and pools with wooded edges or swamps with tall emergent vegetation such as spike-rush, cumbungi, common reed and club-rush. It has been observed frequenting wet meadows, flooded grasslands, shallow freshwater swamps with short emergent and floating vegetation, or shrubby/wooded lakes and swamps dominated by river red gum, oak, lignum and *Melaleuca* species (Corrick and Norman 1980; Fjeldsa 1985; Gosper 1981; Hobbs 1956b; Jaensch *et al.* 1988; Marchant and Higgins 1990; Vestjens 1977b). Foraging occurs at night in still or slow-moving water on exposed shores, in swampy vegetation or near trees for roosting (Gosper 1981; Recher and Holmes 1982). Nesting occurs among dense trees or shrubs to heights of up to 20 m, but may occur on treeless islands on the ground among low shrubby vegetation (Braithwaite and Clayton 1976; Close *et al.* 1982; Marchant and Higgins 1990; Vestjens 1977b). Breeding is commonly in central parts of swamps in areas of up to 20 ha (Braithwaite and Clayton 1976). Nests are typically established on horizontal limbs in crowns of trees, forks under the canopy or on top of shrubs such as lignum and reed beds. Nests consist of a flattish construction of sticks (Marchant and Higgins 1990).

The diet of the rufous night heron primarily consists of aquatic animals, particularly fish, but also frogs, yabbies, crayfish and insects. It is considered to be an opportunistic feeder that will take any suitable prey including eggs and nestlings of other bird species, house mice and human refuse (Hobbs 1976; Marchant and Higgins 1990; Vestjens 1977b).

Breeding and flood requirements for breeding

Breeding in the rufous night heron is reportedly stimulated by flooding (Briggs 1990) and numerous studies support this. For example, studies have found that laying correlated with rainfall and a three-month lag (Halse and Jaensch 1989), breeding was more successful with flooding occurring after drying (Crome 1988), nesting correlated with area of river red gum flooded in excess of four months (Briggs *et al.* 1997), and the number of nests correlated with river flow with a three- to nine-month lag (Kingsford and Auld 2005; Kingsford and Johnsons 1998). Since flooding is the main stimulus for breeding, due to the high incidence of flooding in spring and summer in the Murray-Darling Basin, this period is regarded as the main breeding season.

The lag period required for breeding success can be inferred from Briggs *et al.* (1997), who found that Ciconiiforme nesting correlated with area of river red gum flooded in excess of four months. Similarly, Briggs and Thornton (1999) found a mean lag time of seven months following autumn flooding, and Kingsford and Auld (2005) found that the number of rufous night heron nests correlated with flooding with a three- to nine-month lag. It is therefore suggested that the rufous night heron requires a minimum flood lag time of three months, and that breeding success may be enhanced with a lag time of up to nine months.

Clutch size for the rufous night heron is approximately two to four eggs (Marchant and Higgins 1990). Number of broods per season is unknown, but broods have been observed in summer and winter (Hanscombe 1915). Eggs are laid every two days. Incubation is by both parents for 21–22 days. New chicks are brooded by both parents; chicks start to leave the nest two to three weeks after hatching, but return to the nest for six to seven weeks (Marchant and Higgins 1990). Age of fledging and maturity is unknown but breeding has been observed in juveniles with immature plumage, presumed to be two to three years old (Braithwaite and Clayton 1976). Breeding duration is estimated at three months. The minimum flood duration for successful breeding is six months, but breeding success will increase when flooding occurs for up to 12 months.

The rufous night heron does not exhibit a water depth preference. While a correlation between occurrence of rufous night heron and water depth has been reported (Halse *et al.* 1993), this may be due to the effect of flood duration and recession of waters. Based on the minimum flood duration period of six months, it is estimated that the rufous night heron requires slow recession of floodwaters for successful breeding.

Information about the required flood frequency for maintaining viable breeding populations of rufous night heron is not available and its longevity has not been documented. However, based on the estimated life-span of wild colonial waterbirds in Australia, of eight to 10 years (Scott 1997), it is estimated that the rufous night heron requires a large flood frequency of one in four years and small floods every one to two years to maintain population numbers.

Drying prior to flooding is an important characteristic of the flood regime that promotes successful breeding in the rufous night heron (Crome 1988). It is therefore likely that complete drying for a few months will increase breeding success.

Silver gull: *Larus novaehollandiae*

The silver gull has a slim body, a short slender bill, a sloping forehead, narrow wings and short to medium length legs (Figure 3.16). It has a length of 36–44 cm, a wingspan of 91–96 cm and a weight of 265–315 g. Adults exhibit a distinctive bright red bill and legs, while juveniles exhibit a dull black bill and pink to brown legs with patterned brown markings on their wings. The sexes have a similar appearance. Non-breeding adults exhibit a slightly duller red bill and legs (Higgins and Davies 1996). The silver gull is a gregarious bird, often observed in large flocks of hundreds to thousands. It has been observed in a range of habitats within the coastal zone and inland wetlands and has a widespread distribution in Australia and New Zealand. It occurs along the coast of Australia and is widespread in areas of the Murray-Darling Basin and inland drainage basins of Queensland (Higgins and Davies 1996).

Habitat and diet

The habitat of the silver gull includes coastal settings, offshore islands and inland wetlands (Higgins and Davies 1996). At inland locations the silver gull tends to inhabit permanent wetlands such as lakes, weirs and reservoirs. It may also be observed on floodplains, swamps, billabongs, suitable farmland and near ephemeral wetlands. It may be attracted to flooded areas after flows or rainfall (Higgins and Davies 1996). Foraging of human refuse does occur, but the silver gull may also be observed foraging in pasture or on cultivated land. Foraging may occur in shallow water less than 5 cm deep, in deeper water by taking food from the surface or in the air when consuming swarming insects (Hobbs 1961; Storr 1964a; Wheeler and Watson 1963). Roosting at inland locations occurs around the margins of wetlands or floodwaters (Favaloro 1947; Higgins and Davies 1996; Storr 1964a). Breeding primarily occurs in coastal settings, but at inland locations breeding may occur on islands or islets within lakes, saltponds or floodwaters. Nests may be located in tree stumps or on embankments of wetlands (Hobbs

Figure 3.16: Silver gull, *Larus novaehollandiae*. Photograph: Kerrylee Rogers (DECCW).

1961; Wheeler and Watson 1963). Nests have been observed in broken sprouts of dead river red gums above the waterline at Lake Cowal (Vestjens 1977b). Nesting is reportedly solitarily or in small colonies. Nests consist of a constructed platform of grass and other suitable material; they may be unlined, or lined with grass or feathers (Favaloro 1947).

The silver gull is an opportunistic scavenger largely sustained on domestic and industrial organic waste, but some also feed on pelagic plankton (Higgins and Davies 1996). Excluding human refuse, the diet also consists of crustaceans, arachnids, insects, fish and reptiles, annelids, molluscs and plants (Higgins and Davies 1996; Smith and Carlile 1993; Vestjens 1977b).

Breeding and flood requirements for breeding

Breeding in inland colonies of the silver gull occurs in response to flooding (Murray and Carrick 1964; Wheeler and Watson 1963). Breeding at coastal locations appears to be seasonal (Higgins and Davies 1996). As breeding occurs opportunistically at inland locations (Smith 1991), it is suggested that breeding of the silver gull in the Murray-Darling Basin is primarily stimulated by flooding with season being a secondary stimulus. Correlations between silver gull abundance and water level (Harper 1990), and silver gull nesting and rainfall with a three-month lag (Halse and Jaensch 1989) support this hypothesis. The main breeding season is late August to late February (Higgins and Davies 1996), coinciding with the main flooding season in the Murray-Darling Basin.

The silver gull breeds opportunistically at inland locations and will arrive at a suitable location within hours of rainfall (Hobbs 1961). However, there may need to be some lag time between rainfall and breeding before flood conditions are suitable, and before gonads and eggs are developed for breeding. Halse and Jaensch (1989) found that nesting correlated best with peak rainfall with a three-month lag, but a strong correlation was also found between breeding and rainfall with a two-month lag.

Clutch size ranges from one to three eggs, with laying at intervals of two to four days. Up to four replacement clutches have been observed per season and successive broods may be raised within a season (Higgins and Davies 1996). Both sexes incubate the eggs for 19–26 days (Wheeler and Watson 1963), or an average 24 days (Storr 1964a). Young have been observed leaving the nest on the second day after hatching (MacGillvray 1925), but generally remain within the nest for at least a week (Higgins and Davies 1996). The young are led away from the colony at about three to four weeks to form smaller groups (Wheeler and Watson 1963). Fledging occurs between five and seven weeks (Wheeler and Watson 1963) and young can disperse from the nest at about six weeks old (Carrick *et al.* 1957). Young can breed as early as two years old (Carrick *et al.* 1957), though first breeding generally occurs at a later age (Mills 1973). Breeding duration is estimated at two to three months. A minimum flood duration of four to six months is required for successful silver gull breeding. When the flood duration is longer, breeding success will increase as multiple clutches may be brooded within a season.

The silver gull does not exhibit a water depth preference, however, Wheeler and Watson (1963) indicated that in inland areas breeding generally occurs on islands within large shallow lakes. Extensive flooding may reduce the habitat available for breeding. Breeding may be successful with a moderate recession of water, but will be greatest when the recession of water is slow.

Colonies of silver gull typically nest on offshore islands along the New South Wales coast rather than in inland settings, and the species is highly dispersive (Higgins and Davies 1996). It is therefore not essential for flooding to occur at a prescribed frequency at inland locations for the maintenance of breeding populations. Similarly, it is not necessary to propose an inter-flood dry-period for the silver gull, as breeding may be maintained at coastal locations. There is no evidence whether silver gull breeding is enhanced with drying. The silver gull is an opportunistic feeder and, when a wetland dries, may forage on nearby pasture or other non-wetland habitats (Wheeler and Watson 1963).

Whiskered tern: *Chlidonias hybridus*

The whiskered tern is the largest marsh tern, small with a short slightly forked tail (Figure 3.17). The adult breeding whiskered tern is pale grey above and has mostly grey underparts with a black cap, while the non-breeding whiskered tern has a white head and black band from the eye to the nape. It has an approximate length of 23–25 cm, a wingspan of 74–78 cm and a weight of 80 g. The sexes have a similar appearance, though the bill of the male is slightly heavier (Higgins and Davies 1996). The whiskered tern is gregarious and is commonly observed in small to large flocks in association with other terns. It tends to inhabit a wide variety of habitats including freshwater and brackish wetlands in coastal and central Australia. It has a scattered distribution throughout Australia but a wide distribution throughout the Murray-Darling Basin, where breeding is widespread (Higgins and Davies 1996). The whiskered tern is regarded as near-threatened in Victoria (*Flora and Fauna Conservation Act 1988*).

Habitat and diet

The whiskered tern prefers shallow, terrestrial, freshwater, permanent or ephemeral wetlands including lakes, swamps, billabongs, river pools, reservoirs, large dams, sewage ponds, flooded saltmarsh and farmland. It commonly occurs in wetlands with submerged and emergent vegetation such as grass, sedges, reeds and rushes, and occasionally swamps of lignum (Bourke 1956; Dostine and Morton 1989a; Fjeldsa 1985; Higgins and Davies 1996; McEvey 1965; Watson 1955). Foraging usually occurs on inland wetlands in permanent swamps or floodwaters with

Figure 3.17: Whiskered tern, *Chlidonias hybridus*. Photograph: Dan Herbert (Hunter Bird Observers Club).

vegetated margins or emergent vegetation (Bourke 1956; Dostine and Morton 1989a). Foraging may also occur while in flight and on the ground in wetlands and farmland. Roosting habitats include muddy spits and banks, dense reeds or perches on the branches of dead trees, logs and fence posts (Bourke 1956; Higgins and Davies 1996). Nesting occurs in temporarily flooded inland wetlands among inundated vegetation such as lignum, canegrass, reeds or floating ribbonweed or flooded paddocks in waters up to 1.8 m deep (Bourke 1956; Bryant 1949; Higgins and Davies 1996; Smith and O'Connor 1955). Floating nests consist of a mound of water ribbons, sticks, grass or other suitable vegetation in a saucer shape with a well-shaped cavity in the centre, attached to vegetation on small islets or on submerged or partly submerged vegetation (Bourke 1956; Higgins and Davies 1996; Smith and O'Connor 1955). Colonies of whiskered tern will abandon nests if sites become dry (Bourke 1956).

The diet of the whiskered tern is primarily carnivorous and consists of insects, crustaceans, fish and other small invertebrates such as frogs, centipedes, spiders and small waterbirds. Occasionally seeds are consumed (Higgins and Davies 1996). Gut analyses indicate that the whiskered tern may also consume lizards, waterbird chicks such as grebes, and the house mouse (Crawford 1977; Dostine and Morton 1989a).

Breeding and flood requirements for breeding

Flooding stimulates breeding in the whiskered tern (Belcher 1902; Bourke 1956; Smith and O'Connor 1955). The whiskered tern can nest in colonies of thousands after heavy rains cause flooding of low-lying areas. Nesting often occurs in association with hoary-headed grebes and Australasian grebes, both of which exhibit a strong flood stimulus (Belcher 1902). Similarly, whiskered tern abundance has been correlated with water level (Gosper *et al.* 1983), preferring deep, seasonally flooded wetlands (Halse *et al.* 1993). While the numbers of whiskered tern at Bool Lagoon in New South Wales was found to correlate with season (Harper 1990), this may reflect habitat suitability rather than breeding stimuli. Hobbs (1961) observed the seasonal arrival of whiskered tern at wetlands in south-western New South Wales, but breeding was undertaken only when conditions were suitable. Therefore, it is likely that breeding of the whiskered tern is primarily stimulated by flooding and that season may be a secondary stimulus. The main breeding season in the Murray-Darling Basin is between October and late January (Higgins and Davies 1996), with breeding in New South Wales reportedly ending in mid-November (Bourke 1956).

Identifying a lag time before breeding is difficult, particularly as the whiskered tern is highly nomadic (Chapman and Lane 1997) and the breeding cycle is short (Bourke 1956). Breeding will commence quickly once a suitable site is found (Bourke 1956). This suggests that a lag time is not essential for successful breeding in the whiskered tern.

The clutch size for the whiskered tern is two or three eggs (Bourke 1956). Second clutches have been laid in the wet area of a swamp after recession of water (Higgins and Davies 1996). Both sexes incubate the eggs; the incubation period is unknown, but likely to be approximately 18–20 days (Higgins and Davies 1996). The young leave the nest at six to eight days and fledging occurs at 14–18 days. Dispersal from the nest probably occurs within one week of fledging (Bourke 1956). Breeding has been observed while whiskered tern exhibit immature plumage, indicating that sexual maturity may occur rapidly (Higgins and Davies 1996). Bourke (1956) indicated that the complete breeding cycle is less than 40 days and it is estimated that there is a relatively short breeding duration of approximately one to two months. Flood durations of at least two months are required for successful breeding of whiskered tern.

An ideal water depth for successful breeding has not been documented. However, as nests have been observed floating on water up to 1.8 m deep (Higgins and Davies 1996) and less than

60 cm deep (Bourke 1956), water depth may not be essential. The flood duration and rate of rise and fall of water levels may be important. Nests have been washed away when water levels have increased suddenly (Bourke 1956), indicating that sudden rises in water level can affect breeding success. Prescribing a rate of recession is difficult as the breeding duration is short, and the whiskered tern has been observed nesting and raising young in waters that have almost entirely receded (Bourke 1956). Colonies have deserted after waters receded quickly and swamps became dry (Bourke 1956). It is therefore suggested that the rate of recession should be slow enough to maintain some floodwaters at a site for a minimum of two months.

The flood frequency for breeding of whiskered tern is determined from its longevity, estimated to be approximately 10 years in the wild (Oiseaux 2009b). A large flood frequency of one flood in every five years is prescribed but, due to the short breeding period, it is unlikely that long periods of flooding are essential for the maintenance of breeding populations of whiskered tern.

As nesting occurs on temporarily flooded wetlands (Bourke 1956; Higgins and Davies 1996) it is evident that the whiskered tern does require an inter-flood dry-period for successful breeding events. It is unlikely that a long duration of drying is required; rather, that drying occurs for a short period to promote the productivity of prey items such as insects and crustaceans.

Caspian tern: *Sterna caspia*

The Caspian tern is a large gull-like tern with a dagger-like bill, large angular head, bulky body and long narrow wings and legs (Figure 3.18). It is pale grey above and white below with a black cap and red bill. The forehead, crown and nape of the adult breeding Caspian tern are black, while non-breeding adults exhibit a cap with white streaking throughout. The sexes have a similar appearance and slight seasonal changes in plumage may be evident. The Caspian tern has an average length of 47–54 cm, a wingspan of 130–145 cm and a weight of approximately

Figure 3.18: Caspian tern, *Sterna caspia*. Photograph: Chris Herbert (Hunter Bird Observers Club).

0.680 kg (Higgins and Davies 1996). The Caspian tern is gregarious when breeding, but may occur singly or in small flocks at other times. It inhabits a wide variety of coastal habitats and inland wetlands, especially large lakes and rivers. It has a scattered distribution globally. Within Australia its distribution is largely limited to the coast and major river drainage basins in the Murray-Darling Basin (Higgins and Davies 1996). The Caspian tern is regarded as vulnerable in Victoria (*Flora and Fauna Conservation Act 1988*).

Habitat and diet

The Caspian tern inhabits coastal and inland wetlands. Inland wetlands may be fresh or saline and include permanent, semi-permanent and ephemeral lakes, waterholes, reservoirs, rivers and creeks (Higgins and Davies 1996; Hobbs 1961). It has been observed foraging in open wetlands such as the sheltered shallow waters on the margins of lakes and rivers (Pierce 1983, 1984). Roosting typically occurs near water such as on bare exposed sand, banks or shores of waterways (Higgins and Davies 1996). Breeding commonly occurs in coastal settings, but it breeds occasionally in inland wetlands (Horton 1973; Serventy 1977). Breeding may occur in the open or among low or sparse vegetation such as herbs, tussocks and other prostrate plants (Higgins and Davies 1996).

The diet of the Caspian tern consists almost entirely of fish. However, diet studies indicate that it may also consume crustaceans, including shrimp, and insects (Higgins and Davies 1996; Middleton 1987).

Breeding and flood requirements for breeding

Breeding of the Caspian tern primarily occurs in coastal settings; breeding in inland wetlands occurs opportunistically (Horton 1973; Serventy 1977). Hobbs (1961) indicated that the Caspian tern is an occasional visitor that appears to leave inland settings to breed at coastal settings. Since there are few observations of inland breeding, it is difficult to determine the stimulus for breeding or the Caspain tern's typical breeding season in the Murray-Darling Basin. As breeding occurs in spring and summer in southern parts of Australia and year-round in northern parts, flooding may be the primary stimulus for breeding, with season a secondary stimulus in temperate regions.

In Australia, the clutch size for the Caspian tern is one to three eggs (Tarr 1960). Eggs are laid at two- to three-day intervals and terns can re-lay eggs numerous times if they fail (Jones 1980; Pennycook 1949). Both parents incubate the eggs for approximately 21 days. The young chicks have been observed leaving the nest within days of hatching (Tarr 1960). Fledging takes approximately five weeks and young reach sexual maturity at three to four years old (Higgins and Davies 1996). Breeding duration is estimated at two months.

As few Caspian tern breed in wetlands of the Murray-Darling Basin, further details on its water requirements for breeding within the Basin are not provided.

Gull-billed tern: *Sterna nilotica*

The gull-billed tern is a large gull-like tern with a thick black bill, heavy rounded head, and thick neck and body (Figure 3.19). It is pale silver-grey above and white below, and has a black mask. During the breeding season the gull-billed tern exhibits a black cap. It has a length of 35–38 cm, a wingspan of 100–115 cm and a weight of approximately 265 g. The sexes have a similar appearance, but exhibit seasonal plumage changes (Higgins and Davies 1996). The gull-billed tern is gregarious when it is breeding and can occur in small to large flocks outside the breeding season. It typically inhabits inland and coastal wetlands, croplands and grasslands. It has a scattered distribution globally and within Australia. Its distribution is

Figure 3.19: Gull-billed tern, *Sterna nilotica*. Photograph: Chris Herbert (Hunter Bird Observers Club).

widespread east of the Great Dividing Range, the Riverina and major river drainage basins within the Murray-Darling Basin (Higgins and Davies 1996).

Habitat and diet

The gull-billed tern prefers shallow floodplain wetlands, particularly ephemeral wetlands such as lakes, swamps and lagoons with mudflats or watercourses (Higgins and Davies 1996). Frequented wetlands may be fresh or saline and fringed with vegetation such as grass, reeds, rushes and canegrass (Higgins and Davies 1996). Foraging typically occurs on freshwater wetlands with food taken from mud or sand at or near the water's edge. It has been observed foraging on dry land, ploughed paddocks and in rural towns (Higgins and Davies 1996; Hobbs 1976). The gull-billed tern may be observed roosting on the edges of wetlands on exposed banks, flats, bars or spits with no or low vegetation (Higgins and Davies 1996; Hobbs 1976). The gull-billed tern breeds on large inland lakes and swamps that may often be ephemeral. Nests are constructed on low exposed islands, banks or flats that are bare or with low sparse vegetation of dry grass, reeds and rushes. Nests are constructed by scraping mud or sand into small mounds, which may be bare or lined with a few small twigs, dry grass, pine needles, feathers, dung, stones and plant debris (Higgins and Davies 1996). The diet of the gull-billed tern consists of insects, fish, worms, crustaceans (including freshwater crayfish, prawns and crabs), centipedes, lizards, rodents and young grebes (Bridgewater 1931; Higgins and Davies 1996; Hobbs 1976).

Breeding and flood requirements for breeding

Breeding of the gull-billed tern appears to be stimulated by flooding or rainfall (Higgins and Davies 1996) and is also regulated by the season. This is supported by its preference for 'moderately permanent' wetlands (Halse *et al.* 1993) and its occurrence on inland wetlands during summer months (Hobbs 1961). The main breeding season is between September and January (Higgins and Davies 1996), however, since eggs have been observed in nests as late as June, it is likely that breeding may occur whenever conditions are favourable.

Due to the nomadic nature of the gull-billed tern and in the absence of further information, it is likely that it selects breeding sites in a similar manner to the whiskered tern, requiring little or no lag time to commence breeding.

The clutch size for gull-billed terns is typically two or three eggs. The laying interval is unknown. Both sexes incubate the eggs for approximately 16 days (Higgins and Davies 1996). Young have been observed leaving the nest at about three days old, but are usually dependent until three months old. The time to reach sexual maturity is unknown (Higgins and Davies 1996). Breeding duration is estimated at approximately four months and, on the basis of little or no lag time requirement, it is suggested that the gull-billed tern requires flood durations of approximately four months for successful breeding.

The breeding habitat of the gull-billed tern is commonly large inland lakes and swamps which are usually ephemeral, a similar habitat to that of the whiskered tern. Hobbs (1961) also observed nesting on the open plain away from water (except for small stock tanks), which indicates that wetland habitats may not always be needed for breeding. There is no evidence of failed nesting attempts due to rapid rise or recession of water levels (Higgins and Davies 1996).

The flood frequency for breeding of the gull-billed tern is determined on the basis of its longevity, estimated at approximately 16 years in the wild (Rydzewski 1978). That record is based on wild gull-billed terns in Europe and, as wetland availability is more limited in Australia, it is probable that its longevity is less than 16 years in the Murray-Darling Basin. It is likely that a large flood frequency of one flood in every four years is required for successful breeding. However, as the gull-billed tern is very dispersive, it may be less vulnerable to the effects of reduced flood frequency in the Murray-Darling Basin.

In the absence of further information on breeding and flooding, it is difficult to prescribe an inter-flood dry-period for the gull-billed tern. However, based on its preference for moderately permanent (Halse *et al.* 1993) and ephemeral wetlands (Higgins and Davies 1996), the gull-billed tern may exhibit a preference for wetlands that undergo a drying phase.

Deep-water foragers

Black swan: *Cygnus atratus*

The black swan is a large black waterbird with a distinctive red bill (Figure 3.20). The sexes have a similar appearance; however, females are usually smaller with a shorter neck and bill than males. They do not exhibit seasonal changes in plumage. The black swan has a length of 110–140 cm and a wingspan of 160–200 cm. Males have a weight of 6 kg, while females have a weight of approximately 5 kg (Marchant and Higgins 1990). The black swan may be observed in pairs, small groups or flocks of thousands. As it requires 40–50 m to take off for flight, it commonly frequents permanent large open waterbodies. These may be fresh, brackish or saline and usually have emergent and aquatic vegetation. The black swan is endemic to Australia and has a widespread distribution throughout eastern Australia, and eastern and southern parts of Western Australia (Marchant and Higgins 1990).

Habitat and diet

The black swan occurs in a range of settings, from tropical to temperate wetlands and within freshwater wetlands, estuarine and marine waters. It exhibits a preference for large open

Figure 3.20: Black swan, *Cygnus atratus*. Photograph: Kerrylee Rogers (DECCW).

permanent waterbodies (Marchant and Higgins 1990). Large numbers have been observed in inland areas after flooding and they can use shallow freshwater swamps, deep swamps with emergent vegetation, fresh meadows, lakes, reservoirs, sewage ponds and sheltered margins on rivers (Corrick and Norman 1980; Fjeldsa 1985; Gosper 1981; Marchant and Higgins 1990). Foraging occurs in shallow or deep open water, on wetland margins or exposed mudflats, or near floating or submerged aquatic vegetation or emergent vegetation (Briggs 1979; Frith *et al.* 1969). Breeding may occur in any of these wetland settings; however, the black swan commonly uses ephemeral waters during times of widespread flooding. Breeding sites include ponds, dams, vegetated swamps, wooded lakes and open still estuarine waters (Frith 1982). Nests are established in shallow water at depths of 30–60 cm where there is soft vegetation for nest construction and feeding, and tall emergent vegetation for shelter and protection of nests (Braithwaite 1981a; Marchant and Higgins 1990). Nests are generally positioned as far from the shore as possible, while still being attached to or supported by vegetation such as cumbungi or spike-rush. Nests have been observed on top of flooded stumps, in the bases of trees in wooded swamps, on masses of floating debris and on the ground on islands within lakes (Frith 1982).

The black swan is regarded as herbivorous with its diet consisting almost entirely of aquatic plants; however, insects and molluscs may be incidentally ingested (Marchant and Higgins 1990). Consumed plant species, such as ribbonweed (*Vallisneria* spp.), cumbungi and pond weed (*Potamogten* spp.), are typical of wetlands of the Murray-Darling Basin (Frith *et al.* 1969).

Breeding and flood requirements for breeding

Use of ephemeral waters after flooding (Frith 1982) supports the hypothesis that breeding is stimulated by flooding (Briggs 1990). In addition, correlations have been reported between peak breeding, rainfall and rainfall with a one-month lag (Halse and Jaensch 1989), breeding and maximum area of aquatic plants (Briggs *et al.* 1997) and the black swan's occurrence in deep seasonal waters (Halse *et al.* 1993). The black swan will breed any time that flood conditions are suitable (Braithwaite and Frith 1969a) and nesting will occur at different times in response to heavy rainfall (Frith 1982). Its abundance has been correlated with climatic variables (Kingsford *et al.* 1999; Woodall 1985) and water levels (Harper 1990). Season may also have some effect. While Frith (1982) claimed that breeding may occur whenever conditions are suitable, black swans have been described as 'truly winter-breeding' (Halse and Jaensch 1989) and sexual activity has been observed on a seasonal basis (Braithwaite and Frith 1969a). In addition, abundance was found to vary with season, rather than water level and rainfall (Gosper *et al.* 1983). It is therefore suggested that, as breeding may occur at any time, flooding is the primary stimulus for breeding with season being a secondary stimulus. The black swan breeds in pairs on small waters or locations with constant water levels; however, colonies of thousands of birds may breed on large lakes or swamps that are characterised by fluctuating water levels and abundant food (Marchant and Higgins 1990). Breeding in New South Wales primarily occurs between April and October (Frith 1982; Marchant and Higgins 1990).

The lag time for the black swan was estimated by Halse and Jaensch (1989), who found that peak breeding occurred at the same time as or within one month of the rainfall peak and that breeding generally occurred in proportion to the amount of rainfall. This short lag period may be due to the species' ability to inhabit a range of wetland types, which provides enough food that it can quickly commence breeding when conditions are suitable.

Clutch size typically ranges from four to six eggs (Frith 1982; Marchant and Higgins 1990). Eggs are laid at intervals of one to three days (Braithwaite 1977; Tingay *et al.* 1977) and both sexes incubate the eggs for 35–48 days. Young black swans may begin to fly when the primary feathers have developed, at 150–170 days old (Frith 1982), however, the young remain close to their parents for some time after fledging (Marchant and Higgins 1990). Sexual maturity is not reached until approximately 18–36 months old (Braithwaite 1981b). The breeding duration of the black swan is approximately seven to eight months. The flood duration required for breeding success is estimated at seven to nine months.

Nests are reportedly established at depths of 30–60 cm and as far from the shore as possible (Frith 1982). The establishment of nests at these depths and shore distances may reflect a need to maximise flooding of nests. In addition, abundance of the black swan is reportedly greatest in unvegetated waterbodies with depths in excess of 2 m (Broome and Jarman 1983). It is therefore suggested that waters should recede at a relatively slow pace to maintain flooding of black swan nests for the duration of breeding.

The flood frequency for breeding of the black swan is estimated from the longevity of wild black swans. While this is estimated to be up to 40 years (Jackson 2004), the most reliable estimate of longevity is based on the capture of a banded black swan 24 years and eight months after it was banded as a chick (Australian Bird and Bat Banding Scheme 1995a). The age of this captured black swan is an upper limit and anecdotal evidence suggests that life expectancy in the wild is more likely to be about 10 years. It is therefore suggested that suitable flood conditions should occur at least every five years to maintain population numbers.

Crome (1988) suggests that black swans require inter-flood drying to promote breeding success, based on the observation of two nests over a four-year period. As breeding occurs in response to flooding and ephemeral wetlands are favoured, this suggestion is plausible. However, since breeding commences rapidly in response to flooding, the black swan may not require drying prior to flooding to sustain its reproduction. In addition, the diet of the black swan consists almost entirely of aquatic plants (Marchant and Higgins 1990), which do not exhibit a significant capacity for drying. Therefore, the role of inter-flood drying in promoting breeding success is unknown and requires further study.

Hardhead: *Aythya australis*

The hardhead is a medium-sized chocolate-brown duck with distinctive white under-tail coverts and a pale blue-grey band on its bill (Figure 3.21). The sexes exhibit a similar appearance; however, males have prominent white eyes and females have brown eyes. The hardhead has a length of 45–60 cm, a wingspan of 65–70 cm and a weight of 0.8–0.9 kg (Marchant and Higgins 1990). The hardhead is usually observed in pairs or small flocks, but may be observed in flocks of thousands. It tends to inhabit open waters on permanent freshwater lakes and swamps. It is endemic to Australia and has a wide distribution, absent only from the most arid areas of central Australia (Marchant and Higgins 1990). The hardhead is regarded as vulnerable in Victoria (*Flora and Fauna Conservation Act 1988*).

Habitat and diet

The hardhead tends to favour inland wetlands, though it may also be observed in estuarine and inshore environments. It exhibits a preference for deep waters with abundant aquatic vegetation, such as deep swamps and lakes, billabongs, pools, creeks on alluvial plains and floodwaters (Fjeldsa 1985; Frith 1959a, 1959b), but has also been observed on freshwater meadows, seasonal swamps, reed swamps, wooded lakes and swamps, shallow floodwaters over alluvial plains, rice fields and sewage ponds (Corrick 1982; Frith 1959a; Gosper 1981; Marchant and Higgins 1990; Vestjens 1977b). While it prefers abundant aquatic vegetation, such as water lilies, *Eragrostis* species, cumbungi, spike-rush, lignum, common reed, club-rushes and even wooded lakes and swamps with river red gum, black box and *Melaleuca* species, it has also been observed on water with little aquatic vegetation (Corrick 1982; Frith 1959a; Gosper 1981; Marchant and Higgins 1990; Vestjens 1977b). Roosting occurs in large flocks on the water during the day or on sandbanks and other low perches by night. Foraging occurs on deep water where the hardhead dives for food (Marchant and Higgins 1990). It breeds in densely vegetated freshwater wetlands with permanent or deep floodwaters (Frith 1982) that may contain climax habitats with complex flora, diverse invertebrates and high organic matter (Crome 1988). Nests

Figure 3.21: Hardhead, *Aythya australis*. Photograph: Chris Herbert (Hunter Bird Observers Club).

are commonly constructed among dense vegetation such as reeds or lignum and above the water level along rivers, creeks, channels, billabongs and dams (Marchant and Higgins 1990). Frith (1982) indicated that vegetation may extend 60–150 cm above the nest and that nests may be constructed 10–90 cm above the water level; however, the hardhead has also been observed nesting up to 27 m from water.

The diet of the hardhead consists of a mixture of aquatic plants and animals, with a preference for freshwater shellfish such as mussels (Frith 1959b; Frith *et al.* 1969; Marchant and Higgins 1990).

Breeding and flood requirements for breeding

Breeding of the hardhead appears to be stimulated by flooding and season (Braithwaite and Frith 1969a; Briggs 1990), as its arrival and breeding correlate with extensive flooding (Frith 1957a). The height of breeding in south-western Australia is well-defined in spring and correlates with rainfall with a two- or three-month lag (Halse and Jaensch 1989). Breeding can occur on a seasonal basis irrespective of flooding, but unstable water levels have led to the failure of breeding (Braithwaite and Frith 1969a). Abundance records show similar correlations with rainfall and other climatic factors (Kingsford *et al.* 1999; Woodall 1985). It is therefore suggested that both flooding and season stimulate breeding, with breeding primarily occurring between August and December in the Murray-Darling Basin. It can also occur any time conditions are suitable (Frith 1982; Marchant and Higgins 1990).

The lag time for breeding is estimated at two to three months (Halse and Jaensch 1989). Hardhead clutch size ranges from six to 18 eggs, and eggs are incubated by the female for 25–32 days (Braithwaite 1975; Frith 1982). No literature is available on the period to fledging, sexual maturity or the success of broods (Marchant and Higgins 1990). Due to limited systematic information on breeding, it is difficult to estimate the required breeding duration. The majority of other Australian ducks (e.g. chestnut teals, grey teals, Australasian shoveler, Pacific black duck, Australian shelduck, freckled duck and maned duck) fledge at between two and three months. Therefore, it is estimated that the hardhead requires flooding for a period of four to six months.

The hardhead exhibits a preference for deep permanent waters that are unvegetated and have a depth of over 2 m (Broome and Jarman 1983; Halse *et al.* 1993). This preference may be related to the habitat of preferred prey items. Based on the duration of flooding required for breeding success, it is estimated that flooding should recede at a moderate to slow rate.

The longevity of the hardhead in the wild is approximately three to four years (Australian Bird and Bat Banding Scheme 2000b; Scott 1997). It is therefore estimated that conditions should be suitable for breeding approximately every second year to maintain numbers of breeding adults. Permanent water sources may be used as breeding sites when conditions at semi-permanent settings are not suitable for breeding (Frith 1982).

It is difficult to determine whether the hardhead requires inter-flood drying to promote breeding success. Crome (1988) found that breeding occurred following drying and that breeding did not occur without drying. However, this was based on observations of breeding activity in only two nests. Consumed items do have some capacity for drying and productivity of these items may be enhanced when flooding follows a period of drying. It is therefore likely that inter-flood drying for a few months may promote hardhead breeding success.

Musk duck: *Biziura lobata*

The musk duck is a heavy, wide-bodied, short-winged duck with a triangle-shaped bill (Figure 3.22). The plumage is a uniform dark grey-brown. The male exhibits a distinct lobe beneath its bill and has a length of 66 cm, wingspan of 87 cm and weight of 2.4 kg. During the breeding

Figure 3.22: Musk duck, *Biziura lobata.* Photograph: Marjorie Kibby (Hunter Bird Observers Club).

season the plumage of the adult male develops a characteristic musk odour and an oily sheen that spreads over the water. The lobe on the female musk duck is markedly less developed than that of the male, it has no musk odour and has a smaller build, with a length of approximately 55 cm, wingspan of 72 cm and weight of 1.55 kg. There are no seasonal changes in appearance (Marchant and Higgins 1990). The musk duck is regarded as a solitary bird, though it may congregate on waterbodies. It is rarely observed on land or in flight; instead, it is more commonly seen on well-established, open or well-vegetated water bodies. It is endemic to Australia and has a widespread distribution throughout the Murray-Darling Basin and south-west Western Australia (Marchant and Higgins 1990). The musk duck is regarded as vulnerable in Victoria (*Flora and Fauna Conservation Act 1988*).

Habitat and diet

The musk duck may be observed on inland wetlands or in estuarine and sheltered inshore habitats (Marchant and Higgins 1990). It is rarely observed flying or walking on land and is regarded as almost entirely aquatic, preferring the stable conditions and abundant aquatic vegetation of deep water environments such as permanent swamps, lakes and estuaries (Fjeldsa 1985). Little is known of its preferred roosting habitat, but it has been observed loafing and preening on the water during the day and night, and sleeping on the water with its bill tucked on its back (Marchant and Higgins 1990). The musk duck forages on water and is regarded as an expert diver, reportedly diving 6 m down for food (Frith 1982). Preferred breeding sites include deep freshwater swamps, lakes, billabongs and rivers. Nesting typically occurs among dense vegetation such as cumbungi, spike-rushes, lignum, *Lepidosperma* and *Melaleuca* species, which provides cover for nests (Corrick and Norman 1980; Fjeldsa 1985; Frith 1959a, 1982; Lowe 1965). Nests reportedly occur in an average water depth of 0.9 m and are constructed an average of 0.25 m above the water level. Nests are occasionally observed on low stumps, on the ground, in clumps of grass or branches of trees overhanging water. Nests are constructed from the stems of surrounding vegetation in a rough cup-shape that may be flattened during incubation (Marchant and Higgins 1990).

The diet of the musk duck primarily consists of aquatic invertebrates including insects, freshwater crayfish, snails and mussels, but it may also include some plant material (Frith *et al.* 1969; Marchant and Higgins 1990).

Breeding and flood requirements for breeding

Briggs (1990) indicated that the musk duck is one of two true seasonally breeding waterbirds in the Murray-Darling Basin, the other being the blue-billed duck. Most breeding records indicate very regular breeding seasons in these species (Braithwaite and Frith 1969a; Frith 1957a). Others refer to their inability to travel great distances (Frith 1957a), which suggests that they are unable to disperse in search of suitable flooded habitat for breeding and perhaps explains why they are strongly seasonal breeders. However, Marchant and Higgins (1990) referred to the dispersal of flocks that breed on ephemeral waters and the musk duck's ability to breed outside the primary breeding season when conditions are favourable. In any case, the musk duck appears to prefer deep permanent water sources, where conditions will suit breeding and season is the main factor required to stimulate breeding (Halse *et al.* 1993). Peak breeding has also been correlated with rainfall with a three-month lag at a range of permanent, seasonal and episodic wetlands in south-western Australia (Halse and Jaensch 1989). As this correlation is based on records from only one breeding season, it may not be causal. A strong correlation was also found between photoperiod increment and breeding, and nesting was observed in response to winter rainfalls and absent when water levels at Murray Lagoon, South Australia, did not rise due to prevalent dry climatic conditions (McCracken *et al.* 2000). While season appears to be the main stimulus for breeding, flooding is an essential requirement, particularly for musk ducks breeding in ephemeral settings. The primary breeding season is between September and October, but breeding may occur as early as June and as late as December. It may also occur out of season during wet years (Marchant and Higgins 1990).

When breeding occurs at wetlands without permanent water conditions, it is likely that a lag period is required to enable physical condition to increase before the commencement of breeding. Musk duck breeding is reported to peak three months after peak rainfall, but some breeding may occur with a two-month lag (Halse and Jaensch 1989).

Musk duck clutch size ranges from one to seven eggs (Frith 1982); clutches greater than four eggs may be indicative of laying by another species of waterbird (Lowe 1965). The female incubates the eggs for about 24 days. In captivity, the time from hatching to fledging is reportedly about three to four months. There is no information about fledging period in the wild (Marchant and Higgins 1990). Similarly, the age at which young are independent of females and are sexually mature is unknown in the wild; in captivity, the lobe was not fully developed in a three-year-old male (Fullagar and Carbonell 1986). The total breeding duration is estimated at four to five months and flooding should occur for a minimum period of seven to eight months to promote breeding success.

Nests of the musk duck are constructed over well-defined water depths of 0.4–1.4 m (Marchant and Higgins 1990), and abundance is generally associated with waterbodies over 2 m deep (Broome and Jarman 1983; Halse *et al.* 1993). At seasonal or ephemeral sites, waters should be maintained for the required flood duration of seven to eight months to ensure optimal hydrological conditions and to promote the survival of consumed food items. Floodwaters should recede at a relatively slow pace.

The musk duck has been observed in the wild at an age in excess of six years and two months (Australian Bird and Bat Banding Scheme 2000c). However, this is likely to be a maximum longevity and most ducks have a wild life expectancy of three to four years (Scott 1997). It is therefore suggested that floods should occur every one to two years at temporary wetland settings to maintain resident breeding population numbers.

As the musk duck reportedly prefers deep permanent sites for breeding, inter-flood drying is not an essential requirement. However, at seasonal or ephemeral wetlands, vegetation used for nesting show some capacity for drying and inter-flood drying may promote the productivity of preferred food. Therefore, while inter-flood drying is not essential, it may promote the breeding success of the musk duck in ephemeral wetlands.

Blue-billed duck: *Oxyura australis*

The blue-billed duck is a small compact duck with a large round head and short neck (Figure 3.23). Males exhibit a distinctive blue bill and have rich chestnut plumage on the upper body, a dark head and upper neck, and dark stiff pointed tail feathers. Females have a brown bill, and darker plumage and tail feathers than males. The blue-billed duck has a length of approximately 40 cm, a wingspan of 60 cm and a weight of 850 g. Males exhibit slight seasonal changes in plumage, although these differences are evident year-round (Marchant and Higgins 1990). The blue-billed duck is gregarious and may congregate in large flocks on large lakes. It is commonly found on densely vegetated swamps with cumbungi or lignum. The blue-billed duck is endemic to Australia with its distribution largely limited to the Murray-Darling Basin, south-eastern Australia and south-western Australia. In New South Wales it is mostly observed in southern parts of the Murray-Darling Basin (Marchant and Higgins 1990). The blue-billed

Figure 3.23: Blue-billed duck, *Oxyura australis*. Photograph: Dan Herbert (Hunter Bird Observers Club).

duck is regarded as endangered in Victoria (*Flora and Fauna Conservation Act 1988*) and vulnerable in New South Wales (*Threatened Species Conservation Act 1995*).

Habitat and diet

The blue-billed duck may be observed in inland wetlands in temperate parts of Australia. It exhibits a preference for large deep permanent wetlands with stable water levels and abundant aquatic vegetation (Braithwaite and Frith 1969b; Fjeldsa 1985). It may also gather on large fresh open lakes, swamps, sewage ponds, saline wetlands and salt lakes (Marchant and Higgins 1990). Feeding generally occurs in open water within or beside tall dense vegetation (Frith 1982), and roosting occurs on open water or in small concealed bays. Nesting occurs on deep fresh swamps and lakes that are densely vegetated, at least around the margins, with rushes, sedges or lignum. While nesting, males are likely to loaf near the nest site (Marchant and Higgins 1990). Nesting is usually solitary, but the blue-billed duck may nest close to other nests. Nests may be established in temporary or permanent wetlands, in inland or coastal settings. Nests are usually positioned over water or on islands within lakes, among aquatic vegetation such as cumbungi, lignum, spike-rush, *Eragrostis* and *Chenopodium* species, after flooding has commenced. The nest consists of a deep cup-shaped bowl and is constructed from plant material (Marchant and Higgins 1990).

The diet of the blue-billed duck is herbivorous, comprising seeds and leaves (Marchant and Higgins 1990). Molluscs, crustaceans, copepods and insects may also be consumed (Frith *et al.* 1969).

Breeding and flood requirements for breeding

The blue-billed duck is one of two duck species identified as seasonally breeding (Briggs 1990). Some studies refer to regularity of breeding (Braithwaite and Frith 1969a; Frith 1957a, 1967), while others found greater variability in breeding patterns (Halse and Jaensch 1989). For example, stable water levels in autumn have initiated unseasonable breeding in the blue-billed duck (Braithwaite and Frith 1969a). The reason for this unseasonable breeding was not established but it may be related to the productivity of prey items. In any case, it is evident that the blue-billed duck exhibits habitat preferences similar to those of the musk duck, inhabiting deep permanent waterbodies and breeding in temporary or permanent waterbodies (Braithwaite and Frith 1969b; Broome and Jarman 1983; Frith 1957a; Halse and Jaensch 1989; Halse *et al.* 1993; Hobbs 1961; Marchant and Higgins 1990). It is therefore postulated that the blue-billed duck behaves in a similar manner to the musk duck, in that breeding at permanent sites is stimulated by season and at temporarily flooded sites breeding is stimulated by suitable hydrological conditions. The main breeding season is between September and February (Marchant and Higgins 1990), but this is likely to vary according to water levels and food availability (Frith 1982; Frith *et al.* 1969; Marchant and Higgins 1990).

Based on breeding at a range of wetland settings, a relationship was established between breeding and peak rainfall with a three-month lag (Halse and Jaensch 1989). Blue-billed duck clutch sizes range from three to 12 eggs, with five to six eggs being more typical (Frith 1982). Eggs are laid at 24-hour intervals and are incubated by the female for about 24–27 days (Marchant and Higgins 1990). In captivity, blue-billed duck young have been closely associated with the female parent for 28–35 days, after which they start to gain independence. However, broods have been observed intact for eight weeks after hatching (Wheeler 1960). Sexual maturity is reached by yearlings (Braithwaite and Frith 1969a). Breeding duration is estimated at two to three months, and it is estimated that in temporary wetlands the blue-billed duck requires flooding for a minimum duration of five to six months.

As nesting occurs in vegetation over water, it is evident that the blue-billed duck has some reliance on flood depth for breeding. However, there is no information about the range or average flood depth at which nests are established, except that they are usually established within 1 m of the edge of vegetation in deep water (Marchant and Higgins 1990). Deep permanent waters are reportedly preferred habitats for foraging (Braithwaite and Frith 1969a; Broome and Jarman 1983; Frith 1957a; Halse *et al.* 1993; Hobbs 1961), with foraging and diving recorded at depths of up to 3 m (Frith 1982; Marchant and Higgins 1990). It is therefore suggested that the blue-billed duck prefers deeper waters of 1–3 m and that floodwaters should recede at a moderate to slow rate.

While the blue-billed duck may live for up to 16 years in captivity (delHoyo *et al.* 1992), the life expectancy of wild ducks in the Murray-Darling Basin is reportedly three to four years (Scott 1997). It is therefore estimated that flooding at temporary wetland settings should occur approximately every one to two years to maintain population numbers.

Like the musk duck, the blue-billed duck prefers deep permanent sites for breeding, which indicates that inter-flood drying is not an essential requirement for breeding. However, in seasonal or ephemeral settings, plants used for feeding and nesting exhibit some capacity for drying and inter-flood drying may promote the productivity of consumed items. Therefore, while inter-flood drying is not essential, it may promote breeding success in temporary wetlands.

Eurasian coot: *Fulica atra*

The Eurasian coot is a medium-sized waterbird with a glossy black head and neck, and slate-black body plumage (Figure 3.24). The bill is white and the coot has a large white frontal shield. The sexes have a similar appearance and do not exhibit seasonal variations in plumage. The Eurasian coot has a length of approximately 35–39 cm and a wingspan of 56–64 cm. Females are slightly smaller than males, with a weight of 0.52 kg compared to the male weight of 0.57 kg

Figure 3.24: Eurasian coot, *Fulica atra*. Photograph: Chris Herbert (Hunter Bird Observers Club).

(Marchant and Higgins 1993). It is a gregarious bird that inhabits large open bodies of fresh or brackish water. It has a global distribution. In eastern Australia it may be found throughout New South Wales and Victoria, most of eastern Queensland and eastern and southern South Australia and scattered throughout the Northern Territory. In Western Australia it is mostly limited to the south-west, and there are scattered observations throughout most of coastal Western Australia (Marchant and Higgins 1993).

Habitat and diet

The Eurasian coot is primarily found in inland and estuarine settings, but may occasionally be observed in marine settings (Marchant and Higgins 1993). It exhibits a preference for shallow permanent or ephemeral wetlands with highly diverse submerged or aquatic vegetation, and often with open deep water in excess of 2 m (Briggs 1979; Broome and Jarman 1983; Corrick and Norman 1980; Marchant and Higgins 1993; Martin *et al.* 1979). Favoured wetlands include rivers, creeks, pools, billabongs, swamps, lakes, lagoons, floodwaters, saltpans, claypans, freshwater meadows and grass and reed swamps (Marchant and Higgins 1993). The Eurasian coot may be observed roosting or loafing among dense emergent vegetation in large open areas of water, on the grassy margins of wetlands, on floating mats of vegetation or on grass next to wetlands (Marchant and Higgins 1993; Ross 1978; Small 1960). Foraging may occur in the water or on land. When foraging in water, the Eurasian coot forages in open, deep or shallow water, among or at the edge of emergent or aquatic vegetation, or on floating mats of aquatic vegetation. Foraging on land occurs on grassy areas near wetlands (Marchant and Higgins 1993; Ross 1978; Small 1960).

Breeding occurs on ephemeral and permanent wetlands including floodwaters. Breeding sites include the edge of floating or dense emergent vegetation and sometimes islands within wetlands (Fletcher 1960; Hobbs 1961; Jackson and Lyall 1964; Macdonald 1968; Marchant and Higgins 1993; Masters and Milhinch 1974). Nests are established in rushes and clumped vegetation in open water or at the edges of wetland habitats (Marchant and Higgins 1993). Nests are generally established above the waterline at heights of 25–30 cm or up to 75 cm; the Eurasian coot may reuse old nests for subsequent clutches or seasons (Brown and Brown 1980). Nests are large bulky structures and have been described as a mound or pyramid with a central depression. They are constructed of plant material such as rushes, lignum, sticks, twigs, leaves or bark and are attached to rushes, submerged logs, waterlilies or other anchors (Brown and Brown 1980; Marchant and Higgins 1993).

The Eurasian coot is regarded as herbivorous and its diet consists mostly of aquatic vegetation, seeds and grass. It may also consume insects, molluscs, crustaceans and, occasionally, eggs (Marchant and Higgins 1993).

Breeding and flood requirements for breeding

Breeding of the Eurasian coot appears to be stimulated by flooding (Briggs 1990), and a relationship has been established between breeding and rainfall with a two-month lag (Halse and Jaensch 1989). There are also observations of breeding on floodwaters (Crome 1988; Frith 1957a; Hobbs 1961) and evidence of breeding throughout the year when conditions are suitable (Marchant and Higgins 1993). Abundance reportedly correlates with rainfall (Woodall 1985) and deep permanent waters (Halse *et al.* 1993). However, Gosper *et al.* (1983) found that abundance correlated with season rather than other climatic and water level variables. It is evident that breeding is stimulated by flooding. It primarily occurs between September and November in New South Wales and August to January in Victoria, but may occur whenever conditions are suitable (Halse and Jaensch 1989; Marchant and Higgins 1993).

The lag time for peak breeding is between two to three months after peak rainfall, with some breeding commencing within one month of rainfall (Halse and Jaensch 1989). Breeding of the Eurasian coot is reported to occur relatively quickly on rising floodwaters compared to other rails (Frith 1957a). It is therefore suggested that the ideal lag time for breeding success of the Eurasian coot is two to three months; however, some breeding may be successful with a lag time as short as one month.

In Australia, Eurasian coot clutch size ranges from one to 14 eggs; large clutch sizes may be clutches from two females in one nest (Marchant and Higgins 1993). Eggs are laid at intervals of 24–48 hours and replacement eggs may be laid after loss. Eggs are incubated by both sexes for 23–26 days (Brown and Brown 1980). The chicks are fed by the parents for seven to 10 days and may leave the nest soon after hatching, but return to the nest frequently. Chicks begin to forage at about four weeks old and are dependent on their parents for about five weeks; however, the family unit may stay together for the remainder of the season until breeding recommences (Marchant and Higgins 1993). Breeding duration is estimated at approximately two months and flood duration should extend for a minimum period of three months. Success may be enhanced with flood durations of four to five months.

Nests are typically established in vegetation over water (Marchant and Higgins 1993). No further information is available about water depths, but it can be inferred from nesting locations that Eurasian coot nests are established at water depths in excess of about 30 cm, but not at depths greater than 2 m. Eurasian coot abundance is generally associated with deep permanent waters (Broome and Jarman 1983; Halse *et al.* 1993). Breeding can still be successful when waters recede at a moderate pace.

The Eurasian coot can live longer than 20 years (Staav and Fransson 2008), but the maximum longevity reported in the wild in Australia is seven years and two months (Australian Bird and Bat Banding Scheme 1999e). It is therefore suggested that conditions should be suitable for breeding at a similar rate to ducks, at every one to two years.

Crome (1988) inferred that inter-flood drying may enhance breeding of the Eurasian coot in temporary wetland settings. It is likely that productivity of consumed items may be enhanced by drying and that consumed items have some capacity for inter-flood drying. Therefore, while drying is not an essential requirement for breeding success, flooding following a drying period in temporary wetlands may promote wetland productivity and breeding success.

Dabbling ducks

Chestnut teal: *Anas castanea*

The chestnut teal is a small compact dabbling duck (Figure 3.25). The sexes display marked sexual differences in plumage: males exhibit a glossy bottle-green head, rich chestnut underparts and dark upper parts while females have a dark brown crown, a pale throat that is streaked brown and scalloped dark-brown plumage. Both sexes have deep red eyes, blue to grey bill, and green-grey legs and feet. The male displays duller eclipse plumage outside the breeding season. Males have a length of 40–50 cm and a weight of approximately 0.7 kg. Females are typically smaller, with a length of 35–45 cm and a weight of approximately 0.6 kg (Marchant and Higgins 1990). The chestnut teal is commonly seen in pairs, small groups or large flocks with grey teals. It may be observed foraging in the shallows of permanent open freshwater wetlands, in semi-permanent or permanent saline and estuarine wetlands or on deeper waters on high-altitude wetlands and creeks. It is endemic to Australia and is most commonly observed in south-eastern Queensland, south-western Western Australia, New South Wales, Victoria and Tasmania. It is presumed to migrate between breeding and non-breeding areas within Australia (Marchant and Higgins 1990).

Figure 3.25: Chestnut teal, *Anas castanea*. Photograph: Max Carpenter (DECCW).

Habitat and diet

The chestnut teal typically inhabits freshwater, brackish and estuarine wetlands in the coastal region, but permanent fresh water such as lakes, reservoirs and sewage ponds may support chestnut teals during dry periods (Norman and Brown 1988). Records indicate that the chestnut teal may be observed in areas with and without emergent vegetation, such as fresh meadows, shallow or deep swamps, rivers, river pools and billabongs, farm dams and occasionally rice fields (Corrick and Norman 1980; Frith 1957a, 1957b; Gosper 1981; Marchant and Higgins 1990; Vestjens 1977b). Foraging mostly occurs along the edges of wetlands, over sand and mud flats or among aquatic or emergent vegetation such as couch (*Cynodon dactylon*), water couch and spike-rush (Norman 1983; Norman and Mumford 1982; Norman *et al.* 1979). Chestnut teals may be observed during the day swimming or loafing on banks, logs, dead trees, rocks or short grassland that is partially or completely surrounded by water (Marchant and Higgins 1990; Norman *et al.* 1979). Breeding mostly occurs in coastal regions (Norman and Brown 1988), but may also occur in swamps, lakes and billabongs (Corrick and Norman 1980; Gosper 1981; Hobbs 1956a; Marchant and Higgins 1990). Breeding commonly takes place on the ground or in tree hollows; some chestnut teal have been observed nesting in flooded lignum swamps in New South Wales (Frith 1982). Nesting on the ground usually occurs near water in grassed areas such as water couch. Generally, no additional material is used to construct the nest, but small amounts of nearby plant material may be adventitiously added (Marchant and Higgins 1990).

The diet of the chestnut teal consists of seeds and insects with some other plant material, molluscs and crustaceans (Marchant and Higgins 1990). Norman and Mumford (1982) reported that 74% of the diet of chestnut teals at Gippsland Lakes, Victoria, consisted of plant material and that animal matter comprised the remainder.

Breeding and flood requirements for breeding

Breeding in the chestnut teal is reportedly stimulated by flooding and season (Briggs 1990) and it is known to exploit inland floodwaters for breeding when available (Frith 1957a). For example, chestnut teal abundance correlated with rainfall in the Brisbane region (Woodall 1985). Due to its ability to exploit floodwaters, breeding of chestnut teal may occur at any time, but it exhibits strong seasonal breeding where waters are more permanent (Frith 1967). Breeding generally occurs between July and March, with the majority occurring between August and October (Marchant and Higgins 1990). Breeding also exhibits a weakly latitudinal gradation, generally occurring earlier in Victoria than in New South Wales (Marchant and Higgins 1990).

As breeding information is primarily gained from captive and semi-captive birds (Norman 1982; Norman and Hurley 1984; Norman *et al.* 1979), it is difficult to estimate a lag time before breeding commencement. Breeding behaviour reportedly follows the common patterns of the *Anas* genus (Frith 1967) and the chestnut teal is likely to have similar lag times as the grey teal, with breeding commencing relatively quickly once conditions are suitable. Therefore, it is likely that the chestnut teal requires a minimum lag time of one month and an ideal lag time of two months (Frith 1957a; Halse and Jaensch 1989).

The clutch size for the chestnut teal ranges from five to 17 eggs, even up to 27 eggs (Frith 1982); however, most lay seven to 10 eggs (Marchant and Higgins 1990). The eggs are generally laid daily, with two to three eggs laid each day (Frith 1982). The chestnut teal will lay replacement clutches after loss or successful fledging, with two to three broods laid during a breeding season (Frith 1982). Females incubate eggs once the clutch is complete, for 23–28 days (Frith 1982). Hatching to first flight may occur within 56 days, but the chestnut teal is not regarded as fully grown until 60–80 days after hatching (Frith 1982). Breeding duration, from laying to independence, is estimated at three to four months. The chestnut teal requires minimum flood durations of four to five months and ideal flood durations of six months.

As nesting generally does not occur in the water and the chestnut teal usually forages at the edges of wetlands (Marchant and Higgins 1990), it is likely that the species does not have a flood depth preference. Rather, flooding should be deep enough to provide habitat for the entire breeding duration. Floodwaters should recede at a moderate to slow rate.

The maximum longevity of chestnut teal in the wild is reported to be 6.5 years (Australian Bird and Bat Banding Scheme 1999a). This is likely to be an upper limit, with estimates of lifespan in the Murray-Darling Basin thought to be closer to three to four years (Scott 1997). As the chestnut teal is responsive to flooding and can travel long distances to access suitable breeding habitats, large floods should occur in the network of wetlands within the Murray-Darling Basin on a one- to two-year basis to sustain breeding population numbers.

There is little evidence that the chestnut teal requires inter-flood drying, but its responsiveness to flooding indicates that it can readily exploit increases in prey items once flooding commences. This is likely to be high at wetlands that experience a drying period.

Grey teal: *Anas gracilis*

The grey teal is a small slender grey duck that has a similar appearance to the chestnut teal (Figure 3.26). The crown is darker grey than the chin, throat and upper neck and the body plumage is grey with a mottled pattern formed by the pale edging on body feathers. It has a dark green bill and bright red eyes. It has a length of 42–44 cm and a wingspan of 60–67 cm; males are generally slightly larger than females. The sexes have a similar appearance, and seasonal and eclipse plumage is not evident (Marchant and Higgins 1990). Like the chestnut teal, the grey teal is commonly observed in pairs, small groups or large flocks. It tends to prefer inland wetlands but has a wide distribution throughout most of Australia, except arid central Australia. Large numbers have been observed on the floodplains at the confluence of the Murrumbidgee and Lachlan rivers (Marchant and Higgins 1990).

Figure 3.26: Grey teal, *Anas gracilis*. Photograph: Chris Herbert (Hunter Bird Observers Club).

Habitat and diet

The grey teal has a widespread distribution in inland wetlands, estuarine and marine waters and occasionally farm dams (Marchant and Higgins 1990); however, it appears to favour shallow productive inland waters where food is abundant (Gentilli and Bekle 1983). The grey teal responds opportunistically to flooding on inland floodplains by following the edge of spreading waters in floodplain swamps, channels and pools vegetated with sedges, canegrass, lignum or tall emergent vegetation such as cumbungi (Corrick and Norman 1980; Fjeldsa 1985; Frith 1959a; Gosper 1981). It has also been observed in high numbers in response to wetlands being filled by rain or floodwaters (Gentilli and Bekle 1983; Marchant and Higgins 1990). Frequented habitats may include floodplains, lakes, reservoirs, rivers, creeks, billabongs, pools, wetlands with inundated or fringing timber, wooded wetlands, grasslands, irrigation channels, sewage ponds and artificial wetlands (Fjeldsa 1985; Frith 1959a; Frith *et al.* 1969; Gentilli and Bekle 1983; Gosper 1981; Vestjens 1977b). The grey teal prefers inland wetlands but may be observed in coastal or estuarine settings during the dry season or drought (Gentilli and Bekle 1983; Marchant and Higgins 1990; Norman 1983). Roosting occurs close to feeding grounds in areas partly or entirely isolated by water (Marchant and Higgins 1990). Breeding occurs opportunistically in response to flooding of alluvial floodplains and lakes (Frith 1959a; Gentilli and Bekle 1983); breeding locations must be large enough to persist until ducklings have fledged and shallow enough to allow diving for submerged food (Gentilli and Bekle 1983). Eggs are commonly laid in the cavities or hollows of trees located along watercourses, within flooded forests or near large wetlands. The grey teal commonly uses cavities in river red gum or black box, but nesting has also been observed under saltbush, in scrub such as tea-tree (*Leptospermum* spp.) or lignum (*Muehlenbeckia florulenta*), or on the ground among rushes (Marchant and Higgins 1990).

The diet of the grey teal consists mostly of seeds of aquatic plants, but may include animal material such as molluscs, gastropods, crustaceans, crayfish, insects and fish (Briggs *et al.* 1985; Frith 1959b; Marchant and Higgins 1990).

Breeding and flood requirements for breeding

Breeding of the grey teal is reportedly stimulated by flooding (Briggs 1990), a hypothesis supported by relationships observed between breeding of grey teal, maximum area of live river red gum and maximum area of aquatic plants (Briggs *et al.* 1997). There have been correlations between peak breeding and rainfall with a two-month lag (Halse and Jaensch 1989), observations of influxes of birds and subsequent breeding in response to suitable flood conditions (Frith 1957a), observations of the stimulation of testis and ovaries in response to increased water levels (Braithwaite and Frith 1969a), and breeding in response to rising water levels (Crome 1986, 1988). Frith (1959c, 1962) went to considerable effort to explore the initiation of breeding seasons in the grey teal, and observed increasing numbers of flocks and breeding behaviour as water levels began to rise. Frith (1959c) postulated that breeding of the grey teal is timed to coincide with increasing water levels and that duckling emergence coincides with increased productivity of available food. Abundance studies further support the role of flooding in stimulating breeding with grey teal abundance correlating with river discharge (Gosper *et al.* 1983), rainfall (Woodall 1985) and the presence of deep seasonal wetlands (Halse *et al.* 1993). The primary breeding season is between June and February (Marchant and Higgins 1990), but breeding may occur whenever conditions are favourable (Crome 1986; Frith 1957a; Fullagar *et al.* 1988; Marchant and Higgins 1990).

Commencement of breeding appears to occur rapidly in response to flooding, as early as 23 days after the commencement of water rising (Frith 1957a), and correlations have been observed between peak breeding and peak rainfall with a one- to two-month lag (Halse and Jaensch 1989). The mean duration between flooding under nests and peak number of grey teal broods was found to be four months following winter/spring flooding and five months following autumn flooding on the Murrumbidgee River (Briggs and Thornton 1999). It is therefore cautiously suggested that the grey teal requires a minimum lag of one month and an ideal lag of two to five months before commencement of breeding.

Clutch size ranges from six to 14 eggs, with up to 30 eggs observed within a nest (Frith 1982); seven to eight eggs appears to be the normal clutch size. Eggs are laid at intervals of about 24 hours and re-laying may occur as early as 20 days after loss of a brood. Subsequent broods may be laid after fledging, which may occur as early as 32 days after laying of the first brood (Marchant and Higgins 1990). Eggs are incubated for an average of 28 days (range 25–31 days) (Fullagar *et al.* 1988). Broods appear to fledge at about 55 days old, enabling two successful broods to be raised per breeding season. Maturity is reached quite rapidly, with both sexes capable of breeding within the first year (Marchant and Higgins 1990). Breeding duration is estimated at three to four months. The grey teal requires a minimum flood duration of four to five months and an ideal flood duration of five to nine months to promote breeding success.

While nests may be established in the hollows of trees in water, they may also be established on the ground and among or beneath vegetation (Marchant and Higgins 1990). Hence, the grey teal does not seem to exhibit a requirement for certain water depths for nest establishment. Halse *et al.* (1993) identifed a preference for deep wetland settings, but typical foraging behaviour, which includes up-ending, dabbling at the water's surface, dredging mud at the water's edge and stripping seeds from emergent vegetation (Marchant and Higgins 1990), indicates that food sources may be obtained from deep or shallow wetlands. However, preferred prey items generally (though not always) occur at the margins of wetlands or in relatively shallow water less than 1 m deep. This suggests that the grey teal does not have a water depth requirement; it may be observed at a range of water depths, depending on food availability. It appears that selection of breeding sites is based on the persistence of habitat throughout the breeding season (Gentilli and Bekle 1983). Inundation of breeding habitat should be maintained for the required flood duration and floodwaters should recede at a moderate pace.

Banding studies indicate that the grey teal may survive in the wild in excess of 31 years (Australian Bird and Bat Banding Scheme 1999b). This is a substantial time and could represent an upper limit. The analysis was based on an average life expectancy of three to four years for ducks within the Murray-Darling Basin (Scott 1997), which is more probable for this species. As the grey teal is known to disperse over great distances to access ephemeral and seasonal wetlands for breeding (Frith 1962), it is likely that large floods should occur in the network of wetlands within the Murray-Darling Basin on a one- to two-year basis to sustain breeding populations of grey teals.

Studies by Crome (1986, 1988) found that breeding of grey teal was enhanced when flooding followed an extended period of drying. It is likely that this was in response to increased productivity of chironomids and other aquatic insects that feed on decomposing plant material. Therefore, complete drying for at least one month may promote breeding success.

Australasian shoveler: *Anas rhynchotis*

The Australasian shoveler is a slim short-necked duck with a massive spatulate bill (Figure 3.27). It exhibits sexual and seasonal plumage differences. Males in breeding plumage have a black rump and back, a grey-blue head with a vertical white crescent between the bright yellow eyes, a grey-black bill and orange legs and feet. The underparts are chestnut with white patches. The male eclipse plumage is typically much duller. Female shovelers have mottled brown upper parts, chestnut underparts, a dark brown eye, a dark grey and olive-tinged bill, and brown-orange legs and feet (Marchant and Higgins 1990). The Australasian shoveler has a length of 45–55 cm, a wingspan of 70–80 cm and a weight of approximately 650 g (Marchant and Higgins 1990). It prefers permanent wetlands with fringing vegetation and open water, but may be

Figure 3.27: Australasian shoveler, *Anas rhynchotis*. Photograph: Chris Herbert (Hunter Bird Observers Club).

observed in a range of wetlands including floodwaters. It is endemic to Australia and New Zealand, and its distribution in Australia is largely limited to the eastern states and south-western Australia. Breeding is primarily limited to south-western New South Wales, Victoria and south-eastern South Australia (Marchant and Higgins 1990). Australasian shovelers are regarded as vulnerable in Victoria (*Flora and Fauna Conservation Act 1988*).

Habitat and diet

Due to the structure of the bill, which is suited to filter-feeding, the Australasian shoveler is limited to aquatic habitats and prefers open water or soft mud in inland wetlands. It is largely limited to wetlands in the temperate zone (Marchant and Higgins 1990). These settings include deep permanent lakes and swamps, and temporary swamps with seasonal fluctuations in water level (Braithwaite and Frith 1969b; Fjeldsa 1985). The Australasian shoveler appears to have a preference for freshwater habitats, but has been observed in high numbers on brackish and saline wetlands (Marchant and Higgins 1990). It has been observed frequenting billabongs, watercourses, floodwaters on alluvial plains, freshwater meadows, shallow swamps, reed swamps, wooded lakes, irrigated fields, sewage ponds and farm dams (Braithwaite and Frith 1969b; Corrick 1982; Frith 1959a; Gosper 1981; Marchant and Higgins 1990; Vestjens 1977b). Fjeldsa (1985) suggested that the presence or absence of fringing or emergent vegetation may not be important; however, birds commonly feed away from vegetation cover (Corrick and Norman 1980; Frith *et al.* 1969). Foraging occurs mostly at night or dusk on water at depths of approximately 30 cm, allowing the Australasian shoveler to reach food by up-ending (Marchant and Higgins 1990). During the day it rests on mud banks among dense vegetation cover or floats on deep water. It has been observed perched on logs or other low perches. Breeding occurs in grassy sites near freshwater wetlands (Marchant and Higgins 1990). Crome (1988) suggested that it breeds more readily in swamps that are productive and have a high diversity of flora and invertebrates. Nests are established on the ground in open paddocks, on low embankments, among long tussock grasses or low herbs, such as bulrushes, thistles, clover, reeds or crops (Frith 1982; Hobbs 1956a; Marchant and Higgins 1990).

The diet of the Australasian shoveler primarily consists of animal material although seeds may also be consumed (Frith 1959b; Frith *et al.* 1969; Marchant and Higgins 1990). Diet analyses indicate that it may consume bivalves, gastropods, crustaceans, insects and some plant material, including seeds (Frith 1959b; Frith *et al.* 1969).

Breeding and flood requirements for breeding

Most studies indicate that breeding of the Australasian shoveler is stimulated by flooding (Briggs 1990; Frith 1957a) or factors associated with flooding, such as rainfall (Frith 1982; Halse and Jaensch 1989) and increasing water levels (Braithwaite and Frith 1969a; Crome 1988). Australasian shoveler abundance has been correlated with climatic factors (Kingsford *et al.* 1999) and the presence of deep, moderately permanent wetlands (Halse *et al.* 1993). Despite the strong role of flooding in breeding, the Australasian shoveler appears to exhibit relatively well-defined breeding seasons between August and December, with a maximum breeding season of June to February (Marchant and Higgins 1990).

There is relatively little information about breeding of the Australasian shoveler, making it difficult to determine the required characteristics of flooding. Halse and Jaensch (1989) suggested that peak breeding correlates with rainfall with a two- to three-month lag and that some breeding commences as early as one month after rainfall.

The clutch size of the Australasian shoveler is thought to be nine to 11 eggs (Frith 1982), although this has not been quantified. Duration of laying is unknown; incubation by the female reportedly occurs for 25 days (Marchant and Higgins 1990). Fledging occurs between eight and 10 weeks old and young are sexually mature at one year old (Marchant and Higgins 1990). The Australasian shoveler has a breeding duration of approximately three months and it is estimated that it requires flooding for a minimum period of four months and ideally five to six months.

As nests are not established in or over water, there does not appear to be a water depth requirement for breeding. Rather, nests are established on the ground, close to water or within 200 m of water (Frith 1982; Hobbs 1956a; Marchant and Higgins 1990). Halse *et al.* (1993) indicated that the Australasian shoveler has a preference for deep, moderately permanent wetlands, but foraging apparently occurs in shallow waters of less than 30 cm (Frith *et al.* 1969). When breeding occurs at temporary or ephemeral wetlands, waters should recede at a moderate pace so as to maintain foraging habitats.

Based on the recapture of banded birds, the Australasian shoveler can live at least 10.5 years (Australian Bird and Bat Banding Scheme 2000a). Similar to recaptures of other ducks (e.g. grey teals, Pacific black ducks), this estimate is markedly higher than the three- to four-year average life expectancy for ducks within the Murray-Darling Basin (Scott 1997). Therefore, to take a cautionary approach, large floods should occur in suitable wetlands within the Murray-Darling Basin on a one- to two-year basis.

Crome (1988) implied that breeding of the Australasian shoveler may be enhanced when flooding follows a period of complete drying within wetlands. Although this conclusion was based on only one breeding record, the theory was partly supported by Frith *et al.* (1969), who found that fluctuations in water levels determined the availability of littoral flora and associated fauna, which provide food and habitat for the Australasian shoveler throughout the year. As Frith *et al.* (1969) indicated, the Australasian shoveler has a preference for permanent swamps; it is unknown whether complete drying is necessary. It is evident that a number of preferred prey items exhibit some tolerance for drying and that drying may promote the productivity of those items. It is therefore suggested that inter-flood drying is not essential, but breeding success may be enhanced when drying occurs for a month or so.

Pacific black duck: *Anas superciliosa*

The Pacific black duck is a large heavily-built dabbling duck with a long head and bill (Figure 3.28). Its plumage is mostly mid-brown with buff-coloured edges on the feathers. It has a dark brown crown with a dark line and cream edging above and below, passing through its eye. The Pacific black duck has conspicuous white underwings that are readily observed during flight (Marchant and Higgins 1990). It has a length of 47–60 cm and a wingspan of 80–100 cm. The sexes have a similar appearance and there are only small seasonal changes in appearance (Marchant and Higgins 1990). The Pacific black duck is a gregarious bird and may be found in a range of habitats, but it exhibits a preference for deep permanent swamps with dense vegetation (Marchant and Higgins 1990). It has a widespread distribution throughout Australia and New Zealand, except in arid areas where it may be observed as a vagrant (Marchant and Higgins 1990). It is particularly widespread in the Murray-Darling Basin (Morris *et al.* 1981).

Habitat and diet

The Pacific black duck may be observed in a range of habitats including temperate and tropical wetlands, and estuarine and marine waters. It does not exhibit marked habitat preferences and is generally observed where conditions are moist or flooded. It may be observed using aquatic and terrestrial habitats, shallow or deep water, among vegetation or on wetland shores, mudflats and pastures. It has been observed using almost any pool, puddle, drain, channel, creek or farm dam (Fjeldsa 1985; Frith 1982; Marchant and Higgins 1990). Specific settings in

Figure 3.28: Pacific black duck, *Anas superciliosa*. Photograph: Max Carpenter (DECCW).

which the Pacific black duck has been observed include deep swamps vegetated with species such as cumbungi, lignum and *Melaleuca* species, shallow seasonal swamps, fresh meadows, sewage ponds, rivers and creeks passing through timbered or open country, open or wooded lakes, flooded or irrigated crops and pastures, and floodwaters on alluvial plains (Corrick and Norman 1980; Fjeldsa 1985; Frith 1959a; Frith *et al.* 1969; Marchant and Higgins 1990; Serventy 1947; Vestjens 1977b). It appears to show seasonal changes in habitat use associated with rainfall patterns – flooded meadows and seasonal swamps are frequented during wet periods, while sites with permanent water are frequented during dry periods (Goodrick 1979; Marchant and Higgins 1990). Loafing occurs on the water or on grassy banks and islands, mud banks or low perches of fallen timber, tree branches, posts or other suitable perches (Marchant and Higgins 1990; Norman *et al.* 1979). Foraging occurs in open water and is limited to depths where the Pacific black duck can feed from the substrate by up-ending (Marchant and Higgins 1990). Breeding occurs in a range of sites away from water such as grassland, cropland, scrubland and forest, or beside any wetland (Frith 1982; Marchant and Higgins 1990; Owen and Sell 1985). Nesting commonly occurs in tree hollows, sometimes in nests of other waterbirds. Less commonly, nests may be on the ground and hidden among vegetation (Marchant and Higgins 1990). Nests in tree hollows generally occur in more open settings and are more conspicuous than other dabbling duck nests. Nests constructed in river red gum have a mean height of the entrance above the trunk base of 290 cm, trunk circumference of 310 cm, hollow angle of 54° and depth of hollow to eggs of 39 cm (Marchant and Higgins 1990).

The diet of the Pacific black duck mostly consists of plant material such as seeds, but may include animal matter such as insects and crustaceans. Other material consumed includes molluscs and spiders (Frith 1959b; Frith *et al.* 1969; Goodrick 1979; Marchant and Higgins 1990).

Breeding and flood requirements for breeding

Breeding in the Pacific black duck is primarily stimulated by flooding (Briggs 1990; Crome 1986, 1988; Frith 1959c; Halse and Jaensch 1989) and by season (Briggs 1990; Frith 1959c; Halse and Jaensch 1989). While Frith (1957a) stressed the regular nature of breeding irrespective of climatic conditions, this appears to be in relation to resident populations that confined their breeding to more permanent waters. In contrast, Braithwaite and Frith (1969a) discussed breeding activity in Pacific black duck in temporary habitats. They found relationships between the length of sexual activity and water level fluctuations; greater availability of food following water level fluctuations was important for stimulating and sustaining gonad activity. Briggs *et al.* (1997) found a similar relationship between breeding of Pacific black duck and maximum area of aquatic plants at wetlands on the Murrumbidgee floodplain. Pacific black duck abundance has a similar pattern, with studies showing relationships between abundance, flooding and associated factors (Harper 1990; Kingsford *et al.* 1999; Woodall 1985) and between abundance and season (Gosper *et al.* 1983). Pacific black duck breeding is generally confined to the end of winter but may extend from June to December (Crome 1986; Fullagar *et al.* 1988; Marchant and Higgins 1990).

Peak breeding reportedly occurs one to two months after peak rainfall, with some breeding occurring within the same month as the commencement of rainfall. Inverse relationships were also established with temperature with a one-month lag and photoperiod with a two-month lag (Halse and Jaensch 1989). Similarly, Crome (1986) found peak breeding occurred three months after water levels began to rise. These relationships strongly reflect the winter–spring nature of breeding, but indicate that a lag of one to three months after rainfall or flooding may be required before breeding is initiated in temporary wetlands settings.

Clutch size for the Pacific black duck ranges from seven to 14 eggs, but is usually eight to 10 eggs (Frith 1982). Eggs are laid at daily intervals and the female incubates them for 26–32 days, with a mean of 29 days. The young are capable of flight at 52–66 days. Pair bonds are formed at about six months of age and young reach sexual maturity within their first year (Marchant and Higgins 1990). Breeding duration is estimated at three to four months. In temporary wetlands, flooding should occur for a minimum period of four to five months and breeding success is enhanced when flooding occurs for a period of six to seven months.

Water depths do not appear to be essential for nest establishment as the Pacific black duck will nest in a range of settings, both away from water and in or beside a wetland (Frith 1982; Marchant and Higgins 1990). While feeding generally occurs in open water, food is obtained by up-ending, indicating that floodwaters should be at a depth whereby aquatic vegetation can be accessed from near the surface. Flooding should remain for long enough to maintain prey items and foraging habitats. Waters should recede at a moderate pace.

The longevity of the Pacific black duck may be in excess of 15.5 years (Australian Bird and Bat Banding Scheme 1999c), but this is probably the upper limit and longevity is more likely to be three or four years in the Murray-Darling Basin (Scott 1997). Breeding should occur a number of times within the average life expectancy to maintain breeding population numbers. As the Pacific black duck can disperse in inland areas (Marchant and Higgins 1990), it is suggested that flooding should occur approximately every one to two years in the network of wetlands in the Murray-Darling Basin.

Crome (1986, 1988) observed the relationship between enhanced breeding success and wetland drying at Lake Merrimajeel and Murrumbidgil Swamp near Booligal, New South Wales. This relationship was particularly strong for the Pacific black duck and it was postulated that the consumption of chironomids, which are highly productive in response to decaying plant material, improves the condition of the Pacific black duck and enhances breeding success. It is therefore suggested that inter-flood drying for a few months may enhance breeding success of the Pacific black duck.

Pink-eared duck: *Malacorhynchus membranaceus*

The pink-eared duck is a small duck with a large square-tipped bill, striped flanks, breast and lower neck, a dark eye patch and white face (Figure 3.29). It gets its name from a small distinctive pink patch behind its eye. The sexes have a similar appearance, although females are generally smaller. The pink-eared duck does not exhibit seasonal changes in plumage (Marchant and Higgins 1990). It has a length of approximately 36–45 cm, a wingspan of 57–71 cm and a weight of 270–480 g (Marchant and Higgins 1990). The pink-eared duck may be observed in pairs, small groups or very large flocks within shallow wetlands. It is endemic to Australia and has a widespread distribution, except in arid central parts (Marchant and Higgins 1990). It is especially common in the Murray-Darling Basin (Morris *et al.* 1981).

Habitat and diet

The pink-eared duck has a widespread distribution in inland wetlands. The specialised bill shape forces the pink-eared duck to forage in mud, making it almost entirely aquatic (Marchant and Higgins 1990). It exhibits a preference for shallow stagnant turbid waters with abundant aquatic fauna (Fjeldsa 1985; Frith 1959a; Hobbs 1957b). It has been observed in large numbers on floodwaters on alluvial plains vegetated with saltbush, bluebush, canegrass and lignum. It is rarely observed on deep waters in swamps, lakes and billabongs as it prefers to forage on the margins of wetlands (Frith 1982). Roosting occurs in flocks at the water's edge or on low fallen timber and logs (Frith 1982). Breeding is concentrated at turbid receding floodwaters on alluvial plains (Hobbs 1957b) or on seasonal or permanent lakes and swamps (Marchant and Higgins 1990; Vestjens 1977b). Nests are constructed on any available surface, over water in hollows or forks of trees, on top of posts, logs, stumps and suitable plant material (Marchant and Higgins 1990). Breeding appears to be more successful in settings that have undergone

Figure 3.29: Pink-eared duck, *Malacorhynchus membranaceus*. Photograph: Chris Herbert (Hunter Bird Observers Club).

complete drying before flooding (Crome 1988). Nests are generally constructed at an average height of 1.04 m, but may be at heights of up to 5 m.

The pink-eared duck is carnivorous and mostly consumes aquatic invertebrates, primarily chironomid larvae; some seeds may also be consumed (Marchant and Higgins 1990). Analyses of gizzards collected in New South Wales indicate that the pink-eared duck consumes gastropods, crustaceans, insects, seeds and other plant material (Briggs 1982; Briggs *et al.* 1985; Frith 1959b; Frith *et al.* 1969).

Breeding and flood requirements for breeding

There is strong evidence that flooding is the primary stimulus for breeding in the pink-eared duck (Briggs 1990; Crome 1986, 1988; Frith 1957a, 1959c; Halse and Jaensch 1989). The duck is known for its erratic breeding seasons as it prefers to breed on receding floodwaters (Braithwaite and Frith 1969a; Frith 1957a, 1959c). However, this preference may be related to increases in food supplies after peak flooding (Crome 1986). Due to the pink-eared duck's strong reliance on flooding, breeding may occur whenever conditions are favourable (Marchant and Higgins 1990). The pink-eared duck does not have a main breeding season, but conditions suitable for breeding generally occur between August and February in southern parts of Australia and between March and May in northern parts of Australia (Marchant and Higgins 1990).

The lag time for breeding of pink-eared duck is thought to be approximately one to three months (Crome 1986; Frith 1959c; Halse and Jaensch 1989) with breeding correlating best with a two-month lag after the rainfall peak (Halse and Jaensch 1989). Therefore, it is likely that the pink-eared duck requires a minimum lag time of one month; increasing lag periods of two to three months may enhance breeding success.

Clutch sizes of the pink-eared duck range from three to 10 eggs, and average seven eggs. Laying commences eight to 26 days after floodwaters begin receding. The females incubate the eggs for about 26 days (Frith 1959c). No information is available on the time for young to fledge or to reach sexual maturity. However, the majority of Australian ducks (chestnut teal, grey teal, Australasian shoveler, Pacific black duck, Australian shelduck, freckled duck and maned duck) fledge between two and three months. It is estimated that the pink-eared duck requires a flood duration of approximately three to four months and an ideal flood duration of four to six months to maximise breeding success.

Nests of pink-eared ducks are usually established over water with depths of 0.2–1.63 m (average 0.84 m) (Marchant and Higgins 1990). While the pink-eared duck prefers to breed on receding floodwaters, the breeding site should be inundated for four to six months and recede at a moderate pace.

The maximum age of recaptured banded pink-eared duck is one year and three months (Australian Bird and Bat Banding Scheme 2000d). This is particularly young and may reflect the difficulty in recapturing banded pink-eared ducks. Flooding should occur several times throughout its life to ensure population persistence. Based on the average life expectancy for ducks within the Murray-Darling Basin of three to four years (Scott 1997) and the dispersive nature of the pink-eared duck, it is postulated that flooding should occur in the Murray-Darling Basin approximately every one to two years.

Crome (1986, 1988) observed enhanced breeding success when flooding followed a period of drying at wetlands near Booligal in New South Wales. Crome (1986) postulated that this resulted from the increased productivity of chironomids consuming decaying plant material following a period of drying. A few months of complete drying in a wetland promotes the breeding success of the pink-eared duck.

Freckled duck: *Stictonetta naevosa*

The freckled duck is a dark grey-brown, heavy-bodied duck with a long neck, narrow bill and large head (Figure 3.30). The dark brown plumage is evenly freckled all over the body with white or buff-coloured freckles. The sexes have a similar appearance, however, the female is smaller and its plumage is generally paler (Marchant and Higgins 1990). The freckled duck exhibits slight seasonal plumage changes, with both sexes displaying paler plumage during the non-breeding season. It has an approximate length of 51–56 cm and a wingspan of 77–82 cm. Males weigh 0.98 kg and females weigh 0.84 kg (Marchant and Higgins 1990). The freckled duck is gregarious and may be observed in small groups or large loose flocks on shallow open lakes, among flooded lignum or dense coastal vegetation. It is endemic to Australia and its distribution is limited to south-eastern Australia. It is occasionally seen in coastal regions and the interior parts of south-western Australia (Marchant and Higgins 1990). The freckled duck is regarded as endangered in Victoria (*Flora and Fauna Conservation Act 1988*), and vulnerable in New South Wales (*Threatened Species Conservation Act 1995*) and South Australia (*National Parks and Wildlife Act 1972*).

Habitat and diet

The freckled duck may be observed in coastal regions during droughts, but has a stronghold on inland wetlands throughout temperate Australia (Marchant and Higgins 1990). Due to its filter-feeding and dabbling habit, it is almost entirely aquatic and is found on densely vegetated waters such as floodwaters, swamps and creeks vegetated with lignum or canegrass (Frith 1965; Hobbs 1961). When floodwaters recede, the freckled duck moves from ephemeral habitats

Figure 3.30: Freckled duck, *Stictonetta naevosa*. Photograph: Chris Herbert (Hunter Bird Observers Club).

to large permanent open waters such as lakes and reservoirs. It has also been observed on rivers, river pools, billabongs, farm dams, sewage ponds, saltpans, saline and freshwater meadows, shallow freshwater swamps with emergent vegetation, and open parts of deep freshwater swamps (Corrick 1982; Corrick and Norman 1980; Frith 1982; Marchant and Higgins 1990). Foraging primarily occurs in shallow waters of less than 5 cm where food is collected by filtering bottom sediments. Foraging is therefore limited to shallow waters or the edges of deeper waters, however, the freckled duck can filter feed from the surface in deeper water habitats when food is available (Briggs 1982; Frith 1965; Frith *et al.* 1969). It may be observed loafing during the day on exposed mudbanks, sand spits or headlands, or even among dense vegetation or on emergent posts (Marchant and Higgins 1990). While breeding, males will loaf near the nest. Breeding occurs in floodwaters, particularly among flooded lignum, oak or *Melaleuca* species (Marchant and Higgins 1990). Nest sites are usually exclusively used by freckled duck (Braithwaite 1976; Frith 1965; Marchant and Higgins 1990).

The freckled duck is a specialised filter-feeder and consumes both insects and plant material (Briggs 1982; Frith *et al.* 1969). Briggs (1982) suggested it may prefer insects and some plant material, while Frith *et al.* (1969) found a predominance of plant material with some nematodes and molluscs.

Breeding and flood requirements for breeding

Breeding of the freckled duck is stimulated by flooding or related hydrological factors (Briggs 1990; Halse and Jaensch 1989). While Braithwaite and Frith (1969a) identified a regular breeding season in the freckled duck, they also found that breeding could occur at any time in response to rainfall and flooding. They suggested that freckled ducks may have variable and erratic breeding seasons, but did not identify clear relationships between water level fluctuations and the species' gonad cycle. They concluded that breeding in the freckled duck was more related to food availability, which is influenced by water level, rather than a direct relationship between breeding and water levels. In any case, it is evident that flooding stimulates breeding in the freckled duck. The main breeding season is June to December (Frith 1965), but some breeding has been observed as early as April. Breeding may occur whenever food and flood conditions are suitable (Marchant and Higgins 1990).

The lag time for breeding of freckled duck is reported to be three months after peak rainfall, with some breeding evident two months after peak rainfall (Halse and Jaensch 1989). Clutch sizes of the freckled duck generally range from five to seven eggs, and eggs are usually laid daily. Females incubate eggs for about 28 days and the young are able to fly by about nine weeks old. In captivity, males have become sexually mature at six months and females have laid eggs at two years old (Marchant and Higgins 1990). Breeding duration is thought to be three months and floodwaters should remain at a breeding site for six months to maintain breeding and foraging habitats, and food availability. Floodwaters should recede at a moderate to slow pace.

Nests are usually constructed in flooded vegetation such as lignum, which suggests that the freckled duck has a water depth requirement before commencement of breeding (Marchant and Higgins 1990). However, little information on required water depths is available for this species, except that nests were observed within approximately 1 m of water (Frith 1965).

The freckled duck is regarded as endanged and vulnerable within the Murray-Darling Basin, and few have been observed breeding (Braithwaite 1976; Crome 1988; Frith 1965; Halse and Jaensch 1989). Therefore, there is no information on the longevity of this species. Based on an estimated life expectancy of three to four years for ducks within the Murray-Darling Basin (Scott 1997), it is estimated that large floods that maintain breeding and foraging habitats for five to six months should occur approximately every one to two years.

Crome (1988) inferred that breeding in the freckled duck may be enhanced when flooding follows a period of drying; however, this is based on an observation of one nest in one breeding season. Since the freckled duck consumes both insects and plants (Marchant and Higgins 1990), the hypothesis is probably true: insect populations can increase rapidly in wetlands that have undergone a drying phase and plant species consumed by this species exhibit some capacity for drying.

Grazing waterfowl

Maned duck: *Chenonetta jubata*

The maned (Australian wood) duck is a medium-sized grey and brown duck, with a small head and bill and a goose-like appearance (Figure 3.31). Males have a darker head, small dark mane, speckled brown-grey breast and black lower belly. They are slightly larger than females. Maned ducks are approximately 48 cm in length and have a wingspan of 80 cm and a weight of 0.8 kg. Females have a paler head with two white stripes above and below the eye, white lower belly and undertails. They do not exhibit seasonal plumage. Males exhibit eclipse plumage (Marchant and Higgins 1990). The maned duck may be seen in pairs or flocks. Habitats frequented by the maned duck include timbered areas near water, short pasture and muddy banks of waterways such as dams, swamps, lakes, reservoirs and sewage farms. It exhibits a widespread distribution throughout Australia, except some arid regions. It is particularly common in eastern Australia (Marchant and Higgins 1990).

Habitat and diet

The maned duck has a widespread distribution within grasslands, woodlands and inland wetlands. It prefers freshwater settings, but may occasionally be observed in coastal saltmarsh or brackish settings (Gosper 1981; Marchant and Higgins 1990). Preferred wetlands have open

Figure 3.31: Maned duck, *Chenonetta jubata*. Photograph: Max Carpenter (DECCW).

banks and wide open beaches for loafing and nearby grasslands for grazing. They may include farm dams, creeks, rivers and river pools, irrigation channels, swamps, rainwater pools, bores, springs, freshwater meadows, shallow swamps, sewage ponds, wooded lakes and ornamental ponds (Corrick and Norman 1980; Fjeldsa 1985; Frith 1959a; Gosper 1981; Marchant and Higgins 1990; Serventy 1947; Vestjens 1977b). When frequenting densely vegetated, deep and large swamps or lakes the maned duck generally remains near the edge of the wetlands (Frith 1982). As a grazing waterfowl, it is typically observed foraging in short grass or among herbs either on the land or in shallow water at the edges of wetlands. Loafing is undertaken during the day and night on open banks and wide beaches within wetlands. Breeding is associated with freshwater wetlands; however, the nests may be established some distance from the wetland. Nests are typically established in hollows in live trees, located in or near water in densely timbered areas. However, nests may be established in holes in dead trees or in trees up to 1.5 km from water (Frith 1982; Kingsford 1986; Marchant and Higgins 1990). No material is added to nests except for some down, after the eggs have been laid (Marchant and Higgins 1990).

The maned duck generally consumes grass, clover and other green herbage. Grain and insects may also be consumed when preferred food items are limited (Frith 1959b; Marchant and Higgins 1990).

Breeding and flood requirements for breeding

As nesting occurs some distance from wetlands and within tree hollows (Marchant and Higgins 1990), the role of flooding in stimulating bird breeding is limited. Rainfall and warmer temperatures promote the growth of grasses that are grazed by the maned duck and that would sustain the species throughout the breeding season. Therefore, it is evident that breeding activity is stimulated by rainfall and season (Briggs 1990), and that the maned duck is one of the few waterbirds that do not exhibit a flooding requirement for breeding. Halse and Jaensch (1989) found a correlation between breeding success and rainfall with a two- or three-month lag. Briggs *et al.* (1991) found that the body condition of the maned duck was related to food availability, whether it was breeding, the stage of breeding and (in females) whether it had paired. Interestingly, Kingsford *et al.* (1999) found a significant correlation between the abundance of the maned duck and wetland area. Breeding of the maned duck commonly occurs in simple pairs in solitary. As breeding is largely dependent on rainfall and the growth of grass species, breeding may occur at any time when these conditions are available (Marchant and Higgins 1990). The main breeding season is between July and December in southern parts of New South Wales and Victoria, between January and March in northern parts of New South Wales, and between August and September in inland locations (Frith 1982; Kingsford 1986).

The maned duck requires a lag time of two to three months in which to build up fat reserves (Halse and Jaensch 1989) and stimulate follicle development and gonad activity (Briggs *et al.* 1991). Clutch sizes are large, and generally range from eight to 11 eggs (Marchant and Higgins 1990). No information on the period for laying is available; however, females incubate the eggs for 28–34 days (Frith 1982; Marchant and Higgins 1990). The period until fledging is reportedly 57 days (Kingsford 1986), but young are brooded for at least an additional two weeks (Marchant and Higgins 1990). There is no information on the period until the young are able to breed. Breeding duration is estimated at approximately four months and wetlands should be inundated for six to seven months to provide sufficient foraging and roosting habitat for the lag and breeding periods.

Depth of flooding does not seem to be an essential requirement for breeding; however, the maned duck prefers deep permanent wetlands with fringing trees (Halse *et al.* 1993). There is no indication that nests will be abandoned should water levels recede in nearby wetlands, but it is essential that grazing pastures and herbage be maintained throughout the breeding period when the energy requirements of female birds are higher (Kingsford 1989). It is therefore estimated that floodwaters should recede at a moderate to slow rate to ensure the maintenance of foraging habitats.

The Australian Bird Study Association, which has undertaken regular bird banding of Australian birds, has recaptured maned ducks at the age of nine years and one month (Australian Bird and Bat Banding Scheme 1996). This is likely to represent a maximum value rather than average life expectancy. Scott (1997) indicated that ducks generally have a life expectancy of three to four years. Kear (2005) indicated that swans generally live longer than geese, which generally live longer than ducks. Therefore, it is likely that the maned duck should breed at least every second year with flooding substantial enough to maintain habitats for the lag and breeding periods.

Breeding success is enhanced when it occurs in habitats that have undergone complete drying (Crome 1986, 1988). In fact, Crome (1986) suggested that the cycle of drying is a greater stimulus for breeding of the maned duck than flooding. However, breeding did occur when wetlands did not undergo complete drying. It is likely that complete drying of a wetland for a short time enhances breeding success.

Plumed whistling-duck: *Dendrocygna eytoni*

The plumed whistling-duck is a small, pale, goose-like duck (Figure 3.32). It has an upright stance, long legs and neck, and a pink bill and legs. Its plumage is olive brown on the crown and hind neck, dark brown on the mantle, back and scapulars and black brown on the rump. It has

Figure 3.32: Plumed whistling-duck, *Dendrocygna eytoni*. Photograph: Chris Herbert (Hunter Bird Observers Club).

characteristic elongated buff-coloured flank-plumes. It has a length of approximately 40–60 cm, a wingspan of 75–90 cm and a weight of 600–1400 g. The plumed whistling-duck does not have seasonal changes in its plumage and the sexes have a similar appearance. The males are slightly larger (Marchant and Higgins 1990). The plumed whistling-duck may be observed in small groups or in flocks of thousands at the edges of wetlands located near open short grass plains. It is endemic to Australia and has a widespread distribution throughout northern and eastern Australia (Marchant and Higgins 1990).

Habitat and diet

The plumed whistling-duck is strongly associated with grasslands, including native grassland or pasture, or dry or moist land. It may also be observed dabbling in the shallow edges of wetlands or roosting on bare open banks (Frith 1982; Marchant and Higgins 1990). The plumed whistling-duck can be observed in a range of freshwater wetland settings; as the wetlands are not used for foraging, the type of wetland is reportedly not important (Marchant and Higgins 1990). It has been observed in billabongs, pools, watercourses, swamps, flooded depressions, lakes, farm dams, floodwaters, sewage ponds, permanent dams and waterholes (Gosper 1981; Hobbs 1956a; Marchant and Higgins 1990; Vestjens 1977b). Foraging occurs in grasslands that are located near the wetlands utilised for roosting (Marchant and Higgins 1990). Roosting occurs on open banks; scrubland and forests are generally avoided as roosting sites (Frith 1982). Breeding generally occurs at inland sites in tall grasslands near permanent or temporary wetlands (Marchant and Higgins 1990). The nests are established on the ground some distance from the wetland so that they are not flooded (Frith 1982; Hobbs 1956a, 1961; Marchant and Higgins 1990). Nests are commonly established approximately 1 km from water to escape the risk of flooding (D'Ombrain 1944; Morse 1922). Surrounding vegetation is used to construct nests and there are obvious tracks, in the long grass or bushes, between the nest and the water (Marchant and Higgins 1990).

The diet of the plumed whistling-duck consists almost entirely of plant material. Gizzard analyses from northern Queensland indicate that it may consume a range of plant species typical of floodplain wetlands, such as rushes (*Cyperus*, *Potamogeton* and *Eleocharis* species), grasses (*Paspalidium flavidum*) and nardoo (*Marsilea* species) (Marchant and Higgins 1990).

Breeding and flood requirements for breeding

Relatively little information on the flood requirements for breeding in the plumed whistling-duck is available. It is probable that breeding is influenced by rainfall and water availability (Marchant and Higgins 1990), as the diet of the plumed whistling-duck can include many floodplain plants. Anecdotal evidence indicates a correlation between floodwaters and species occurrence in south-western New South Wales (Hobbs 1961). No correlations were found between abundance of plumed whistling-duck, rainfall or dam water levels in the Brisbane region (Woodall 1985). In the absence of further information, the view that 'the movements (and most likely the breeding of wild ducks) are probably dictated by their food supply as it fluctuates with rainfall and flooding' is considered appropriate (Frith 1957a, p. 24). The plumed whistling-duck breeds in pairs in solitary, primarily between September and January in south-eastern Australia, but breeding can occur as late as May in northern parts of Australia (Marchant and Higgins 1990).

Information on the lag time required to stimulate breeding of the plumed whistling-duck is not available. It is likely that some lag period is required to stimulate the growth of plants and to enable the species to build fat reserves for the breeding season. In the absence of further information, it is cautiously suggested that the plumed whistling-duck requires a lag of two to three months after significant rainfall or flooding to stimulate plant productivity.

Clutch sizes are not well known, but may range from eight to 14 eggs (Frith 1982), which are laid at intervals of one to six days in captivity (D'Ombrain 1945). Both parents incubate the eggs after laying is complete for 28–30 days (D'Ombrain 1945; Marchant and Higgins 1990). The period until fledging, independence and reproductive maturity is unknown. As there is no information on time until fledging or time for young birds to become completely independent of the parents, it is difficult to estimate the required breeding duration. The majority of Australian ducks (e.g. chestnut teal, grey teal, Australasian shoveler, Pacific black duck, Australian shelduck, freckled duck and maned duck) fledge at between two and three months. Based on this, it is likely that the breeding duration of the plumed whistling-duck is approximately three to five months. Habitats should be maintained for approximately five to eight months; the species does not necessarily require flooding for the entire period, as foraging habitats may be maintained for some time under drawdown conditions. As nesting occurs away from water and roosting occurs on open banks, significant amounts of standing water are not essential and waters can recede at a moderate pace.

The required depth of flooding is unknown, but should be at a level to maintain the wetland plant communities on which the plumed whistling-duck feeds. These communities generally grow in shallow water depths of about 60 cm above and below the waterline (see Chapter 2).

Accounts of breeding do not indicate the flood frequency required to maintain populations. This may be partly related to breeding being stimulated in response to rainfall and corresponding increases in plant growth. There is no information on the life expectancy of the plumed whistling-duck in captivity or in the wild. Scott (1997) indicated that ducks in Australia generally have a life expectancy of three to four years. Therefore, breeding in the plumed whistling-duck should occur at least every second year with flooding or rainfall substantial enough to promote the growth of wetland plants.

There is no information to suggest that successful breeding in the plumed whistling-duck is dependent on the drying of wetlands. However, many plant species consumed by the plumed whistling-duck either require some drying, or have their growth enhanced by a period of drying (see Chapter 2). Therefore, it is likely that a drying phase of a few months will promote wetland productivity and enhance the breeding success of the plumed whistling-duck.

Australian shelduck: *Tadorna tadornoides*

The Australian shelduck is a large-bodied duck with a small black head, black bill and orange breast (Figure 3.33). The head is black brown and is tinged iridescent green or blue and the breast and mantle are cinnamon brown. The species has a black back, scapulars and rump. The sexes have a similar appearance, but females have a white ring around the eye and base of the bill and are smaller in size, with a length of 56–58 cm and a wingspan of 94–116 cm. Males have a length of approximately 59–72 cm and a wingspan of 96–132 cm. The Australian shelduck weighs 1.3–1.6 kg and moults into a duller plumage during the non-breeding season (Marchant and Higgins 1990). The Australian shelduck can be seen in pairs, small groups or flocks of thousands on deep fresh waters, large saline lakes, billabongs, lagoons, estuaries, sand spits and islands. It is endemic to Australia and its distribution is primarily limited to the temperate south-eastern and south-western parts of Australia (Marchant and Higgins 1990).

Habitat and diet

The Australian shelduck may be observed on grasslands, croplands, inland wetlands, estuarine waters and wooded grasslands. It prefers open areas with large expanses of water, mudflats and wide beaches, but may also occur in small wetlands with short sparse emergent vegetation (Marchant and Higgins 1990). Favoured habitats include large lakes, large shallow swamps on

Figure 3.33: Australian shelduck, *Tadorna tadornoides*. Photograph: Chris Herbert (Hunter Bird Observers Club).

alluvial plains and deep freshwater swamps (Corrick and Norman 1980; Fjeldsa 1985), but it may also frequent tidal and inland rivers, river pools, billabongs, fresh meadows, shallow swamps, saltmarsh, rush-dominated swamps, sewage ponds, sheltered inland waters and large dams (Corrick 1982; Corrick and Norman 1980; Fjeldsa 1985; Frith 1982; Marchant and Higgins 1990). Foraging occurs by grazing at the edge of wetlands or by up-ending in the water within wetlands. The greatest numbers of Australian shelduck are associated with waters near feeding grounds of short grass, pasture or crops of seed or grain. Roosting occurs by day on the muddy margins of wetlands or on the water. Breeding occurs in wetland settings and nests are usually established in holes in live or dead trees (Marchant and Higgins 1990). Nests have also been observed on the bare ground or among grass. The location of nests with respect to water is variable. Usually only natural debris, occurring in tree hollows, is used to construct the nest (Marchant and Higgins 1990).

The diet of the Australian shelduck is not well known, but may consist of a range of plant material and invertebrates (Frith 1982; Marchant and Higgins 1990).

Breeding and flood requirements for breeding

Breeding of the Australian shelduck is stimulated by flooding and season (Briggs 1990). This hypothesis is supported by Halse and Jaensch (1989), who found strong positive correlations between the time of egg laying and rainfall, and rainfall with a one-month lag, and strong negative correlations between time of laying, temperature and photoperiod with a one-month lag. Breeding can occur between July and November in south-western New South Wales, irrespective of flooding, and usually occurs within dead trees standing in water (Hobbs 1961). Halse and Jaensch (1989) regarded the Australian shelduck, black swan and great cormorant as the only true winter-breeding waterbirds, but observations from New South Wales indicated that breeding is more likely to occur in late winter and spring (Hobbs 1961). Breeding of the Australian shelduck is solitarily and regularly occurs between mid-winter and spring (Frith 1982).

When flooding stimulates breeding in the Australian shelduck, egg laying correlates with rainfall or with rainfall and a one-month lag (Halse and Jaensch 1989). This indicates that the Australian shelduck responds rapidly to this cue and that the lag time required to commence breeding is approximately one month or less. The clutch size ranges from eight to 14 eggs, with eggs laid at daily intervals. The eggs are incubated by the females for 30–33 days (Marchant and Higgins 1990). Fledging reportedly occurs at 70 days (Riggert 1977) and parents desert the young just before or after they can fly (Storr 1964b). Juveniles establish pairs quickly, but breeding does not occur until at least 22 months old (Frith 1982). It appears that the Australian shelduck maximises its use of inundated areas by being one of the last species to leave a wetland (Chapman and Lane 1997). The breeding duration of the Australian shelduck is estimated at three to four months and it is estimated that it requires flooding for a period of three to five months.

Studies do not provide an ideal water depth, however, preferred food items (Frith 1982), with the exception of ribbonweed, generally occur at relatively shallow depths of approximately 0.6 m. The rate of water recession should be moderate.

Recaptures of banded birds by the Australian Bird Study Association indicated that Australian shelducks may live in excess of 13 years and one month (Australian Bird and Bat Banding Scheme 1994). This estimate may represent an upper limit rather than average life expectancy. It is estimated that large breeding events should occur at least every five to six years. As the Australian shelduck is hunted in parts of the Murray-Darling Basin and it does not reach sexual maturity until two years of age (Marchant and Higgins 1990), it is suggested that flooding should occur approximately every two to three years.

Breeding success of the Australian shelduck is enhanced when breeding sites undergo inter-flood drying (Crome 1986, 1988). This is likely to be a result of the life-cycle of wetland plants (see Chapter 2), which have greater productivity following a drying phase. Therefore, it is likely that inter-flood drying of a few months (up to six months) may enhance the breeding success of the Australia shelduck.

Shoreline foragers

Black-tailed native-hen: *Gallinula ventralis*

The black-tailed native-hen is a large dark fleet-footed rail with an erect narrow tail (Figure 3.34). Its plumage is slate grey on the sides of its head and throat, slate grey with a blue tinge on the foreneck and breast and olive brown on the remainder of its upper parts. Its bill and frontal shield are light green, the iris is yellow and the legs and feet are coral pink. It does not exhibit any seasonal changes. Males and females have a similar appearance, but females are slightly smaller and their plumage is generally duller and paler than that of males. It has an approximate length of 30–38 cm, a wingspan of 55–66 cm and a weight of 0.4 kg (Marchant and Higgins 1993). The black-tailed native-hen is gregarious and may be seen singly, in pairs or large groups. It frequents a range of settings that are open, dry and close to water and vegetation cover. It has a widespread distribution throughout mainland Australia, particularly west of the Great Dividing Range and south-western parts of Western Australia (Marchant and Higgins 1993).

Habitat and diet

Habitat selection for the black-tailed native-hen is opportunistic. It appears to favour permanent or ephemeral inland wetlands in low-rainfall areas, particularly fresh or brackish wetlands. These wetlands commonly include dense clumps of vegetation such as lignum, canegrass and saltbush, or they may be sparsely wooded. Frequented wetlands include shallow

Figure 3.34: Black-tailed native-hen, *Gallinula ventralis*. Photograph: Chris Herbert (Hunter Bird Observers Club).

lakes, swamps, pools, floodplains and flats of rivers and creeks and inundated depressions (Marchant and Higgins 1993). Foraging by the black-tailed native-hen occurs at the edge of water, such as the grassy or muddy margins of waterholes, lakes, swamps, rivers and creeks, or on open ground near wetlands including scrubs and dry pastures or crops (Marchant and Higgins 1993). Roosting is undertaken by night and among long-grass or dense vegetation (Marchant and Higgins 1993) and the black-tailed nataive-hen may be observed resting during the day near the roots of trees (Nicholls 1942). Breeding may occur at a range of wetlands including swamps, waterholes, dams, lakes, river flats or temporarily inundated claypans, farmland and floodplains with receding waters. Breeding is also undertaken near water and often among clumps of dense vegetation such as lignum, nitre bush, bluebush, saltbush and grass (Marchant and Higgins 1993; Stone 1912). Nests are cup-shaped and constructed of plant material such as stalks and leaves of lignum, reeds, twigs, leaves, canegrass, long strips of bark and grass. The nests are lined with grass, reeds, feathers and other suitable material (Chaffer 1940; Howe 1909; Marchant and Higgins 1993).

The diet of the black-tailed native-hen consists of seeds, other plant material and insects (Marchant and Higgins 1993). Vestjens (1977b) found that it consumed *Polygonum* seeds and a range of insects in New South Wales. Other studies referred to the consumption of plant material and some insects (Marchant and Higgins 1993).

Breeding and flood requirements for breeding

Breeding of the black-tailed native-hen appears to occur opportunistically in response to heavy rainfall with nesting occurring as waters from flooding dry-up (Dingle 2004; Halse and

Jaensch 1989; Marchant and Higgins 1993). Breeding occurs in New South Wales continuously between winter and summer (Hobbs 1961). Marchant and Higgins (1993) suggested that breeding primarily occurs between August and December in eastern Australia, but it may occur whenever rainfall is sufficient.

Breeding of the black-tailed native-hen appears to be greatest two to three months after rainfall, with some breeding occurring as early as one month after rainfall (Halse and Jaensch 1989). The clutch size for the black-tailed native-hen is usually five to seven eggs. The period taken to lay eggs is unknown; eggs are incubated for 19–20 days. The young appear to remain in the nest for some time, but the period until fledging, parental independence and reproductive maturity is unknown (Marchant and Higgins 1993). Due to minimal information about breeding of the black-tailed native-hen it is difficult to establish an approximate breeding duration. Based on references to breeding early in the drying cycle following flooding (Chapman and Lane 1997), it is suggested that breeding duration may not be more than one to two months. It is cautiously suggested that the black-tailed native-hen requires minimum flood durations of two to three months, with breeding success enhanced following flooding of five to six months or more.

There is little indication that the black-tailed native-hen requires a specific flood depth for breeding, although it reportedly exhibits a preference for moderately permanent deep September water depths (Halse *et al.* 1993). It is likely that the species prefers shallow flooding that maintains foraging habitats yet does not flood nest sites. Floodwaters should recede at a moderate pace.

There is no information about the recovery or longevity of the black-tailed native-hen, which makes it difficult to infer a flood frequency requirement. To be cautious, and due to the species' preference for inland wetlands in low-rainfall areas, it is suggested that flooding should occur every one to two years to maintain breeding population numbers and habitats.

Similarly, there is little information about the need for inter-flood drying. Other rails appear to breed well following a period of drying (Crome 1988), and great influxes were observed in south-western New South Wales when flooding followed a period of drying (Hobbs 1961). As the black-tailed native-hen favours ephemeral inland settings, it can be inferred that it breeds well following some drying.

Purple swamphen: *Porphyrio porphyrio*

The purple swamphen is a large rail with a large triangular red bill and frontal shield, and distinctive indigo chin, throat, neck, breast, upper belly and flanks (Figure 3.35). It has long legs with thick knees and long slender toes. The remainder of the plumage on its crown, nape and face is glossy black, except for its white undertail coverts. The purple swamphen has a length of 44–48 cm and a wingspan of 70–88 cm. Males are slightly larger, with a weight of 1.05 kg compared to the female weight of 0.86 kg. The sexes have a similar appearance, but the frontal shield on males is bigger. Only slight seasonal variations are observed, with the frontal shield swelling during the breeding season (Marchant and Higgins 1993). The purple swamphen may be observed singly or in loose groups. It frequents margins of wetlands with considerable vegetation and is commonly observed in urban areas. It has a widespread global distribution and in Australia has a widespread distribution in the eastern states. It may also be observed in temperate south-western Western Australia and parts of the Northern Territory (Marchant and Higgins 1993).

Habitat and diet

The purple swamphen frequents inland wetlands and nearby grasslands, but may also be observed in estuarine or littoral settings. Frequented wetland types are usually fresh or brackish and may be permanent, semi-permanent, seasonal or ephemeral (Brown and Brown

Figure 3.35: Purple swamphen, *Porphyrio porphyrio*. Photograph: Kerrylee Rogers (DECCW).

1977; Corrick 1982; Corrick and Norman 1980; Goodsell 1990; Marchant and Higgins 1993; Norman and Mumford 1985). This includes a range of wetlands such as ponds, swamps and artificial wetlands, and even open habitat next to wetlands such as grasslands, meadows, forest margins, agricultural land, lawns, parks, gardens, golf courses and other artificial open grassy areas (Bell *et al.* 1961; Bravery 1970; Carroll 1966, 1969; Hobbs 1961; Marchant and Higgins 1993; McKenzie 1967). Foraging occurs among aquatic vegetation at the fringes of wetlands or among sedges, rushes, reeds, shrubs and trees (Bravery 1970; Corrick and Norman 1980; Marchant and Higgins 1993; Norman and Mumford 1985). It may also occur while perched and roosting in trees (Norman and Mumford 1985). Nesting generally occurs among reeds in swamps, dams and other suitable wetlands, but may occasionally occur in grass tussocks some distance from water (McLean 1901). The height of reeds around the nest ranges from 10–225 cm, with an average height of 131 cm (Marchant and Higgins 1993). The depth of water around nests ranges from 30–120 cm (Fletcher 1908; Stidolph 1938). Nests are constructed from the grass or reeds in which they are established. Nest material is beaten down to form a platform with a shallow depression in the centre (Hall 1903; McLean 1901).

The diet of the purple swamphen primarily consists of aquatic plants but may include seeds, fruits, insects, frogs, lizards, fish, young birds, eggs and small mammals (Balasubramaniam and Guay 2009; Carroll 1966; Marchant and Higgins 1993; McLean 1901; Norman and Mumford 1985; Vestjens 1977b). Other consumed items include crustaceans, spiders and insects (Norman and Mumford 1985).

Breeding and flood requirements for breeding

Breeding of the purple swamphen is reportedly stimulated by flooding and season (Briggs 1990), which is supported by relationships between breeding, rainfall, photoperiod and temperature (Halse and Jaensch 1989). Abundance has been correlated with deep water depths (Halse *et al.* 1993) and other seasonal factors (Gosper *et al.* 1983). Norman and Mumford (1985) found no relationship between rainfall and gonad dimensions, but did identify a relationship between ovarian follicle diameters and season. The breeding season varies throughout Australia, with observations of breeding between mid-August and December in Victoria (Marchant and Higgins 1993), between September and November in south-western New South Wales (Hobbs 1961) and between January and November in northern parts of Australia (Hall 1903; Lavery 1986; Le Souef 1902). Breeding may be highly dependent on climatic conditions (Halse and Jaensch 1989). Breeding of the purple swamphen is described as communal, monogamous, polygamous and promiscuous (Marchant and Higgins 1993) and typically occurs in groups of two to five adults (Dow 1980).

Breeding of the purple swamphen was reportedly highly correlated with rainfall with a two- to three-month lag, with some breeding commencing as early as one month after rainfall (Halse and Jaensch 1989). The purple swamphen has been observed building a number of trial nests in the month prior to the commencement of breeding (Craig 1980). Nests have been flooded when constructed on rising floodwaters (Marchant and Higgins 1993). The purple swamphen therefore requires a minimum lag time of one month and an ideal lag time of two to three months before breeding commencement.

In Australia, the clutch size ranges from two to six eggs with an average of 4.2 eggs. Eggs are laid daily and are incubated by all adults in the breeding group (McLean 1901) for 25–27 days (Marchant and Higgins 1993). The chicks are brooded by all adults in the breeding group for up to two months and the young are able to breed within their first year (Brown and Brown 1977). Breeding duration is estimated at up to three months. The purple swamphen requires a minimum flood duration of four months and an ideal flood duration of five to six months. Waters should recede at a moderate to slow pace.

The purple swamphen requires specific water depths at nest sites, ranging from 30–90 cm (Fletcher 1908; Stidolph 1938). This is likely to relate to the need to protect eggs from predators, such as swamp harriers, when nests are exposed (Marchant and Higgins 1993) or the need for nest sites to be flooded for the duration of breeding.

The oldest recovered purple swamphen is reportedly nine years old and was recovered from New Zealand (Dakota 2009). The oldest recovered specimen from Australia was five years and five months old (Australian Bird and Bat Banding Scheme 1979). Estimates of flood frequency based on longevity indicate that the purple swamphen is likely to require flooding at least every two years to maintain breeding populations. However, breeding may be maintained at settings with permanent water during dry periods.

There is no explicit information about the species' requirement for an inter-flood dry-period. As the purple swamphen frequents a range of permanent, semi-permanent, seasonal or ephemeral wetland settings it probably exhibits some capacity to survive an inter-flood dry-period, though that is not likely to be an essential requirement. The aquatic plants consumed by the purple swamphen, such as cumbungi, common reed, ribbonweed and spike-rush (Norman and Mumford 1985), exhibit an adaptive capacity to survive periods of drying, but drying is not essential for their maintenance.

Masked lapwing: *Vanellus miles*

The masked lapwing is a large conspicuous lapwing (Marchant and Higgins 1993). Its crown and nape are black with a white collar extending to the lower hind neck, the sides of the breast and the face. The majority of the remaining plumage is uniform pale grey brown (Figure 3.36). It has a length of 30–37 cm, a wingspan of 75–85 cm and a weight of 230–400 g. The sexes have a similar

Figure 3.36: Masked lapwing, *Vanellus miles*. Photograph: Max Carpenter (DECCW).

appearance and do not exhibit seasonal plumage changes (Marchant and Higgins 1993). The masked lapwing may be observed in pairs, family groups, small parties or large flocks during the non-breeding season. It frequents a range of open habitats and is a common bird in urban areas. It has a widespread distribution throughout the eastern states and northern parts of Australia and may be observed in all parts of Victoria and New South Wales (Marchant and Higgins 1993).

Habitat and diet

The masked lapwing frequents a range of habitats and is commonly observed in urban areas and modified habitats. It exhibits a preference for short-grassed areas that are fringing wetlands. Frequented wetlands may include permanent or temporary swamps, marshes, billabongs, lakes, reservoirs, receding floodwaters, farm dams, saltmarshes, waterholes, lagoons, waterlogged paddocks, bore drains, river flats and river beds (Favaloro 1943; Hobbs 1961; Marchant and Higgins 1993; van Tets *et al.* 1967). Foraging and roosting occurs on the ground among short grass, bare mud or gravel or among sparse vegetation (Loyn 1978; Marchant and Higgins 1993). Breeding also occurs on the ground among short grass less than 12 cm high (Favaloro 1943; Fletcher 1924; Marchant and Higgins 1993). Nests are commonly located at a short distance from water (Barlow *et al.* 1972), are established in depressions in the ground and may be unlined or lined with nearby material (Dove 1937; Favaloro 1943; Thomas 1969). There is some indication that if the nest becomes waterlogged, additional material will be added to raise the eggs (Thomas 1969).

The masked lapwing typically consumes molluscs, worms, millipedes, centipedes, insects, crustaceans, gastropods, bivalves, frogs and some plant material (Favaloro 1943; Marchant and Higgins 1993; van Tets *et al.* 1969; Vestjens 1977b).

Breeding and flood requirements for breeding

The primary breeding season for masked lapwings is mid-June to late November in New South Wales (Marchant and Higgins 1993), early July to November in Victoria (Dann 1981; Favaloro

1943; Wheeler 1955) and July to early December in northern parts of the Murray-Darling Basin (Marchant and Higgins 1993). However, as breeding occurs following rainfall (Dann 1981; Thomas 1969), it may occur at almost any time (Marchant and Higgins 1993). Thomas (1969) established a correlation between egg laying, mean temperature and rainfall and indicated that photoperiodism may not be the primary factor controlling breeding. It has been proposed that egg laying is timed so that hatching occurs when there is peak food availability (Dann 1981; Thomas 1969). It is therefore likely that flooding or rainfall is the primary stimulus for breeding in the masked lapwing, with season a secondary stimulus. Breeding of the masked lapwing occurs in pairs.

Studies of masked lapwing breeding have not considered the lag time before breeding commencement; however, peak breeding has been observed approximately one month following heavy rainfall (Dann 1981) and approximately two weeks after copulation and/or nest laying (Barlow *et al.* 1972; Thomas 1969). Masked lapwing clutch size generally ranges from one to four eggs (Thomas 1969), laid at one- to two-day intervals (Bourke 1953; D'Ombrain 1928). Eggs will be re-laid if they are lost after an interval of six to 28 days (Barlow *et al.* 1972; Favaloro 1943) and multiple clutches may be laid within a season (Barlow *et al.* 1972). The eggs are incubated by both sexes for 28–30 days (Bourke 1953; D'Ombrain 1928). The young usually fledge at about six to seven weeks old, but this may occur as early as five weeks and as late as eight weeks (Thomas 1969). The young may remain dependent on parents for up to six months after fledging (Marchant and Higgins 1993) and breeding may occur within the first year (Barlow *et al.* 1972). Breeding duration is estimated at approximately three months. Therefore, the masked lapwing requires a minimum flood duration of approximately four months. Second clutches may be laid and reared if flooding occurs for longer periods.

The masked lapwing establishes nests on mounds surrounded by or near water (Favaloro 1943), but there does not appear to be a water depth requirement. The eggs must remain out of the water; nests may be built up to protect eggs and young from submergence (Thomas 1969). Floodwaters should recede at a moderate rate.

The oldest banded masked lapwing recovered was 13 years and one month old (Australian Bird and Bat Banding Scheme 2002). It is therefore estimated that suitable floods should occur at least every six years. As the masked lapwing exhibits relatively rapid breeding and fledging, and can use non-wetland habitats for breeding, suitable floods are likely to occur more frequently. Small habitat-maintenance floods should occur approximately every other year.

There is no specific literature pertaining to a requirement for inter-flood drying; however, as flooding promotes the productivity of dietary items (Dann 1981; Thomas 1969) it is probable that some wetland drying enhances breeding success. Inter-flood drying of a few months is compatible with the 11-month breeding cycle of masked lapwings (Barlow *et al.* 1972).

Banded lapwing: *Vanellus tricolor*

The banded lapwing is a medium-sized lapwing with narrow pointed wings, a long tail and shorter legs (Figure 3.37). It has a distinctive appearance with a glossy black cap and white stripe from behind the eye to the side of the nape, as well as a noticeable black U-shaped breast band on white underparts. The hind neck and upper mantle are grey brown and grade into a darker colour on the lower mantle, back, scapulars and tertials. It has an approximate length of 25–29 cm, a wingspan of 61–67 cm and a weight of 0.15–0.2 kg. The sexes have a similar appearance and do not exhibit seasonal changes in plumage (Marchant and Higgins 1993). The banded lapwing is territorial but gregarious and can be observed in parties or large flocks among very short grass or similar unprotected settings. The banded lapwing is endemic to

Figure 3.37: Banded lapwing, *Vanellus tricolor*. Photograph: Chris Herbert (Hunter Bird Observers Club).

Australia and occurs throughout eastern, southern and western parts of mainland Australia and Tasmania (Marchant and Higgins 1993).

Habitat and diet

The banded lapwing prefers open areas in dry or semiarid regions (Marchant and Higgins 1993). Frequented settings include sparsely vegetated open plains with short grass, saline herb fields, grasslands modified by rabbits, low open woodlands, agricultural land and urban areas with mown grass, such as golf courses (Brooker *et al.* 1979; Favaloro 1943; Ford and Sedgwick 1967; Hobbs 1961; McEvey 1965; McEvey and Middleton 1968; van Tets *et al.* 1967, 1969; Whittell 1932). Its reliance on wetlands is unknown, as there is some indication that it avoids wetlands (Favaloro 1943; Hobbs 1961). However, there are observations of banded lapwing within flooded paddocks (Thomas 1968) and in Western Australia it is commonly observed near water (Boehm 1955; Marchant and Higgins 1993). Foraging occurs among short sparse grass and in shallow water (Boehm 1955). Roosting reportedly occurs on hot days in the shade provided by fence posts, trees or short grass (Favaloro 1943; Marchant and Higgins 1993; Whittell 1932). Breeding occurs on the ground among short grass (Favaloro 1943; van Tets *et al.* 1969; Whittell 1932). Nests are generally established in a hollow in the ground or in a depression (Favaloro 1943) among dry grassland, pastures, crops, on roadsides, claypans and dry salt lakes on bare stony ground (Favaloro 1943; Marchant and Higgins 1993; Norton 1922; Whittell

1932). Nests may be unlined or lined with nearby material such as grass, roots, twigs, animal droppings, dirt and stones (McGilp 1922; Whitlock 1920).

The diet of the banded lapwing consists of seeds, leaves, molluscs, worms, insects and spiders (Marchant and Higgins 1993).

Breeding and flood requirements for breeding

Breeding of the banded lapwing appears to be similar to that of the masked lapwing. The primary breeding season is between July and November (Marchant and Higgins 1993). However, as breeding appears to occur in response to rainfall (Dann 1981), it may occur any time conditions are suitable (Hindwood and Hoskin 1954; Marchant and Higgins 1993). Rainfall appears to be the primary stimulus for breeding with hatching reported to coincide with peak food availability (Dann 1981). The banded lapwing breeds in simple pairs, and sometimes with many pairs in the same areas (Marchant and Higgins 1993).

The lag time between rainfall and commencement of breeding appears to be relatively short. Hatching occurs only a few weeks after rainfall, several weeks earlier than the emergence of masked lapwings (Dann 1981). This may be related to the earlier availability of the surface insects consumed by the banded lapwing, compared to worms consumed by the masked lapwing (Dann 1981).

The banded lapwing usually has a clutch of four eggs, but clutch size may range from three to six eggs (Favaloro 1943; Marchant and Higgins 1993). Eggs are laid at daily intervals except the last egg, which takes longer (Marchant and Higgins 1993). The eggs are incubated by both sexes for 26–28 days (Favaloro 1943; Lord 1951). Estimating the breeding duration is difficult due to a lack of information on the time until fledging; however, a pair in captivity bred at two years old (Favaloro 1943; Marchant and Higgins 1993). It is probable that the breeding duration is similar to that of the masked lapwing, estimated at three months. However, as rainfall is the primary stimulus for breeding and nesting generally occurs among dry grassland and similar habitats, flooding may not be necessary for the entire breeding duration. The duration of rainfall and/or flooding need only be long enough to promote food availability for the breeding duration.

Similarly, as nest sites are not associated with water, the banded lapwing does not exhibit a required water depth or a rate of water recession for breeding success. Rainfall and/or flooding should occur for long enough to promote food availability, irrespective of water depths and rates of water recession.

The oldest recovered banded lapwing was three years and six months old (Australian Bird and Bat Banding Scheme 1980). Based on this relatively short period, it is estimated that conditions for breeding should occur at least every two years. Annual rainfall or flood conditions suitable for breeding would promote the long-term viability of banded lapwing populations.

As the banded lapwing does not exhibit a strong requirement for flooding in wetlands, breeding may occur after periods of drying. Much of the plant material consumed by the banded lapwing exhibits a tolerance to drying, and drying is likely to promote an increase of insect prey items consumed by the banded lapwing (Marchant and Higgins 1993). Therefore, a period of inter-flood drying will most likely promote the breeding success of banded lapwing.

Large waders

Yellow-billed spoonbill: *Platalea flavipes*

The yellow-billed spoonbill is a large white or creamy white waterbird with a long pale yellow spatulate bill and yellow legs (Figure 3.38). It has a length of 76–92 cm and a weight of

Figure 3.38: Yellow-billed spoonbill, *Platalea flavipes*. Photograph: Chris Herbert (Hunter Bird Observers Club).

approximately 1.9 kg. The sexes have a similar appearance, although males tend to be larger. During the breeding season, a black line extends around the bare facial skin, lacy outer wing plumes are tipped black and plumes at the base of the lower neck and upper breast are well developed (Marchant and Higgins 1990). The yellow-billed spoonbill may be observed in solitary or in small flocks or colonies in association with other herons, egrets, ibises or spoonbills. It is mainly observed on fresh or brackish inland waters. It is endemic to Australia and has a widespread distribution throughout the eastern states and along the south-western coastline of Western Australia. Southern New South Wales and Victoria are regarded as a breeding stronghold (Marchant and Higgins 1990).

Habitat and diet

The yellow-billed spoonbill may be observed in inland wetlands, wet grasslands and, rarely, sheltered marine locations. It mainly occurs at inland fresh or brackish wetlands with sparse or low vegetation. It exhibits a preference for shallow swamps with abundant aquatic flora, pools, watercourses, billabongs, farm dams and channels and may be observed on pastures flooded by rain or irrigation, wet meadows and shallow parts of lakes or deeper swamps. The yellow-billed spoonbill has been observed in open water or among tall emergent vegetation, shrubs or trees such as common reed, cumbungi, lignum, river red gum and black box (Marchant and Higgins 1990). The yellow-billed spoonbill forages in open water or near tall emergent vegetation or submerged logs. Due to the shape and structure of the bill, foraging is limited to water depths of less than 0.4 m over a substrate of sand, mud or clay (Vestjens 1975b). It roosts at night on live or dead trees or stumps, or on the ground on banks and shores (Lowe 1983a). Breeding mainly occurs in the southern range of its distribution at freshwater wetlands such as lakes, swamps, watercourses and flooded pasture. Highly productive wetlands may be favoured (Crome 1988) and habitats are usually vegetated with trees, lignum or reeds so that nests can be built in vegetation (Jaensch *et al.* 1988; Lowe 1983a; Vestjens 1975b). Nests have been observed on the vertical side-branches or in the crowns of trees over water and are generally established 2–8 m above the water level. Nests may be reused in subsequent years. They comprise a large shallow platform of loosely woven sticks and twigs with a central depression and may be lined with bark. Live and dead stems of reeds, rushes and other vegetation may be utilised if the nest site is near reed beds (Marchant and Higgins 1990).

The diet of the yellow-billed spoonbill primarily consists of aquatic insects but may include freshwater crayfish, freshwater shrimps and fish (Marchant and Higgins 1990). Gut analyses indicate that it may also consume molluscs, spiders and plant material (Vestjens 1975b). Consumed fish species include Australian smelt (*Retropinna semoni*), goldfish (*Carassius auratus*), gambusia (*Gambusia holbrooki*) and flat-headed gudgeon (*Philypnodon* species). Consumed crustaceans and molluscs include freshwater prawns (*Macrobrachium*), western yabby (*Cherax destructor*), Murray crayfish (*Euastacus armatus*), pond snail (*Lymnaea* species) *Glyptophysa* and *Isidorella* species (Hobbs 1957a; Marchant and Higgins 1990; Vestjens 1975b).

Breeding and flood requirements for breeding

Breeding of the yellow-billed spoonbill is reportedly stimulated by season and flooding (Briggs 1990); however, only Halse *et al.* (1993) supports a strong seasonal stimulus. They found that the yellow-billed spoonbill prefers to frequent permanent wetlands, rather than intermittent wetlands that would be characterised by flooding and drying. However, breeding has been correlated with water depth and rainfall with a two-month lag (Halse and Jaensch 1989) and the area of live river red gum flooded in excess of four months (Briggs *et al.* 1997). Breeding success is reportedly greater in response to flooding at an intermittent wetland (Crome 1988). Yellow-billed spoonbill abundance was found to correlate with water level (Harper 1990), and abundance in eastern Australia was found to correlate with far-ranging climate issues such as rainfall in the western region and the Southern Oscillation Index (Kingsford *et al.* 1999). The yellow-billed spoonbill generally breeds between September and April (Vestjens 1977b) but it is suggested that the role of flooding in breeding may be greater than the role of season at semi-permanent and intermittent settings, particularly as breeding may occur any time conditions are suitable (Marchant and Higgins 1990). Breeding occurs in solitary or in small colonies, sometimes in association with other waterbirds such as ibises, cormorants, egrets, herons and other spoonbills.

The yellow-billed spoonbill requires a substantial lag time to promote breeding success, estimated at two to five months (Briggs and Thornton 1999; Briggs *et al.* 1997; Halse and Jaensch 1989) or up to six months following autumn flooding (Briggs and Thornton 1999). The clutch size of the yellow-billed spoonbill ranges from two to four eggs (Vestjens 1977b). There is no information on the interval at which eggs are laid, number of broods per season and whether replacement laying occurs. Incubation is by both parents for 26–31 days. The period until fledging is approximately five weeks, and the young remain dependent on parents for several weeks before departure from the nest (Marchant and Higgins 1990). Therefore, the yellow-billed spoonbill has a breeding duration of two to three months and it is likely that it requires a minimum lag time of two months. Breeding success may increase with a greater lag time. The upper limit of the lag time is unknown. The minimum required flood duration is four to five months and flooding around nests should ideally occur for up to nine months.

Yellow-billed spoonbill breeding and abundance appears to correlate with water depth, perhaps indicating a water depth requirement for breeding (Halse *et al.* 1993; Harper 1990). The correlations may be an artefact of the relationship between water depth and flood duration, but it is evident that nesting occurs on deep waters (Halse *et al.* 1993) and that shallow waters are required for foraging (Marchant and Higgins 1990). The recession of floodwaters should be moderate to slow to promote breeding success.

As there are no records of the longevity of yellow-billed spoonbill in the wild, the captive longevity of 28.6 years (Brouwer *et al.* 1994) is used to estimate the required flood frequency for the maintenance of breeding populations. It is suggested that large floods should occur in the mosaic of wetlands within the Murray-Darling Basin at a frequency of at least one in seven years, with small maintenance floods every other year.

The yellow-billed spoonbill is one of the waterbird species that benefit most from the boom of productivity that follows drying in intermittent swamps (Crome 1988). Inter-flood drying is likely to promote its breeding success; hence, it is suggested that drying should occur at breeding sites for a few months to promote the release of chironomids and other invertebrates that may be consumed.

Royal spoonbill: *Platalea regia*

The royal spoonbill is a large white waterbird with a black spatulate bill and black facial skin, legs and feet (Figure 3.39). It has a length of 74–81 cm, a wingspan of approximately 120 cm and a weight of 1.5–1.8 kg. The sexes are similar in appearance; males are slightly larger, with slightly longer legs and bills than females (Marchant and Higgins 1990). The royal spoonbill is a gregarious bird, occurring in small flocks. It is commonly observed in inland wetlands, sheltered marine locations and wet pasture. It has a widespread distribution throughout eastern and northern Australia but is more commonly observed in coastal locations than inland wetlands (Marchant and Higgins 1990). The royal spoonbill is regarded as vulnerable in Victoria (*Flora and Fauna Conservation Act 1988*).

Habitat and diet

In inland areas the royal spoonbill utilises permanent and ephemeral waters when available (Marchant and Higgins 1990). It prefers freshwater wetlands, including swamps with

Figure 3.39: Royal spoonbill, *Platalea regia*. Photograph: Chris Herbert (Hunter Bird Observers Club).

semi-aquatic herbs or tall emergent vegetation such as rushes (*Eleocharis* species, *Typha* species, *Phragmites australis*, *Scirpus* species and *Eragrostis* species), wet meadows, flooded pasture, rivers, billabongs, pools, the shallow margins of deeper lakes and swamps, and inundated woodlands vegetated with river red gum, black box, oak and *Melaleuca* species (Corrick 1981, 1982; Jaensch *et al.* 1988; Lowe 1982; Marchant and Higgins 1990; Vestjens 1977b). Due to the structure of the bill, foraging is limited to areas with waters less than 0.4 m deep and a soft substrate of sand, mud or clay (Vestjens 1975b). Foraging often occurs among aquatic or emergent vegetation or submerged logs. Roosting occurs communally in trees over or near water. Nesting usually occurs in trees, shrubs, reeds or rushes, such as river red gum, lignum and common reed, located over or near water (Marchant and Higgins 1990; Vestjens 1977b). Nest sites are usually located over water 0.5–1.5 m deep, and usually occur 1–15 m above the water level. When nests are constructed within reeds, they usually occur 0.5–1.5 m above the water level. Nests are bowl-shaped, constructed from interwoven sticks and small twigs and lined with leaves and water weed (Marchant and Higgins 1990; Vestjens 1977b).

The diet of the royal spoonbill consists mainly of fish, shrimp and other crustaceans and insects (Marchant and Higgins 1990; Vestjens 1975b). Gizzard analyses from the royal spoonbill indicated that it had consumed a range of fish species including Australian smelt, goldfish, gambusia, redfin perch and flat-head gudgeon (*Philypnodon* species), insects, crustaceans including shrimp and freshwater crayfish, molluscs and plant seeds of the genera *Medicago* (Vestjens 1975b). Other records indicate that it may also consume amphipods, crabs, tadpoles and nardoo (*Marsilea* species) (Lowe 1982; Marchant and Higgins 1990; Mathews 1910).

Breeding and flood requirements for breeding

Breeding of the royal spoonbill appears to be influenced by flooding and season (Briggs 1990). As nesting trees or emergent vegetation are usually located in or over water, flooding is an essential requirement. Breeding typically occurs between October and March. The role of flooding is supported by established relationships between colony size and water flow (Kingsford and Johnsons 1998), ciconiiforme breeding and area of river red gum flooded in excess of four months (Briggs *et al.* 1997), and increased breeding success after complete drying of an intermittent wetland (Crome 1988). In the Northern Territory, abundance of royal spoonbills reportedly correlates with season (Morton *et al.* 1993). Others have identified relationships between royal spoonbill abundance and season (Marchant and Higgins 1990) and between abundance, season and flooding (Harper 1990). Some researchers found no relationship between abundance, season and/or flooding (Gosper *et al.* 1983; Woodall 1985). As there is no information about ovum and gonadal development in the species it is difficult to establish the relative importance of flooding and season, but it is likely that flooding is the primary stimulus for breeding. Breeding occurs in pairs with other colonial nesting waterbirds, such as egrets, herons, ibises and other spoonbills. Breeding in New South Wales reportedly occurs between November and May, with some nests established as early as October (Marchant and Higgins 1990; Vestjens 1977b).

No information is available on the lag time required for successful breeding of royal spoonbills. However, based on the relationship between ciconiiforme breeding and flooding in excess of four months (Briggs *et al.* 1997), it is probable that it requires some lag time before the commencement of breeding. Clutch sizes of royal spoonbill are generally three or four eggs (range of two to five eggs). While there is no information on whether replacement laying occurs when broods are unsuccessful, it is evident that the species lays only one brood per season. The interval at which laying of eggs occurs is unknown; both sexes incubate the eggs for 20–25 days (Marchant and Higgins 1990). The earliest reported breeding is at three years of age (Schweigman 1999). Very little is known of the breeding duration of the royal spoonbill, but it is likely to be two to three months. Applying this breeding duration to the relationship observed between ciconiiforme breeding and area of live river red gum flooded in excess of four months (Briggs *et al.* 1997), it is suggested that the royal spoonbill requires a lag time of at least one to two months. Breeding success is likely to increase with a greater lag time. The upper limit of the lag time is unknown. The minimum flood duration is four to five months.

Water depth may be a significant factor in successful royal spoonbill breeding. A correlation has been found between abundance and water level (Harper 1990), while nesting reportedly occurs over water 0.5–1.5 m deep (Marchant and Higgins 1990). It is possible that the favoured water depth may be due to a correlation between deep water and long flood durations, thereby indicating that water recession should occur at a moderate to slow pace. Slower water recession may enhance breeding success.

Very little information, including data on maximum life-span, is available to establish a minimum flood frequency for the maintenance of breeding populations of royal spoonbill. Based on wild colonial nesting waterbirds' life expectancy of eight to 10 years (Scott 1997), it is estimated that large floods should occur at least every four to five years with small maintenance floods every other year.

The diet of the royal spoonbill consists of fish, crustaceans and insects; productivity of some of these species would increase in response to flooding following drying. The role of drying in successful breeding is supported by Crome (1988). It is suggested that drying of a few months may promote the productivity of chironomids and other insects and enhance the breeding success of the royal spoonbill.

Glossy ibis: *Plegadis falcinellus*

The glossy ibis is a dark brown medium to slender ibis with a height of 55–65 cm, a wingspan of 80–95 cm and a weight of approximately 0.5 kg (Figure 3.40). It has a curved bill and has a glossy green and purple iridescence. The sexes appear similar and their plumage varies seasonally

Figure 3.40: Glossy ibis, *Plegadis falcinellus*. Photograph: Chris Herbert (Hunter Bird Observers Club).

(Marchant and Higgins 1990). The glossy ibis is a gregarious species, but is commonly seen alone or with other ibis species. The glossy ibis inhabits a range of wetlands including fresh and brackish swamps, lakes, irrigated agriculture, tidal flats and mangrove. It has a widespread distribution and has been observed in significant numbers in floodplain wetlands in the Murray-Darling Basin (Marchant and Higgins 1990). The international importance of glossy ibis habitats is recognised through its inclusion in the China-Australia Migratory Bird Agreement (CAMBA). The glossy ibis is regarded as near-threatened in Victoria (*Flora and Fauna Conservation Act 1988*).

Habitat and diet

The glossy ibis inhabits inland wetlands and, occasionally, wet pasture and sheltered marine locations. In inland areas, it prefers freshwater wetlands including permanent and ephemeral wetlands on floodplains, watercourses, billabongs, pools and shallow swamps with abundant aquatic vegetation. It also frequents shores of large lakes and deeper swamps in open water or among tall emergent vegetation (*Phragmites australis*, *Scirpus*, *Typha* and *Eleocharis* species), shrubs (*Muehlenbeckia* species) and woodland (*Eucalyptus* and *Melaleuca* species). It may be observed on flooded pasture, agricultural land, wet meadows and sewage ponds (Corrick 1982; Corrick and Norman 1980; Dostine and Morton 1989b; Fjeldsa 1985; Gosper 1981; Jaensch *et al.* 1988; Lowe 1957; Marchant and Higgins 1990; Sansom *et al.* 1954; Vestjens 1977b). Foraging occurs in shallow water over a soft substrate or on grassy or muddy margins of wetlands. Roosting occurs in dead or live trees located in or near water (Marchant and Higgins 1990). Breeding is most common in fresh or brackish wetlands that are vegetated with reeds, rushes, shrubs or trees in which to build nests. Nests are positioned 10–50 cm above the water level and are constructed from interwoven sticks that may be lined with eucalyptus leaves or bunches of leafy twigs. Nests are usually reconstructed each season, as other colonial-nesting birds that have earlier breeding seasons may demolish them (Marchant and Higgins 1990; Vestjens 1977b).

The diet of the glossy ibis consists mostly of aquatic invertebrates and insects, such as snails, mussels, crayfish, mites, spiders, crickets, grasshoppers, beetles and frogs. Gut content analyses indicate that it may also consume rice seeds and fish (Marchant and Higgins 1990; Mathews 1910; Vestjens 1977b).

Breeding and flood requirements for breeding

Breeding of the glossy ibis appears to be stimulated by flooding (Briggs 1990). Number of nests has been linked to flows with a lag of three to 12 months and flood extent (Kingsford and Auld 2005), and glossy ibis colony size and water flows (Kingsford and Johnsons 1998). Nesting has been observed following significant rainfall and flooding (Bailey 1933). The glossy ibis breeds in pairs colonially and in association with other colonial-nesting birds such as ibises, spoonbills, herons and egrets. Breeding in New South Wales reportedly occurs between October and February (Vestjens 1977b), but laying during one season at Balranald was limited to a short period between 10 October and 6 November (Lowe 1983a).

Breeding of glossy ibis in the Macquarie Marshes correlated best with flow with a three- or six-month lag, but significant relationships were also observed with flow with a nine- and 12-month lag (Kingsford and Auld 2005). It is therefore estimated that breeding success is greatest with a lag of three to six months; however, success may not decline significantly when the water remains for up to 12 months.

The clutch size of the glossy ibis is approximately two to six eggs (Vestjens 1977b). It is likely that the glossy ibis has only one brood per season; replacement laying has not been observed. Based on observations from Russia, eggs are likely to be laid daily and incubated by both parents for approximately 21 days (Dementiev and Gladkov 1968, cited in Marchant and Higgins 1990). At 10 days old, young may intermingle and they become more active by 15 days old. Observations of the period from hatching to fledging in Russia indicate that this lasts approximately 25 days and the young remain dependent on parents for some time after fledging (Dementiev and Gladkov 1968, cited in Marchant and Higgins 1990). Breeding duration is approximately two months (Marchant and Higgins 1990) and it is estimated that the glossy ibis requires a minimum flood duration of five months, with an ideal flood duration of eight months. It is unlikely that breeding success will be affected with flood durations of up to 12 months.

As nests are established over water there is likely to be a flood depth requirement (Marchant and Higgins 1990); however, no information is available about the specific depth of water required for successful breeding. Water depth is likely to be substantial enough to enable waters to recede at a slow or moderate to slow pace.

The glossy ibis may live in captivity for up to 26.8 years (Brouwer *et al.* 1994); longevity in the wild in North America is significantly less, at 14.3 years (Clapp *et al.* 1982). Glossy ibis life expectancy in inland Australia is likely to be significantly less. Using the life expectancy of eight to 10 years, typical for colonial-nesting waterbirds in the Murray-Darling Basin (Scott 1997), it is estimated that the glossy ibis may require large floods at least every four to five years to boost population numbers.

No information is available about the role of drying in the breeding success of glossy ibis. As its diet primarily consists of aquatic invertebrates and insects (Vestjens 1977b), some of which would exhibit productivity boosts in response to drying, glossy ibis breeding success may be enhanced with drying for a few months.

Australian white ibis: *Threskiornis molucca*

The Australian white ibis is a mainly white ibis with a very long dark down-curved bill and dark head (Figure 3.41). It has a length of 65–75 cm, a wingspan of 110–125 cm and a weight of 1.4–2.5 kg. Males and females have a similar appearance, but females are generally smaller in

Figure 3.41: Australian white ibis, *Threskiornis molucca*. Photograph: Kerrylee Rogers (DECCW).

stature and weight. During the breeding season the skin beneath the wing changes from pink to dark scarlet (Marchant and Higgins 1990). The Australian white ibis is gregarious or solitary; it breeds in large colonies of up to 20 000 pairs. It inhabits wetlands of almost any kind and occurs in urban areas among rubbish dumps and along rivers and creeks. It exhibits a widespread distribution throughout eastern, northern and south-western Australia (Marchant and Higgins 1990).

Habitat and diet

The Australian white ibis inhabits inland wetlands, sheltered marine habitats and pasture. In inland areas, the Australian white ibis inhabits shallow swamps with abundant aquatic vegetation and open fresh water such as watercourses, billabongs, pools, floodwaters, shallow or deep parts of freshwater lakes, freshwater meadows, swamps with tall emergent vegetation such as reeds (*Eleocharis*, *Phragmites*, *Typha*, *Scirpus* and *Eragrostis* species), open water or among shrubs such as lignum or trees (*Melaleuca* and *Eucalyptus*) (Carrick 1959; Corrick 1981, 1982; Corrick and Norman 1980; Fjeldsa 1985; Gosper 1981; Jaensch *et al.* 1988; Marchant and Higgins 1990; Vestjens 1977b). Foraging primarily occurs in shallow water over a soft substrate or on muddy flats or shores. Roosting occurs in trees that are near water, while breeding occurs in fresh, brackish or saline vegetated wetlands with reeds, shrubs or trees (Marchant and Higgins 1990). Breeding is suggested to be more successful in highly productive wetlands that have undergone previous drying (Crome 1988). Nests are built in forks or on large branches of trees, such as tea-tree, weeping willow (*Salix* species) or *Eucalyptus*. Nests in trees are located over water, but they have occasionally been observed on the bare ground, on artificial surfaces, on flattened vegetation such as reeds, rushes, lignum and cumbungi at or near water level, or in trees away from water (Jaensch *et al.* 1988; Marchant and Higgins 1990). Nests may be located at heights of up to 30 m. Nests may be reused in subsequent years, but they are likely to be used by a new pair. Nests are constructed from a platform of sticks, twig and leaves or reeds and rushes, and are lined with ribbonweed (Marchant and Higgins 1990).

The diet of the Australian white ibis includes small aquatic animals such as insects, freshwater crayfish, fish and frogs, and terrestrial animals such as lizards, snakes, rats, mice, carrion, crickets, beetles and earthworms (Carrick 1959; Marchant and Higgins 1990). Consumed crayfish include the yabby (*Cherax albidus*) and freshwater crab (*Paratelphusa leichardti*) (Carrick 1959). The Australian white ibis has populated urban areas and it now exhibits a more non-specific diet (Martin *et al.* 2007).

Breeding and flood requirements for breeding

The stimulus for breeding of the Australian white ibis is flooding (Carrick 1962; Marchant and Higgins 1990) and strong relationships between colony size and water flow have been established (Kingsford and Auld 2005; Kingsford and Johnsons 1998). Breeding may not occur annually. It is strongly influenced by flooding and drought (Marchant and Higgins 1990) and may occur whenever flood conditions are suitable (Carrick 1962). The Australian white ibis breeds in pairs colonially and is commonly associated with other colonial-nesting waterbirds such as ibises, cormorants, spoonbills, herons and egrets. Breeding in central New South Wales reportedly occurs between September and April (Marchant and Higgins 1990; Vestjens 1977b).

The Australian white ibis exhibits a strong relationship between number of nests and water flow with a three- to nine-month lag at the Macquarie Marshes (Kingsford and Auld 2005). The relationship is strongest with a six-month lag, and weakens only slightly with a nine-month lag. The Australian white ibis is reportedly more responsive to flooding than the straw-necked ibis, with egg laying commencing three to four weeks after flooding (Carrick 1962).

The Australian white ibis clutch size ranges from one to six eggs, but averages two or three eggs (Corben and Munro 2006). Lost clutches have been observed to be replaced within 18 days and multiple clutches may be laid per season. Eggs are laid at night at intervals of 48 hours and both sexes incubate the eggs for 20–23 days. Chicks are brooded for approximately three weeks by both parents. After this time, chicks may be left unattended and completely abandon the nest approximately 48 days, or as early as 30 days, after hatching. Young Australian white ibis remain dependent on parents for an average of 21.5 days after fledging. However, some may remain within the colony and dependent on parents for months. The period until young are able to breed is unknown. Breeding duration is estimated at two to three months (Marchant and Higgins 1990). It is therefore estimated that the Australian white ibis requires a minimum flood duration of five to six months. Breeding success increases when flooding occurs for eight to nine months, or even up to 12 months.

Breeding commences in response to a rise in water level and nests are usually built over a water depth of 0.5–1 m. There is evidence of nest abandonment when floodwaters receded quickly (Carrick 1962). Based on the long flood duration required for successful breeding, it is suggested that the recession of waters should be slow.

Available literature does not prescribe a required flood frequency for maintaining Australian white ibis populations. It is a partial migrant and will move throughout the Australian landscape in response to flooding and water availability (Marchant and Higgins 1990). While the Australian white ibis is reported to have survived for 33.2 years in captivity (Brouwer *et al.* 1994), no information is available about its longevity in the wild. Using the life expectancy of eight to 10 years, typical of colonial-nesting waterbirds in the Murray-Darling Basin (Scott 1997), it is suggested that the Australian white ibis requires large floods at least every four to five years and small floods every other year to maintain population numbers.

Breeding success in relation to wetland drying has not been documented. On the basis of its consumption of a range of insects (Carrick 1959), it is probable that inter-flood drying may be required to build fat reserves and stimulate egg laying in the Australian white ibis in semi-permanent and intermittent wetlands. Breeding at both inland and coastal locations occurs at wetlands that undergo seasonal changes in water level (Marchant and Higgins 1990). However, due to the non-specific diet of the Australian white ibis in urban areas (Martin *et al.* 2007), prey items may not be a limiting factor for breeding success in those situations. It is likely that inter-flood drying for at least a few months may enhance breeding success of the Australian white ibis; the ideal duration is unknown.

Straw-necked ibis: *Threskiornis spinicollis*

The straw-necked ibis is a white-necked black-backed ibis with a length of 60–70 cm, a wingspan of 100–120 cm and a weight of 1.1–1.5 kg (Figure 3.42). Males and females exhibit a similar appearance; however, males are larger and have a larger bill than females. The straw-necked ibis does not exhibit breeding plumage (Marchant and Higgins 1990). The straw-necked ibis is gregarious and occurs in small flocks on wet or dry ground in pasture, or in groups in the shallows of wetlands or along wetland margins. It is endemic to Australia, with a widespread distribution except in the driest parts of central Australia. The greatest concentrations of the straw-necked ibis occur in Victoria and southern parts of the Murray-Darling Basin (Marchant and Higgins 1990).

Habitat and diet

The straw-necked ibis is commonly observed in inland wetlands, pasture, cultivated land and, rarely, sheltered marine habitats. Although it is uncommon throughout arid central Australia, it has been observed using permanent and ephemeral wetlands when water is available (Marchant and Higgins 1990). Wetlands utilised include meadows, shallow swamps with semi-aquatic herbs and abundant aquatic vegetation or tall emergent vegetation, shallow open parts

Figure 3.42: Straw-necked ibis, *Threskiornis spinicollis*. Photograph: Chris Herbert (Hunter Bird Observers Club).

of deeper swamps, lakes, watercourses and areas vegetated with shrubs and trees such as lignum, river red gum and *Melaleuca* species (Carrick 1959; Corrick 1982; Corrick and Norman 1980; Fjeldsa 1985; Gosper 1981; Jaensch *et al.* 1988; Marchant and Higgins 1990; Vestjens 1977b). Foraging occurs in damp or dry grassland, often away from wetlands or in aquatic shallows where the vegetation is short and patchy, thereby enabling unimpeded movement (Carrick 1959; Marchant and Higgins 1990). Roosting occurs in trees in wetlands or farmland (Marchant and Higgins 1990; McKilligan 1975). Breeding occurs in fresh, brackish or saline wetlands that are generally vegetated with reeds, shrubs or trees so that nests can be built in them; nests may be built on the ground on islands or along the margins of wetlands or temporary floodwaters (Carrick 1962; Cowling and Lowe 1981; Jaensch *et al.* 1988; Marchant and Higgins 1990). In eastern Australia, breeding conditions are usually associated with flooding but permanent wetlands with stable water levels may be utilised (McKilligan 1975). Nests are commonly established over a water depth of less than 1 m (Carrick 1962). Nests are constructed from a platform of reeds, rushes and lignum that are broken and woven together; they are sometimes lined with grass or similar vegetation (Marchant and Higgins 1990).

The diet of the straw-necked ibis includes various small animals such as freshwater crayfish, frogs, fish, insects, spiders and freshwater snails (Marchant and Higgins 1990). Gut content analyses indicate they may also consume earthworms, molluscs, crabs, isopods, centipedes, lizards, snakes, rats and mice (Carrick 1959).

Breeding and flood requirements for breeding

Breeding of the straw-necked ibis has been related to flooding (Briggs 1990; Carrick 1962; Halse and Jaensch 1989; Hobbs 1961; Kingsford and Johnsons 1998). However, most studies indicate that nesting primarily occurs on a seasonal basis in spring and summer, with one study suggesting that low temperatures may inhibit breeding at locations where water levels are more stable (Serventy and Marshall 1957). This may reflect the seasonal nature of flooding in unregulated rivers within the Murray-Darling Basin (Carrick 1962; Halse and Jaensch 1989; Hobbs 1961; McKilligan 1975). McKilligan (1975, p. 210) suggested that the stimulus for breeding of the straw-necked ibis may include a factor that 'completes the set of conditions necessary for its success'. Studies indicated that straw-necked ibis abundance correlates with flood-related factors such as water level and rainfall (Gosper *et al.* 1983; Halse *et al.* 1993; Woodall 1985), while others indicated the importance of season (Harper 1990). Therefore, breeding of the straw-necked ibis is primarily stimulated by flooding but there may also be a secondary seasonal stimulus. Breeding usually occurs in spring and summer, but may occur whenever conditions are suitable (Marchant and Higgins 1990). The straw-necked ibis breeds in pairs colonially and in association with other colonial-nesting waterbirds such as other ibises, spoonbills, herons and egrets. Drought, flooding and water levels influence the breeding season and therefore the straw-necked ibis may not breed annually at a location (Marchant and Higgins 1990).

The lag time before commencement of breeding for the straw-necked ibis appears to be substantial, with breeding commencement reported to take twice as long as that of the Australian white ibis (Carrick 1962). The number of nests of the straw-necked ibis correlates with flow with a three- to nine-month lag and a strong relationship occurs with a flood lag of six months; the relationship weakens only slightly at nine months (Kingsford and Auld 2005).

Straw-necked ibis clutch size is generally two to five eggs. No information is available about whether replacement clutches are laid or the number of broods reared per season (Marchant and Higgins 1990). Eggs are laid at intervals of 48 hours and are incubated by both parents for about 3.5 weeks (Carrick 1962). No reliable information is available about the period from hatching to first flight, however, fledging was observed four weeks after hatching. Chicks feed independently of parents at two weeks after fledging (Carrick 1962). No information is available about the period until young are able to breed. Breeding duration of the straw-necked ibis is estimated at three months (Marchant and Higgins 1990). It is therefore estimated that successful breeding requires a minimum flood duration of six months, and that flood durations of up to 12 months will enhance breeding success.

The straw-necked ibis exhibits a water depth requirement: breeding commences in response to a rise in water level, and nests are usually built over a water depth of 0.5–1 m (Carrick 1962). Nest abandonment has been observed at the Macquarie Marshes in response to rapid recession of floodwaters that were present for only two to three months (Carrick 1962). Based on the long flood duration required for successful breeding, it is suggested that the recession of waters should be fairly slow.

Required flood frequency for maintaining breeding populations of straw-necked ibis is unknown. The captive longevity of the straw-necked ibis is 39.4 years (Brouwer *et al.* 1994). As no detailed information is available about the life expectancy of wild straw-necked ibis, the typical life expectancy of eight to 10 years for wild colonial waterbirds in the Murray-Darling Basin (Scott 1997) is used to estimate required flood frequencies. It is suggested that large floods should occur at least every four to five years with small floods every one to two years.

It is unknown whether inter-flood drying is an essential requirement for breeding success. While breeding of the straw-necked ibis primarily occurs in settings that exhibit flooding and a corresponding abundant increase in food (Carrick 1962), McKilligan (1975) referred to two wetlands in South Australia where water levels remained fairly constant and breeding occurred on a fairly regular basis. It is therefore suggested that, in wetlands that are used for breeding and do not have abundant food, inter-flood drying may be an essential requirement for stimulating wetland productivity and enhancing the breeding success of straw-necked ibis. The period of drying required is unknown, but is likely to be at least a couple of months.

Brolga: *Grus rubicunda*

The brolga is a large, light grey crane with a long bill, small red head and long neck and legs (Figure 3.43). The sexes have a similar appearance, however, females are slightly smaller. The brolga has a wingspan of 1.7–2.4 m. Males have a length of 105–134 cm and a weight of 4.7–8.7 kg, females have a length of 77–113 cm and a weight of 3.7–7.3 kg. There are no marked seasonal changes in plumage (Marchant and Higgins 1993). The brolga may be seen in flocks during the breeding season, but it pairs off during the non-breeding season. It frequents open swamps, coastal mudflats, grassland, crops and water within deserts and is widely distributed throughout northern Australia and New Guinea. In New South Wales and Victoria it is limited to the major river systems within the Murray-Darling Basin and the western plains (Marchant and Higgins 1993). The brolga is regarded as vulnerable in Victoria (*Flora and Fauna Conservation Act 1988*), New South Wales (*Threatened Species Conservation Act 1995*) and South Australia (*National Parks and Wildlife Act 1972*).

Habitat and diet

The brolga primarily occurs in inland wetlands, grasslands and woodlands throughout tropical and temperate Australia and less commonly in arid and semiarid regions. Populations in southern Australia are generally observed in shallow freshwater marshes less than 0.5 m deep and with emergent vegetation, during the breeding season. The brolga may also be observed in freshwater meadows dominated by annual herbs and rushes such as *Carex*, *Juncus*, *Poa*, *Glyceria australis* and *Eleocharis* (Marchant and Higgins 1993). It may occasionally be observed in shallow freshwater meadows dominated by lignum (Corrick 1982; Marchant and Higgins 1993; Meine and Archibald 1996). During the non-breeding season, the brolga may congregate near deep permanent freshwater marshes and nearby fields and pastures to forage. Foraging is

Figure 3.43: Brolga, *Grus rubicunda*. Photograph: Chris Herbert (Hunter Bird Observers Club).

undertaken in moist mud located in marshes or pasture and roosting occurs in shallow water (Marchant and Higgins 1993). Nesting occurs in shallow wetlands or in the shallow waters of deep wetlands in water depths of up to 40 cm, but usually less than 30 cm. Nests are established on the ground, among vegetation such as tussock grass, sedge or canegrass and occasionally on floating vegetation (Herring 2005; Marchant and Higgins 1993). The brolga may use the same nesting site for up to 20 years or nest nearby (Elliott 1938; Marchant and Higgins 1993). Nests may comprise a raised knoll, mound, platform, trampled grass, floating structure or small depression. Nests are constructed from grasses, sedges, rushes, sticks and leaves from near the nest site. In the absence of construction materials, nests have been constructed from clay and unearthed roots from marsh beds (Dennis 1932; Marchant and Higgins 1993).

The brolga is an omnivore and consumes mostly crops and tubers. Its diet may include a variety of insects, spiders, molluscs, crustaceans, small mammals, reptiles and frogs (Marchant and Higgins 1993). There are no detailed studies of the food consumed by the brolga.

Breeding and flood requirements for breeding

Breeding of the brolga reportedly occurs in response to flooding (Briggs 1990), with nesting preferentially occurring in ephemeral or semi-permanent wetlands (Herring 2005). Breeding occurs after the wet season in northern parts of Australia and after winter rainfall in southern parts (Marchant and Higgins 1993). Despite the strong role of flooding in successful breeding, breeding primarily occurs between December and February. Breeding may occur as early as August and as late as June in northern parts of Australia. In southern parts of Australia, breeding primarily occurs between July and November, but may occur as early as May and as late as March when conditions are favourable (Marchant and Higgins 1993). The brolga breeds in solitary. Pair bonds are formed in its third year and breeding commences soon afterwards (Marchant and Higgins 1993).

Recent research highlighted the role of habitat and water regime in successful breeding of the brolga, particularly the isolated populations in southern parts of Australia (Herring 2005). The brolga reportedly began nesting at ephemeral wetlands approximately 20 to 60 days after they filled.

The typical clutch size is two eggs, but ranges from one to three eggs (Marchant and Higgins 1993). Eggs are laid at two-day intervals and eggs will be re-laid up to two times in the wild if they are lost or infertile. The eggs are incubated by both sexes for approximately 31 days. The young may leave the nest one to two days after hatching, but body feathers are not fully developed until 80–90 days old. Flight is possible by approximately 14 weeks old. The period from fledging to maturity is unclear, but juveniles have been observed with parents for two years, pair bonds are formed in the third year and breeding territories are sought in the fourth year (Marchant and Higgins 1993). Breeding duration for the brolga is estimated at three to four months. Based on the breeding duration and lag time required for successful breeding, it is estimated that wetlands used for breeding should be flooded for four to six months. This flood duration is supported by Herring (2005), who found that breeding occurred more frequently at ephemeral sites inundated for two to six months than at semi-permanent sites inundated for six to 12 months. Breeding did not occur at dry ephemeral sites inundated for zero to two months, or at permanently inundated wetlands.

Depth of flooding and recession of floodwaters are important aspects of successful breeding (Herring 2005). Nesting within floodwaters protects eggs from predators. The mean depth of flooded wetlands where the brolga breeds successfully was 24–72 cm (Herring 2005), and the majority of breeding occurred at depths less than 30 cm (Marchant and Higgins 1993). A decline in water supply to ephemeral wetlands has been associated with the loss of breeding sites (Herring 2005), indicating that the supply of floodwaters to breeding sites should be maintained for the entire flood duration to promote breeding success. However, the retention of water levels for too long may promote the establishment of plant species adapted to more permanent flood conditions, such as cumbungi. Water recession in a timely manner maintains preferred plant species used for nesting, such as common spike-rush and common nardoo, and enables access to tubers that are consumed by the brolga (Herring 2005). Therefore, floodwaters should recede at a moderate to slow rate to promote successful breeding.

The frequency at which flooding should occur to maintain breeding populations of the brolga is difficult to determine. The captive longevity of the brolga is in excess of 41 years, but no information is available on the longevity of wild brolga (Veynet 2006). Large floods should occur at least every 10 years. However, opportunities for maintaining brolga population numbers are limited. Brolga breeding success appears to be low, with usually only one chick raised successfully (Du Guesclin 2003), and reproduction occurring at the earliest in the fourth year (Marchant and Higgins 1993). There is considerable discussion about the viability of southern populations of the brolga (Du Guesclin 2003), where population numbers have not changed for some time. It is estimated that the age of the southern populations has increased, fecundity has decreased and, with few opportunities for recruitment, the populations may be susceptible to collapse. Brolga breeding preferentially occurs at ephemeral wetlands that are characterised by shorter floods (Herring 2005). Large-scale floods, such as those required for successful breeding of some colonial-nesting waterbirds, may promote unsuccessful breeding (Herring 2005) as they are associated with increased flood depths and durations in excess of the required four- to six-month period. It is therefore suggested that moderate sized floods should occur relatively frequently (approximately every five years) to maintain reproductive brolga population numbers. Smaller floods should occur more frequently to maintain breeding and foraging habitats.

Inter-flood drying is an important aspect of brolga breeding as it maintains essential breeding and feeding habitats and limits the expansion of wetland plants adapted to permanent water conditions (Herring 2005). Based on the brolga's preferential selection of ephemeral wetlands, inter-flood drying should occur for approximately six to eight months to promote breeding success in the brolga.

Small waders

Black-winged stilt: *Himantopus himantopus*

The black-winged stilt is a medium sized and extremely long-legged pied wader (or shorebird), with a long fine bill and a length of 33–37 cm (Figure 3.44). Its wings are long and narrow with pointed tips and its legs trail behind the tail during flight. Males are mostly white with a black hind neck and black upper back, scapulars and upper wings and a light grey tinge on the tip of the tail. The plumage is similar on both sexes, except females have duller black mantle, scapulars and upper wing (Marchant and Higgins 1993). Outside the breeding season, the black-winged stilt commonly gathers in small to large flocks with other waders. It gathers in pairs or small family groups during the breeding season. The black-winged stilt has been observed feeding at the edge of wetlands by wading, strutting gracefully through the water and occasionally swimming. It is widespread throughout mainland Australia (Marchant and Higgins 1993).

Habitat and diet

The black-winged stilt frequents shallow open freshwater wetlands, particularly those with dense short coverage of grass or other emergent vegetation. Wetland types that it frequents include swamps, marshy ponds, billabongs, lakes, lagoons, saltmarsh, closed tidal wetlands, sewage farms and saltworks, flooded paddocks and waterholes (Fjeldsa 1985; Hobbs 1961;

Figure 3.44: Black-winged stilt, *Himantopus himantopus*. Photograph: Chris Herbert (Hunter Bird Observers Club).

Marchant and Higgins 1993; Pierce 1980). Roosting occurs in shallow water, on banks, on islets in sheltered wetlands or sand flats within estuaries. Foraging primarily occurs in shallow water or saturated mud, often close to emergent vegetation (Marchant and Higgins 1993; Powlesland and Robertson 1987). Breeding reportedly occurs in a range of habitats including swamps, rivers, flooded saltmarsh, paddocks, sewage ponds, rice paddies, farm dams and waterholes, even in mangrove hollows over water and platforms over tidal flats. Nesting occurs in a range of vegetation types including grassy banks, sedge or tussock hummocks, mats of *Cotula* species, floating vegetation, grass tussocks, lignum, canegrass, rushes, samphire and other suitable vegetation. Nesting occurs on the ground or in damp depressions made from flood debris, driftwood, rootlets, grass, twigs, rushes, algae or other vegetation (Marchant and Higgins 1993). Nests may be built up when threatened with submergence in floodwaters (Carnaby 1933; Hindwood 1939; Marchant and Higgins 1993; Stidolph 1931; Wheeler 1955).

The black-winged stilt has a diet comprising aquatic and terrestrial invertebrates, including molluscs, crustaceans, insects, arachnids and annelids, and occasionally seeds and diatoms (Marchant and Higgins 1993).

Breeding and flood requirements for breeding

Breeding of the black-winged stilt appears to be stimulated by flooding with positive relationships established between breeding and rainfall with a two- to three-month lag (Halse and Jaensch 1989). The black-winged stilt is regarded as a colonial or semi-colonial nesting bird and nests in colonies of up to 500 nests. Breeding occurs in southern and eastern Australia between August and December, but may occur at any time conditions are suitable. Some breeding within the Murray-Darling Basin occurred as late as January to March (Halse and Jaensch 1989).

The lag time between rainfall and breeding of the black-winged stilt is two to three months (Halse and Jaensch 1989). Clutch size ranges from three to six eggs (Carnaby 1933). Eggs are laid at intervals of 24–48 hours, with replacement clutches laid quickly after loss. Second clutches have been observed after successful incubation and hatching (Stokes 1949). Both sexes incubate the eggs, once the clutch is complete, for approximately 25 days (Marchant and Higgins 1993; Stokes 1949). New chicks forage for food near the nest and are brooded by both parents for 28–38 days until fledging is complete. Fledged stilts may remain with parents and can successfully breed within two years of fledging (Marchant and Higgins 1993). Breeding duration is estimated at two to three months and flood duration is estimated at four to six months, with second clutches brooded under prolonged flood conditions.

There does not appear to be a specific flood depth requirement for successful breeding in the black-winged stilt. Shallow waters are used for foraging (Marchant and Higgins 1993) and floods of great depth may cause abandonment of submerged nests (Pullen 1967). Floodwaters should recede at a moderate rate to ensure breeding success.

Based on recapture of banded birds, the longevity of wild black-winged stilt is three years and 11 months (Australian Bird and Bat Banding Scheme 1981). Therefore, floods that induce successful breeding of the black-winged stilt should occur at least every second year to maintain population numbers and habitats.

There is little information on a need for inter-flood drying to promote breeding success of the black-winged stilt. However, as it is frequently observed around floodwaters and ephemeral wetlands and it consumes a range of invertebrates whose populations may respond to wetting following a period of drying (Marchant and Higgins 1993), it is likely that some drying will increase food availability and promote the breeding success of black-winged stilt.

Red-necked avocet: *Recurvirostra novaehollandiae*

The red-necked avocet is a large shorebird with striking chestnut head and neck, long bluish legs and a long slender upcurved bill (Figure 3.45). The body is white with two dark stripes along its back. The red-necked avocet has a length of approximately 40–48 cm, a wingspan of 70–80 cm and a weight of 310 g. The sexes have a similar appearance and do not exhibit seasonal plumage changes (Marchant and Higgins 1993). The red-necked avocet is gregarious and may be observed in flocks at freshwater, brackish or saline wetlands. It is endemic to Australia and may be observed in all regions of New South Wales and Victoria, although it is most common in inland wetlands (Marchant and Higgins 1993).

Habitat and diet

The red-necked avocet may frequent a range of wetlands, but it is primarily observed in shallow inland ephemeral wetlands (Marchant and Higgins 1993). Frequented wetlands may include waterholes, flooded claypans, mud springs, bore drains, dams, lagoons, shallow swamps and margins of rivers (Bryant 1946; Chaffer 1940; Hobbs 1961; Marchant and Higgins 1993; McGilp 1922). Foraging primarily occurs in shallow waters on soft mud, and roosting is undertaken on bare islands or banks near water. Roosting also commonly occurs on water or among low vegetation (Bryant 1946; Chaffer 1940; Marchant and Higgins 1993). Breeding is most common in inland salt lakes, but may also be undertaken in swamps, lakes, flooded paddocks, crops, saltworks, flooded gravel pits and other suitable habitats. Breeding may occur in newly flooded wetlands or areas where waters are receding (Bryant 1946; Hobbs 1961; Marchant and Higgins 1993). Nests are established on mud or vegetated islands, mudbanks, sandbars, ridges and

Figure 3.45: Red-necked avocet, *Recurvirostra novaehollandiae*. Photograph: Chris Herbert (Hunter Bird Observers Club).

mounds. The nests are constructed in a depression on the ground and scantily lined with twigs, grass, water weed and other plant material. They may occasionally be built from mud or gravel (Bryant 1946; Hobbs 1961; Marchant and Higgins 1993).

The diet of the red-necked avocet is mostly carnivorous, comprising insects and crustaceans. It may also include seeds and other plant material (Chaffer 1940; Marchant and Higgins 1993). There have been no detailed studies of the food consumed by the red-necked avocet.

Breeding and flood requirements for breeding

Breeding of the red-necked avocet appears to be stimulated by flooding (Halse and Jaensch 1989; Hobbs 1961). Breeding occurs in simple pairs individually or in small colonies of up to 150 pairs. Breeding primarily occurs from mid to late winter until mid to late summer (Bryant 1946; Marchant and Higgins 1993). Some records indicate that the red-necked avocet may breed as early as May during wet years (Chapman and Lane 1997; Hobbs 1961).

Breeding of the red-necked avocet appears to occur relatively quickly after rainfall, with eggs laid after a one- to two-month lag (Halse and Jaensch 1989). The clutch size of red-necked avocets is usually four eggs, but it ranges from two to five eggs (Marchant and Higgins 1993). The eggs are laid at daily intervals and are incubated for a minimum period of 23 days. No information on the fledging time or period until maturity is available (Marchant and Higgins 1993). Chicks reportedly leave the nest as soon as they hatch (McGilp 1922). Breeding duration is estimated at two months, comprising one month of egg laying and incubation, and one month for fledging to occur. Flooding should occur for three to four months to increase breeding success.

As the red-necked avocet nests in depressions in the ground, rather than on floodwaters, there is no indication that it requires floodwaters of a certain depth. Shallow floodwaters should remain for the duration of breeding to ensure available food to sustain breeding and young red-necked avocets. Floodwaters should recede at a moderate rate.

Based on the recovery of banded red-necked avocet aged approximately seven years and three months (Australian Bird and Bat Banding Scheme 1999f), it is estimated that the red-necked avocet requires flooding conditions suitable for breeding at least every three years, with smaller habitat maintenance floods every other year.

As the red-necked avocet reportedly breeds in newly flooded wetlands or on receding floodwaters (Marchant and Higgins 1993), there is some evidence that it prefers wetlands which experience a period of inter-flood drying. In these sites the red-necked avocet may respond opportunistically to the increase in wetland productivity that follows drying. Productivity of items consumed by the red-necked avocet may increase in response to flooding (Jenkins and Boulton 2003).

Red-capped plover: *Charadrius ruficapillus*

The red-capped plover is a small plover. Its plumage is mostly grey brown and white, with a chestnut red crown and nape (Figure 3.46). It has an incomplete black band running from the nape to the white breast and a black line from the bill through the eye to a chestnut crown. It has a length of 14–16 cm, a wingspan of 27–34 cm and a weight of 35–40 g. The sexes have a similar appearance, but the plumage is slightly duller in females. There are no seasonal changes in plumage (Marchant and Higgins 1993). The red-capped plover is gregarious and very active. It may be observed singly, in pairs, in small groups or, when breeding, in flocks of hundreds. It generally inhabits coastal settings but may be observed in great numbers on bare

Figure 3.46: Red-capped plover, *Charadrius ruficapillus*. Photograph: Chris Herbert (Hunter Bird Observers Club).

areas in and around inland wetlands, particularly salt lakes. It is endemic to Australia and has a widespread distribution, but is more common in the southern parts of Australia (Marchant and Higgins 1993).

Habitat and diet

The red-capped plover inhabits a range of settings from littoral to estuarine and terrestrial wetlands. Due to a preference for saline or brackish waters it commonly occurs in inland salt lakes, and coastal and estuarine wetlands (Marchant and Higgins 1993). It may frequent permanent or ephemeral wetlands that have wide bare mudflats with sparse vegetation (Close and McCrie 1986; Hobbs 1961, 1972; Marchant and Higgins 1993). It has been observed on other inland waters such as rivers, brackish and freshwater lakes, waterholes and dams, springs and artesian bore drains, and swamps (Favaloro 1949; Marchant and Higgins 1993). Foraging is undertaken on mudflats or sand flats, usually away from water, but the red-capped plover may be observed near water and, very rarely, wading in the water (Hobbs 1972; Marchant and Higgins 1993). Roosting appears to be undertaken by day in flocks away from the water, while breeding is undertaken in the open or among low or sparse vegetation (Favaloro 1949; Hindwood and Hoskin 1954; Hobbs 1972; Marchant and Higgins 1993). Nests are constructed on the ground in sand, mud, grit and stony areas. They are positioned on flats along the foreshore of frequented habitats (Green 1956; Gwynne 1932; Hobbs 1972; Wheeler 1955). Nests are not very close to water, generally at distances over 40 m (Hobbs 1972; McGilp 1922). However, they have been observed within approximately 10 m of water (Green 1956).

The diet of the red-capped plover primarily consists of annelids, molluscs, small crustaceans and some plant material (Marchant and Higgins 1993). No detailed studies have been undertaken of the food consumed by the red-capped plover in inland locations.

Breeding and flood requirements for breeding

Breeding of the red-capped plover is reported to commence following rainfall and flooding (Favaloro 1949; Hobbs 1961, 1972). Breeding within the Murray-Darling Basin may occur at any time due to the increase in available food that follows flooding (Favaloro 1949). The importance of flooding in stimulating breeding is supported by the timing of peak breeding three months after a high-rainfall event (Halse and Jaensch 1989). Breeding of the red-capped plover is undertaken in simple pairs, singly or in flocks (Marchant and Higgins 1993). It predominantly occurs between July and continues until January, but may continue until March or any time conditions are suitable (Green 1956; Marchant and Higgins 1993; Wheeler 1955).

The lag time before breeding commencement is up to three months (Halse and Jaensch 1989). The clutch size is generally two eggs, with three- or four-egg clutches occasionally observed (Hobbs 1972; Wheeler 1955). Eggs are laid at intervals of one to three days and eggs will be re-laid if they fail to hatch (Marchant and Higgins 1993). The eggs are incubated mostly by the female for 30–31 days (Gwynne 1932; Marchant and Higgins 1993). The period until fledging and maturity is unknown. Based on a fledging period of approximately 1.5 months, which is similar to that of other plovers (Marchant and Higgins 1993), breeding duration is estimated at 2.5 months. Flooding should occur for a period of five to six months.

As nests are established some distance from water (Hobbs 1972; McGilp 1922), there is no indication that the red-capped plover requires flooding of a particular depth for nesting. Floodwaters should recede at a moderate rate and be shallow enough to provide suitable foraging habitat.

The oldest recovered red-capped plover was 20 years and 11 months old (Australian Bird and Bat Banding Scheme 1999d). This is the longest life-span of all recaptured Australian breeding shorebirds and may not represent a reliable estimate for the species, particularly in inland settings. It is therefore estimated that flooding suitable for breeding of red-capped plovers should occur at least every five years and that habitat-maintenance floods should occur every second year.

There is little indication that the red-capped plover requires inter-flood drying to promote breeding success. However, since consumed items may exhibit increased productivity when flooding follows some drying, and red-capped plovers have been observed on temporary and ephemeral wetlands (Marchant and Higgins 1993), it is evident that an increase in wetland productivity that follows a period of drying is likely to favour this species.

Black-fronted dotterel: *Elseyornis melanops*

The black-fronted dotterel is a small slim plover (Figure 3.47). It has a horizontal stance, with a distinctive black facemask and breastband and chestnut scapulars. It has an approximate length of 16–18 cm, a wingspan of 33–35 cm and a weight of 30–35 g. The sexes have a similar appearance and there are no seasonal changes in plumage (Marchant and Higgins 1993). The black-fronted dotterel may be observed in solitary or in pairs, or in groups of up to 100 birds. It tends to inhabit mud or gravel areas near fresh water and its distribution is limited to Australia and New Zealand. It has a widespread distribution throughout Australia and may be observed in nearly all regions of the Murray-Darling Basin (Marchant and Higgins 1993).

Habitat and diet

The distribution of the black-fronted dotterel is primarily limited to terrestrial freshwater wetlands, but it may occasionally be observed in brackish or saline wetlands. Preferred habitats

Figure 3.47: Black-fronted dotterel, *Elseyornis melanops*. Photograph: Chris Herbert (Hunter Bird Observers Club).

include the margins of inland wetlands, particularly freshwater wetlands with muddy bottoms and margins (Marchant and Higgins 1993). Frequented habitats include swamps, lakes, pools, waterholes, dams and tanks, reservoirs, soaks and springs, billabongs, inundated claypans and paddocks, drainage channels, rivers, bores and sewage farms (Emison and Porter 1978; Hindwood and Hoskin 1954; Hobbs 1961; Maclean 1977; Marchant and Higgins 1993; Masters and Milhinch 1974; McGarvie and Templeton 1974). The black-fronted dotterel also appears to prefer areas with sparse low vegetation, fringing or emergent vegetation (Favaloro 1943; Hindwood and Hoskin 1954; Marchant and Higgins 1993) and occasionally wooded forests (Hobbs 1961). Foraging is undertaken in the soft fine wet sediments at the edge of wetlands, and may occasionally occur in shallow water (Child and Child 1984; Heather 1977; Hindwood and Hoskin 1954; Maclean 1977; Powlesland and Robertson 1987; Sibson *et al.* 1972). Roosting is undertaken along rivers and other frequented wetlands, while breeding occurs along rivers, shallow creeks, lakes, lagoons, natural pans and farm dams (Child and Child 1984; Heather 1973; Maclean 1977; Marchant and Higgins 1993; McGarvie and Templeton 1974; Sharland 1942). Nests are established 4–300 m from the water and are situated on open stony ground or on banks of sand, gravel, pebbles and similar materials (Maclean 1977; Marchant and Higgins 1993). Nests are constructed in depressions in the ground. They may be unlined or lined with pebbles, pellets of dry mud, dry leaves, bark, short grass and stalks, sticks and animal dung (Marchant and Higgins 1993).

The diet of the black-fronted dotterel consists of molluscs, crustaceans, insects and occasionally seeds (Marchant and Higgins 1993). There are no detailed studies of its food consumption.

Breeding and flood requirements for breeding

Breeding of the black-fronted dotterel appears to be in response to flooding and consequently may occur whenever conditions are suitable (Marchant and Higgins 1993; McGilp 1922). There is some indication that season may also be important (Halse and Jaensch 1989; Hindwood and Hoskin 1954). The black-fronted dotterel breeds in simple pairs in solitary. Breeding may occur in Australia between August and February, but may occur at other times when conditions are suitable (Hindwood and Hoskin 1954; Marchant and Higgins 1993; McGilp 1922).

While breeding of black-fronted dotterels was significantly correlated with photoperiod with a one-month increment, a strong correlation was also evident between breeding and rainfall with a three-month lag in south-western Australia (Halse and Jaensch 1989). It is therefore proposed that the black-fronted dotterel requires a lag time of approximately three months.

The black-fronted dotterel usually has a clutch of two or three eggs that are laid at intervals of 48 hours. Multiple clutches may be laid within a season and eggs may be re-laid after failure (Marchant and Higgins 1993; Sharland 1942). The eggs are incubated by both parents for 22–26 days (Marchant and Higgins 1993). Fledging occurs at 27–40 days old, but fledglings remain with parents for at least 57 days after hatching (Child and Child 1984). Breeding duration is estimated at two to three months and flooding should occur for a duration of five to six months.

Depth of flooding is not pertinent to successful breeding, but the black-fronted dotterel reportedly prefers moderately permanent deep wetlands (Halse *et al.* 1993). Floodwaters should recede at a moderate to slow pace and provide shallow waters suitable for foraging.

Based on the longevity of black-fronted dotterel in the wild of six years and 11 months (Australian Bird and Bat Banding Scheme 1990), it is estimated that flooding suitable for breeding should occur at least every three years. Smaller floods should occur at least every second year to ensure the maintenance of habitats.

There is no explicit information indicating that the black-fronted dotterel requires inter-flood drying for breeding success. However, consumed items may increase in response to flooding following a period of drying in inland wetlands (Marchant and Higgins 1993). Some drying may increase the availability of food items and therefore promote the breeding success of black-fronted dotterel.

Red-kneed dotterel: *Erythrogonys cinctus*

The red-kneed dotterel is a medium plum-bodied plover with a large head, high forehead, short tail and moderately long fine bill (Figure 3.48). It has a black cap and breastband, that is striking against its white chin and throat. The chest band grades into chestnut flanks and greenish brown upper plumage. The red-kneed dotterel has an approximate length of 17–19.5 cm, a wingspan of 33–38 cm and a weight of 40–55 g. The sexes are similar and do not exhibit seasonal changes in plumage (Marchant and Higgins 1993). The red-kneed dotterel may be seen singly, in pairs or family groups, or in small to large flocks. It is mostly observed within freshwater wetlands and rarely in brackish to saline settings. Its distribution is mostly confined to Australia and it has been observed in all states (Marchant and Higgins 1993).

Habitat and diet

The red-kneed dotterel prefers temporary or permanent freshwater wetlands, particularly those filled by rain or floodwaters (Bright and Taysom 1932; Fjeldsa 1985; Hobbs 1961; Maclean 1977; Marchant and Higgins 1993; McGill 1943). This includes swamps, lakes, waterholes, dams, reservoirs, claypans, billabongs, creeks, bore drains, inundated ditches and sewage ponds (Marchant and Higgins 1993). Frequented wetlands typically have exposed margins and

Figure 3.48: Red-kneed dotterel, *Erythrogonys cinctus*. Photograph: Chris Herbert (Hunter Bird Observers Club).

soft fine bottom sediments, or scattered fringing or emergent vegetation (Bright and Taysom 1932; Fjeldsa 1985; Maclean 1977; McGill 1943). Foraging is undertaken along the muddy margins of wetlands or by wading in shallow water (Maclean 1977; McGill 1943). There is little information about the roosting behaviour of the red-kneed dotterel, but it has been observed in the shade of bushes (Marchant and Higgins 1993) or floating and resting with grey teal (Maclean 1977). Breeding is undertaken on the shoreline or small islets within freshwater wetlands such as swamps, bore streams, lakes and flooded paddocks, or among vegetation such as belah and lignum (Marchant and Higgins 1993). The nests are commonly established on the ground on sand, soil, mud, mounds, sandbars or islets (Marchant and Higgins 1993). Nests are concealed among dense or dead shrubs, constructed in depressions and scantily lined with twigs, grass and other plant material (Boehm 1963; Bright and Taysom 1932; Hobbs 1961; Maclean 1977; McGill 1943; McGilp 1922).

The diet of the red-kneed dotterel primarily consists of seeds, molluscs, annelids, spiders and insects (Marchant and Higgins 1993), and may include freshwater snails and a range of insects (Vestjens 1977b).

Breeding and flood requirements for breeding

Red-kneed dotterel breeding appears to be in response to flooding, but season may play some role as breeding does not seem to occur throughout the winter months (Halse and Jaensch 1989; Hobbs 1961; Marchant and Higgins 1993). Breeding of the red-kneed dotterel is undertaken in simple pairs, sometimes in groups or colonies of up to 30 pairs. Breeding in the Murray-Darling Basin usually occurs between August and January, but nesting has been observed after rainfall (Marchant and Higgins 1993). Peak breeding in New South Wales occurs during October and November, and ceases in January (Maclean 1977).

The period between rainfall and the commencement of breeding is three months (Halse and Jaensch 1989). The clutch size is usually four eggs, but may be two to five eggs (Maclean 1977; Marchant and Higgins 1993). Eggs are likely to be laid daily and incubated by both sexes. No information is available on the period of incubation, duration to fledging, and period from fledging to maturity. Based on a breeding duration of three months, it is proposed that the red-kneed dotterel requires flooding for six months to ensure breeding success. However, further study of the breeding behaviour is required to confirm the breeding duration.

As nesting occurs beside floodwaters (Marchant and Higgins 1993), there is no indication that the red-kneed dotterel requires a particular flood depth. Rather, it requires flood waters to recede at a moderate to slow pace to ensure breeding success.

There have been no recaptures of banded red-kneed dotterels, so there are no estimates of longevity. Prescribing a flood frequency to ensure the maintenance of red-kneed dotterel populations is therefore difficult. It is cautiously suggested that flooding should occur every other year to ensure the maintenance of red-kneed dotterel populations and habitats.

There is no explicit information indicating that inter-flood drying promotes the breeding success of the red-kneed dotterel. As consumed items at inland wetlands often increase when flooding follows a period of drying (Jenkins and Boulton 2003), red-kneed dotterel breeding success may increase in response to elevated food availability.

Summary of water requirements

The role of season in stimulating reproduction is minimal for waterbirds in the Murray-Darling Basin. With the exception of a few species that may reproduce in the absence of flooding, namely the maned duck, black-tailed native-hen, masked lapwing and banded lapwing, waterbirds rely on flooding and season to reproduce. There has been considerable debate about the role of flooding in stimulating reproduction, but it is now generally accepted that reproduction occurs in response to the flux in wetland productivity and increased food availability associated with flooding (Kingsford and Norman 2002). Adequate food supply increases the condition of waterbirds and ensures they can undertake the demanding task of reproduction.

While flooding is generally identified as a prerequisite to successful waterbird breeding, the specific nature of flooding required for successful breeding varies among waterbirds. Analysis of the water flow requirements for successful breeding of species (Table 3.1) illustrates how the required flood timing, duration, depth, rate of fall and frequency varies considerably between species. This is largely due to physiological differences between waterbird species. For example, flood duration, which is a combination of the lag time before breeding commencement and breeding duration, is driven by the physiological capacity of waterbirds to metabolise food and increase their condition rapidly, the reproductive cycle of individual species and the physiological ability of new chicks to fledge rapidly. Similarly, the minimum required frequency of flooding is largely driven by the success of waterbird reproduction and the life-cycle of waterbirds, but it is also driven by environmental factors such as habitat and food availability.

Variation in flood requirements is also partly explained by the physiological response of habitats and food items to flooding. For example, lag times are likely to be shorter for species that consume food items which are highly responsive to flooding and are available relatively quickly after the commencement of a flood. Lag times may be longer for waterbirds that consume food items which require some time to reach peak productivity. An example is the fish-eaters functional group, which generally exhibits long lag times due to the time required for fish production to reach levels that may sustain waterbirds throughout the breeding period. Duck species, however, consume chironomids which can respond rapidly to flooding (Crome 1986).

The net effect of these drivers (food and habitat response to flooding) is that waterbird breeding exhibits some plasticity. Waterbird breeding plasticity is largely driven by climatic patterns and flood variability. The plasticity is evident in the ability of many waterbirds to commence breeding whenever flood conditions are suitable, irrespective of photoperiod and temperature. This ability is not shared by their northern hemisphere counterparts (Kingsford and Norman 2002).

Waterbird breeding is generally timed to coincide with maximal food availability following flooding; as such, waterbird breeding commonly reflects the spatial and temporal availability of habitats and food within and between wetlands of the Murray-Darling Basin. Within a wetland, the result may be that waterbirds breed in a species-specific sequence. Waterbirds that exhibit short lag times and breeding durations fledge rapidly after the commencement of a flood, while other waterbird species that require long flood durations to fledge their young will complete the sequence of waterbird breeding for a flood. Similarly, waterbird breeding within a wetland will be distributed on the basis of habitat requirements. Species that prefer to breed in temporary or ephemeral settings establish nests at high elevations in a floodplain wetland gradient when conditions are suitable; those that require floodwaters around nests for some time may be established in core wetland areas.

Spatial and temporal variability in waterbird breeding is influenced by changes in wetland availability in the Murray-Darling Basin and other parts of Australia. This variability is largely dependent on the dispersal capability of species and the fluctuations in space and time of wetlands suitable for breeding (Roshier *et al.* 2001a). Waterbirds may travel some distance to wetlands where flooding meets their requirements for successful breeding. The result is that abundance of some waterbird species at a wetland may be high despite flooding being unsuitable for successful breeding (Roshier *et al.* 2002). The ability of some waterbirds to interact with spatially distributed wetlands helps to ensure their survival in the highly variable flow

Table 3.1: Summary of the water requirements for successful breeding of selected waterbird species that breed in the Murray-Darling Basin

Species	Stimulus	Breeding season (flood timing)	Maximum flood timing	Minimum lag time (months)	Ideal lag time (months)	Breeding duration (months)	Minimum flood duration (months)	Ideal flood duration (months)	Ideal flood depth (m)	Rate of fall	Longevity (years)	Large flood frequency (years)	Small flood frequency (years)	Inter-flood dry-period (months)
Great crested grebe	Flood, season	Nov–Feb	Aug–Mar	1	1–3	1–2	2	2–5	1–2	Moderate–fast	19.2 (wild)	n/a	1–2	n/a
Hoary-headed grebe	Flood, season	Oct–Jan	Aug–Mar	2	5–6	1–2	3–4	6–8	Deep	Slow	Unknown	n/a	1–2	n/a
Australasian grebe	Flood, season	Nov	Aug–Apr	3	5–7	3	6	8–10	Deep	Slow	Unknown	n/a	1–2	n/a
Australian pelican	Flood	Sept–Feb	Aug–Mar	2–3	3	4–5	7–8	8	n/a	Slow	16.8 (wild)	5	1–2	n/a
Darter	Flood	Nov–Apr		2	2–3	3	5	5–6	Deep (0.3–3.5)	Moderate	16 (captivity)	4	1–2	1–3
Great cormorant	Flood, season	Sept–Jan, Mar–Aug		3	9	4	7	12	Deep (up to 30)	Slow	18 (wild)	5	1–2	1–3
Little pied cormorant	Flood, season	Sept–Mar		3	9	3–4	6	12	Shallow	Slow	11.7 (wild)	4	1–2	1–3
Little black cormorant	Flood, season	Dec–May (northern), Sept–May (southern)	Any time	3	9	3–4	6	12	1–2	Slow	10 (wild)	4	1–2	1–3
Pied cormorant	Flood, season	Sept–May	Any time	Unknown	Unknown	3–4	Unknown	Unknown	Unknown	Unknown	27.8 (wild)	Unknown	Unknown	Unknown
Great egret	Flood, season	Nov–May	Sept–May	3	9	3–4	6–7	12	Deep	Slow	7.1 (wild)	3	1–2	1–6
Intermediate egret	Flood, season	Dec–Mar (northern), Nov–Apr (southern)		3	9	3–4	6–7	12	Deep	Slow	7 (wild)	3	1–2	1–6
Little egret	Flood	Oct–Mar		<3	3	3–4	4–6	6	Deep	Moderate	8–10 (wild)	4	1–2	Unknown
Pacific heron	Flood, season	Aug–Mar		1–2	5–7	3	4–5	8–10	Deep	Moderate–slow	8–10 (wild)	4	1–2	1–3
White-faced heron	Flood, season	Aug–Oct	June–Mar	1–4	>4	3	4–7	>7	Deep	Moderate–slow	8–10 (wild)	4	1–2	1–3
Rufous night heron	Flood	Sept–Feb		3	9	3	6	12	Deep	Slow	8.9	4	1–2	1–3
Silver gull	Flood, season	Aug–Feb		2	3	2–3	4–6	>6	Shallow	Moderate–slow	28.8 (wild)	n/a	n/a	n/a
Whiskered tern	Flood, season	Oct–Jan		Nil	Nil	1–2	1–2	2	<0.6–1.8	Fast	10 (wild)	5	1–2	>1

Species	Stimulus	Breeding season (flood timing)	Maximum flood timing	Minimum lag time (months)	Ideal lag time (months)	Breeding duration (months)	Minimum flood duration (months)	Ideal flood duration (months)	Ideal flood depth (m)	Rate of fall	Longevity (years)	Large flood frequency (years)	Small flood frequency (years)	Inter-flood dry-period (months)
Caspian tern	Flood, season	Any time (northern), Sept–Feb (southern)		Unknown	Unknown	2	Unknown	Unknown	Unknown	Unknown	23 (wild)	Unknown	Unknown	Unknown
Gull-billed tern	Flood, rain-fall, season	Sept–Jan		Nil	Nil	4	4	4	Unknown	Unknown	16 (wild)	5	1–2	Unknown
Black swan	Flood, season	Apr–Oct		<1	1	7–8	7–9	9	0.3–0.6 (<2)	Slow	24.8 (wild)	5	1–2	Unknown
Hardhead	Flood, season	Aug–Dec	Any time	2	2–3	3–5	5	5–8	Deep (>2)	Moder-ate–slow	3–4 (typical)	1–2	1–2	1–3
Musk duck	Flood, season	Sept–Oct	June–Dec	2	3	4–5	6	6–8	Deep (>2)	Slow	6.2 (wild)	1–2	1–2	1–3
Blue-billed duck	Flood, season	Sept–Feb	Any time		3	2–3	5	5–6	1–3	Moder-ate–slow	3–4 (typical)	1–2	1–2	1–3
Eurasian coot	Flood	Sept–Nov (NSW), Aug–Jan (Vic)	Any time	1	2–3	2	3	4–5	0.3–2	Moderate	7.2 (wild)	3	1–2	1–3
Chestnut teal	Flood, season	Aug–Oct	July–Mar	1	2	3–4	4–5	6	Unknown	Moder-ate–slow	6.5	3	1–2	Unknown
Grey teal	Flood	June–Feb	Any time	1	2–5	3–4	4–5	5–9	Unknown	Moderate	>31 (wild), 3–4 (typical)	1–2	1–2	1–3
Australasian shoveler	Flood	Aug–Dec	June–Feb	1	2–3	3	4	5–6	Deep	Moderate	10.6 (wild), 3–4 (typical)	1–2	1–2	Unknown
Pacific black duck	Flood, season	July–Sept	June–Dec	1	2–3	3–4	4–5	5–7	Unknown	Moderate	15.5 (wild), 3–4 (typical)	1–2	1–2	1–3
Pink-eared duck	Flood	Mar–May (northern), Aug–Feb (southern)	Any time	1–2	2–3	2–3	3–4	4–6	0.2–1.63	Moderate	1.3 (wild), 3–4 (typical)	1–2	1–2	1–3
Freckled duck	Flood	June–Dec	Any time	2	3	3	5	6	Unknown	Moder-ate–slow	3–4 (typical)	1–2	1–2	1–3

Species	Stimulus	Breeding season (flood timing)	Maximum flood timing	Minimum lag time (months)	Ideal lag time (months)	Breeding duration (months)	Minimum flood duration (months)	Ideal flood duration (months)	Ideal flood depth (m)	Rate of fall	Longevity (years)	Large flood frequency (years)	Small flood frequency (years)	Inter-flood dry-period (months)
Maned duck	Rainfall, season	July–Dec (southern), Jan–Mar (northern)	Any time	2	3	4	6	7	Deep	Moder-ate–slow	9.1 (wild), 3–4 (typical)	1–2	1–2	1–3
Plumed whistling-duck	Flood, rainfall	Sept–Jan	Sept–May (northern)	2	3	3–5	5	8	±0.60	Moderate	3–4 (typical)	1–2	1–2	1–3
Australian shelduck	Flood, season	July–Nov		<1	1	3–4	3	5	0.60	Moderate	13.1 (wild), 3–4 (typical)	1–2	1–2	1–3
Black-tailed native-hen	Rainfall	Aug–Dec	Any time	2	3	1–2	3–4	4–5	Shallow	Moderate	Unknown	1–2	1–2	1–3
Purple swamphen	Flood, season	Aug–Dec (southern), Jan–Nov (northern)	Any time	1	2–3	3	4	5–6	0.3–0.90	Moder-ate–slow	5.5 (wild)	2	1–2	1–3
Masked lapwing	Flood, rainfall, season	June–Nov (NSW), July–Dec (Vic, north-ern MDB)	Any time	1	1	3	4	>4	Unknown	Moder-ate–fast	13.1 (wild)	6	1–2	1–3
Banded lapwing	Rainfall	July–Nov	Any time	<1	<1	3	<4	<4	Unknown	Moder-ate–fast	3.6 (wild)	2	1–2	1–3
Yellow-billed spoonbill	Flood, season	Sept–Apr	Any time	2	2–6	2–3	4–5	4–9	Deep	Moder-ate–slow	28.6 (cap-tive)	7	1–2	1–3
Royal spoonbill	Flood, season	Nov–May	Oct–May	1–2	>2	2–3	3–5	>5	0.5–1.5	Moder-ate–slow	8–10	4–5	1–2	1–3
Glossy ibis	Flood	Oct–Feb		3	6–12	2	5	8–12	Deep	Moder-ate–slow	26.8 (cap-tive), 14.3 (wild), 8–10 (typical)	4–5	1–2	1–3
Australian white ibis	Flood	Sept–Apr		1–3	6–9	2–3	3–6	6–12	0.5–1	Slow	33.2 (cap-tive), 8–10 (typical)	4–5	1–2	1–3
Straw-necked ibis	Flood, season	Sept–Feb	Any time	3	6–9	3	6	9–12	0.5–1	Slow	39.4 (cap-tive), 8–10 (typical)	4–5	1–2	1–3

Species	Stimulus	Breeding season (flood timing)	Maximum flood timing	Minimum lag time (months)	Ideal lag time (months)	Breeding duration (months)	Minimum flood duration (months)	Ideal flood duration (months)	Ideal flood depth (m)	Rate of fall	Longevity (years)	Large flood frequency (years)	Small flood frequency (years)	Inter-flood dry-period (months)
Brolga	Flood	Dec–Feb (northern), July–Nov (southern)	Aug–June (northern), May–Mar (southern)	1–2	2	3–4	4	6	0.24–0.72	Moderate–slow	41 (captive)	5		
Black-winged stilt	Flood	Aug–Dec	Any time	2	3	2–3	4	6	Unknown	Moderate	3.11 (wild)	2	1–2	1–3
Red-necked avocet	Flood	July–Feb	May–Feb	1	2	2	3	4	Unknown	Moderate	7.3 (wild)	3	1–2	1–3
Red-capped plover	Flood, rainfall	July–Jan	Any time	3	3	2.5	5	6	Unknown	Moderate	20.11 (wild)	5	1–2	1–3
Black-fronted dotterel	Flood, season	Aug–Feb	Any time	3	3	2–3	5	6	Unknown	Moderate	6.11 (wild)	3	1–2	1–3
Red-kneed dotterel	Flood, season	Aug–Jan	Any time	3	3	3	6	6	Unknown	Moderate–slow	Unknown	Unknown	1–2	1–3

conditions of parts of the Murray-Darling Basin. It also limits our capacity to establish a required flood frequency within a wetland to ensure the long-term survival of waterbird populations.

Despite the difficulty in establishing the specific flood requirements for some species, the plasticity in waterbird breeding ensures that waterbirds readily respond to floods. However, it creates challenges for the supply of water and conservation of habitats to ensure the long-term survival of waterbirds within the Murray-Darling Basin. Since wetland availability is essential for waterbird breeding, conservation efforts should focus on the supply of water rather than on the conservation of spatially distributed reserves (Roshier *et al.* 2002). In addition, as water flow needs are species-specific, the supply of water to wetlands should aim to maximise reproductive output for a range of species, rather than being based on the needs of individual species. This approach is likely to maximise the recruitment of many waterbird species and maintain the ecological integrity of the wetland habitats on which they depend.

References

Andrew IG (1963) White heron invasion, 1957. *Notornis* **10**, 311–315.

Australian Bird and Bat Banding Scheme (1979) *Porphyrio porphyrio* – Purple swamphen. *Corella* **3**, 18.

Australian Bird and Bat Banding Scheme (1980) *Vanellus tricolor* – Banded lapwing. *Corella* **4**, 63.

Australian Bird and Bat Banding Scheme (1981) *Himantopus himantopus* – Black-winged stilt. *Corella* **5**, 39.

Australian Bird and Bat Banding Scheme (1990) *Elseyornis melanops* – Black-fronted dotterel. *Corella* **14**, 77.

Australian Bird and Bat Banding Scheme (1994) *Tadorna tadornoides* – Australian shelduck. *Corella* **8**, 26.

Australian Bird and Bat Banding Scheme (1995a) *Cygnus atratus* – Black swan. *Corella* **19**, 106.

Australian Bird and Bat Banding Scheme (1995b) *Pelecanus conspicillatus* – Australian pelican. *Corella* **19**, 106.

Australian Bird and Bat Banding Scheme (1996) *Chenonetta jubata* – Maned duck. *Corella* **20**, 147.

Australian Bird and Bat Banding Scheme (1999a) *Anas castanea* – Chestnut teal. *Corella* **24**, 51.

Australian Bird and Bat Banding Scheme (1999b) *Anas gracilis* – Grey teal. *Corella* **23**, 23.

Australian Bird and Bat Banding Scheme (1999c) *Anas superciliosa* – Pacific black duck. *Corella* **23**, 23.

Australian Bird and Bat Banding Scheme (1999d) *Charadrius ruficapillus* – Red-capped plover. *Corella* **23**, 94.

Australian Bird and Bat Banding Scheme (1999e) *Fulica atra* – Eurasian coot. *Corella* **23**, 94.

Australian Bird and Bat Banding Scheme (1999f) *Recurvirostra novaehollandiae* – Red-necked avocet. *Corella* **23**, 52.

Australian Bird and Bat Banding Scheme (2000a) *Anas rhynchotis* – Australasian shoveler. *Corella* **24**, 51.

Australian Bird and Bat Banding Scheme (2000b) *Aythya australis* – Hardhead. *Corella* **24**, 71.

Australian Bird and Bat Banding Scheme (2000c) *Biziura lobata* – Musk duck. *Corella* **24**, 51.

Australian Bird and Bat Banding Scheme (2000d) *Malacorhynchus membranaceus* – Pink-eared duck. *Corella* **24**, 71.

Australian Bird and Bat Banding Scheme (2002) *Vanellus miles* – Masked lapwing. *Corella* **26**, 116.

Bailey RF (1933) New nesting records of glossy ibis. *Emu* **33**, 279–291.

Balasubramaniam S and Guay P-J (2009) Purple swamphens (*Porphyrio porphyrio*) attempting to prey upon black swan (*Cygnus atratus*) eggs and preying upon a cygnet on an urban lake in Melbourne, Australia. *Wilson Journal of Ornithology* **120**, 633–635.

Barlow MI, Muller PM and Sutton RR (1972) Breeding data on the spur-winged plover in Southland, New Zealand. *Notornis* **19**, 212–249.

Barnard HG (1913) Birds of the Brunnette Downs (N.T.). *Emu* **13**, 205–210.

Baxter GS (1994) The location and status of egret colonies in coastal New South Wales. *Emu* **94**, 255–262.

Belcher CF (1902) Marsh terns and grebes. *Emu* **2**, 33.

Bell BD, McKenzie HR, Sibson RB, Hogg MJ and Wiblin R (1961) Field study course at Farewell Spit 22–29/1/1961. *Notornis* **9**, 145–156.

Boehm EF (1955) Banded plover as a wader. *Emu* **55**, 296.

Boehm EF (1963) Notes on some South Australian waders. II. *Emu* **63**, 276–282.

Bourke PA (1953) Winter nesting of the spur-winged plover. *Emu* **53**, 105–108.

Bourke PA (1956) Notes on the whiskered tern. *Emu* **56**, 339–348.

Braithwaite LW (1975) Managing waterfowl in Australia. *Proceedings of the Ecological Society of Australia* **8**, 107–128.

Braithwaite LW (1976) Notes on the breeding of the freckled duck in the Lachlan River Valley. *Emu* **76**, 127–132.

Braithwaite LW (1977) Ecological studies of the black swan. I. The egg, clutch and incubation. *Australian Wildlife Research* **4**, 59–79.

Braithwaite LW (1981a) Ecological studies of the black swan. II. Colour and plumage changes, growth rates, sexual maturation and timing and frequency of breeding. *Wildlife Research* **8**, 121–133.

Braithwaite LW (1981b) Ecological studies of the black swan. III. Behaviour and social organisation. *Wildlife Research* **8**, 135–146.

Braithwaite LW and Clayton M (1976) Breeding of the nankeen night heron *Nycticorax caledonicus* while in juvenile plumage. *Ibis* **118**, 584–586.

Braithwaite LW and Frith HJ (1969a) Waterfowl in an inland swamp in New South Wales. III. Breeding. *CSIRO Wildlife Research* **14**, 65–109.

Braithwaite LW and Frith HJ (1969b) Waterfowl in an inland swamp in New South Wales. I. Habitat. *CSIRO Wildlife Research* **14**, 1–16.

Bravery JA (1970) The birds of Atherton Shire, Queensland. *Emu* **70**, 49–63.

Bridgewater AE (1931) A list of the birds of Mansfield, Victoria. *Emu* **31**, 280–285.

Briggs SV (1979) Daytime habitats of waterbirds at four swamps on the Northern Tablelands of New South Wales. *Emu* **79**, 211–214.

Briggs SV (1982) Food habits of the freckled duck and associated waterfowl in north-western New South Wales. *Wildfowl* **33**, 88–93.

Briggs SV (1990) Waterbirds. In *The Murray*. (Eds N MacKay and D Eastburn) pp. 337–344. Murray-Darling Basin Commission: Canberra.

Briggs SV and Maher MT (1985) Limnological studies of waterfowl habitat in south-western New South Wales. II. Aquatic macrophyte productivity. *Australian Journal of Marine and Freshwater Research* **36**, 707–715.

Briggs SV and Thornton SA (1999) Management of water regimes in river red gum *Eucalyptus camaldulensis* wetlands for waterbird breeding. *Australian Zoologist* **31**, 187–197.

Briggs SV, Maher MT and Palmer RP (1985) Bias in food habits of Australian waterfowl. *Wildlife Research* **12**, 507–514.

Briggs SV, Lawler WG and Thornton SA (1991) Effects of breeding and environment on body condition of maned ducks, *Chenonetta jubata*. *Wildlife Research* **18**, 577–587.

Briggs SV, Thornton SA and Lawler WG (1997) Relationships between hydrological control of river red gum wetlands and waterbird breeding. *Emu* **97**, 31–42.

Bright J (1934) Notes on a few birds of the Rochester District. *Emu* **34**, 293–302.

Bright J and Taysom AR (1932) Birds of Lake Cooper, Victoria, and surroundings. *Emu* **32**, 42–48.

Brooker MG, Ridpath MG, Estbergs JA, Bywater J, Hart DS and Jones MS (1979) Bird observations on the north-western Nullarbor Plain and neighbouring regions, 1967–1978. *Emu* **79**, 176–190.

Broome LS and Jarman PJ (1983) Waterbirds on natural and artificial waterbodies in the Namoi Valley, New South Wales. *Emu* **83**, 99–104.

Brouwer K, Schifter H and Jones ML (1994) Longevity and breeding records of ibises and spoonbills Threskiornithidae: in captivity. *International Zoo Yearbook* **33**, 94–102.

Brown RJ and Brown MN (1977) Observations on swamphens breeding near Manjimup, W.A. *Corella* **1**, 82–83.

Brown RJ and Brown MN (1980) Eurasian coots breeding on irrigation dams near Manjimup, Western Australia. *Corella* **4**, 33–36.

Bryant CE (1946) Notes on avocets breeding near Melbourne. *Emu* **46**, 241–245.

Bryant CE (1949) Marsh terns nesting. *Emu* **49**, 267.

Buller KG (1942) Breeding of the great crested grebe in southern Western Australia. *Emu* **42**, 54–55.

Campbell J and Sonter C (1985) Unusual food items of pelicans. *Australian Birds* **20**, 1–3.

Carnaby JC (1933) The birds of the Lake Grace District, W.A. *Emu* **33**, 103–109.

Carrick R (1959) The food and feeding habits of the straw-necked ibis, *Threskiornis spinicollis* (Jameson), and the white ibis, *T. molucca* (Cuvier) in Australia. *CSIRO Wildlife Research* **4**, 69–92.

Carrick R (1962) Breeding, movements and conservation of ibises (Threskiornithidae) in Australia. *CSIRO Wildlife Research* **7**, 71–88.

Carrick R, Wheeler WR and Murray MD (1957) Seasonal dispersal and mortality in the silver gull, *Larus novae-hollandiae* Stephens, and crested tern, *Sterna bergii* Lichstein, in Australia. *CSIRO Wildlife Research* **2**, 116–144.

Carroll ALK (1966) Food habits of pukeko (*Porphyrio melanotus* Temminck). *Notornis* **13**, 133–141.

Carroll ALK (1969) The pukeko (*Porphyrio melanotus*) in New Zealand. *Notornis* **16**, 101–120.

Chaffer N (1940) Photographing the avocet – and other birds – at Lake Midgeon. *Emu* **40**, 126–128.

Chapman A and Lane JAK (1997) Waterfowl usage of wetlands in the south-east arid interior of Western Australia 1992–93. *Emu* **97**, 51–59.

Child P and Child M (1984) The Alexandra black-fronted dotterels: 1982/83 season. *Notornis* **31**, 31–39.

Christidis L and Boles WE (2008) *Systematics and Taxonomy of Australian Birds*. CSIRO Publishing: Melbourne.

Clapp RB, Klimkiewicz MK and Kennard JH (1982) Longevity records of North American birds: Gaviidae through Alcidae. *Journal of Field Ornithology* **53**, 81–124.

Cleland JB (1912) Examination of contents of stomachs and crops of Australian birds. *Emu* **12**, 8–18.

Close DH and McCrie N (1986) Seasonal fluctuation of waders in Gulf St Vincent, 1976–85. *Emu* **86**, 145–154.

Close DH, Bonnin JM, Waterman MH and Connell DJ (1982) Breeding waterbirds on the salt lagoon islands, South Australia. *Corella* **6**, 25–36.

Corben DT and Munro U (2006) Clutch size and fledgling success of the Australian white ibis *Threskiornis molucca* in an urban environment. *Corella* **30**, 55.

Corrick AH (1981) Wetlands of Victoria. II. Wetlands and waterbirds of South Gippsland. *Proceedings of the Royal Society of Victoria* **92**, 187–200.

Corrick AH (1982) Wetlands of Victoria. III. Wetlands and waterbirds between Port Phillip Bay and Mount Emu Creek. *Proceedings of the Royal Society of Victoria* **94**, 69–87.

Corrick AH and Norman FI (1980) Wetlands of Victoria. I. Wetlands and waterbirds of the Snowy River and Gippsland Lakes catchment. *Proceedings of the Royal Society of Victoria* **91**, 1–15.

Cowling SJ and Lowe KW (1981) Studies of ibises in Victoria. I. Records of breeding since 1955. *Emu* **81**, 33–39.

Craig JL (1980) Pair and group breeding behaviour of a communal gallinule, the pukeko, Porphyrio *P. Melanotus*. *Animal Behaviour* **28**, 593–603.

Crawford DN (1972) Birds of Darwin area, with some records from other parts of Northern Territory. *Emu* **72**, 131–148.

Crawford DN (1977) Notes on the feeding of two *Chlidonias* terns. *Emu* **77**, 146–147.

Crome FHJ (1986) Australian waterfowl do not necessarily breed on a rising water level. *Wildlife Research* **13**, 461–480.

Crome FHJ (1988) To drain or not to drain? Intermittent swamp drainage and waterbird breeding. *Emu* **88**, 243–248.

Dakota A (2009) *Porphyrio porphyrio*. Animal Diversity Web. University of Michigan: Michigan. <animaldiversity.ummz.umich.edu>.

Dann P (1981) Breeding of the banded and masked lapwings in southern Victoria. *Emu* **81**, 121–127.

delHoyo J, Elliot A and Sargatal J (1992) *Handbook of Birds of the World, Vol. 1*. Lynx Edicions: Barcelona.

Dementiev GP and Gladkov NA (1968) *The Birds of the Soviet Union*. Israel Program for Scientific Translations: Jerusalem.

Dennis RF (1932) Nest of brolga. *Emu* **32**, 223.

Dickinson P (1951) Stomach contents of New Zealand inland shags. *Marine and Freshwater Research* **2**, 245–253.

Dingle H (2004) The Australo-Papuan bird migration system: another consequence of Wallace's Line. *Emu* **104**, 95–108.

D'Ombrain EA (1928) Nesting of the spur-winged plover in captivity. *Emu* **28**, 110.

D'Ombrain AF (1944) Notes on young of plumed tree-duck. *Emu* **44**, 324–327.

D'Ombrain AF (1945) Nesting notes on plumed tree ducks. *Emu* **45**, 80–82.

Dostine PL and Morton SR (1988) Notes on the food and feeding habits of cormorants on a tropical floodplain. *Emu* **88**, 263–266.

Dostine PL and Morton SR (1989a) Feeding ecology of the whiskered tern, Chlidonias Hybrida, in the Alligator Rivers region, Northern Territory. *Wildlife Research* **16**, 549–562.

Dostine PL and Morton SR (1989b) Food of the darter *Anhinga melanogaster* in the Alligator Rivers region, Northern Territory. *Emu* **89**, 53–54.

Dove HS (1937) Two interesting nests. *Emu* **37**, 184.

Dow DD (1980) Communally breeding Australian birds with an analysis of distributional and environmental factors. *Emu* **80**, 121–140.

Du Guesclin P (2003) Brolga (*Grus rubicunda*) Action Statement no. 119. Arthur Rylah Institute, Dept of Sustainability and Environment: Melbourne.

Edwards HV (1920) The black-listed shag. *Emu* **20**, 106.

Elliott AJ (1938) Birds of the Moonie River District adjacent to the border of New South Wales with Queensland. *Emu* **38**, 30–49.

Elliott AJ (1940) Birds at Lake Midgeon. *Emu* **40**, 123–125.

Emison WB and Porter JW (1978) Summer surveys of birds in the Mt Cobberas-Snowy River area of Victoria, Australia. *Emu* **78**, 126–136.

Favaloro N (1943) Notes on the two resident Victorian plovers. *Emu* **43**, 145–153.

Favaloro NJ (1947) Silver gulls in north-west Victoria. *Emu* **47**, 352–356.

Favaloro NJ (1949) Notes on the red-capped dotterel in inland localities. *Emu* **49**, 13–18.

Favaloro NJ (1952) An inland record of the pied cormorant. *Emu* **52**, 161–162.

Fjeldsa J (1983) Social behaviour and displays of the hoary-headed grebe *Poliocephalus poliocephalus*. *Emu* **83**, 129–140.

Fjeldsa J (1985) Classification of waterbird communities in south-eastern Australia. *Emu* **85**, 141–149.

Fjeldsa J (1988) 'Comparative ecology of the Australasian grebes (Aves: Podicipedidae)'. Royal Australasian Ornithologists Union: Melbourne.

Fletcher JA (1908) Bird notes from Cleveland, Tasmania. I. *Emu* **8**, 210–214.

Fletcher JA (1924) Birds of the steppes. *Emu* **24**, 107–117.

Fletcher SE (1960) Coots nesting in Tasmania. *Emu* **60**, 154.

Ford J and Sedgwick EH (1967) Bird distribution in the Nullarbor Plain and Great Victoria Desert region, Western Australia. *Emu* **67**, 99–124.

Frith HJ (1957a) Breeding and movements of wild ducks in inland New South Wales. *CSIRO Wildlife Research* **2**, 19–31.

Frith HJ (1957b) Wild ducks and the rice industry in New South Wales. *CSIRO Wildlife Research* **2**, 32–50.

Frith HJ (1959a) The ecology of wild ducks in inland New South Wales. I. Waterfowl habitats. *CSIRO Wildlife Research* **4**, 97–107.

Frith HJ (1959b) The ecology of wild ducks in inland New South Wales. III. Food habits. *CSIRO Wildlife Research* **4**, 131–155.

Frith HJ (1959c) The ecology of wild ducks in inland New South Wales. IV. Breeding. *CSIRO Wildlife Research* **4**, 156–181.

Frith HJ (1962) Movements of the grey teal, *Anas gibberifrons* Muller (Anatidae). *CSIRO Wildlife Research* **7**, 50–70.

Frith HJ (1965) Ecology of the freckled duck, *Stictonetta naevosa* (Gould). *CSIRO Wildlife Research* **10**, 125–139.

Frith HJ (1967) *Waterfowl in Australia*. Angus and Robertson: Sydney.

Frith HJ (1982) *Waterfowl in Australia*. 2nd edn. Angus and Robertson: Sydney.

Frith HJ and Davies SJJF (1961) Breeding seasons of birds in subcoastal Northern Territory. *Emu* **61**, 97–111.

Frith HJ, Braithwaite LW and McKean JL (1969) Waterfowl in an inland swamp in New South Wales. II. Food. *CSIRO Wildlife Research* **14**, 17–64.

Fullagar PJ and Carbonell M (1986) The display postures of the male musk duck. *Wildfowl* **37**, 142–150.

Fullagar PJ, Davey CC and Rushton DK (1988) Is it true that Australian ducks are different? In *Proceedings of International Symposium on Wetlands*. Shortlands Wetland Centre: Newcastle. (Eds B Gilligan, M Maddock and K McDonald) pp. 81–98.

Gentilli J and Bekle H (1983) Modelling a climatically pulsating population: grey teal in south-western Australia. *Journal of Biogeography* **10**, 75–96.

Goodrick GN (1979) Food of the black duck and grey teal in coastal northern New South Wales. *Wildlife Research* **6**, 319–324.

Goodsell JT (1990) Distribution of waterbird broods relative to wetland salinity and pH in south-western Australia. *Australian Wildlife Research* **17**, 219–229.

Gosper DG (1981) Survey of birds on floodplain-estuarine wetlands on the Hunter and Richmond rivers in northern N.S.W. *Corella* **5**, 1–18.

Gosper DG, Briggs SV and Carpenter SM (1983) Waterbird dynamics in the Richmond Valley, New South Wales, 1974–77. *Wildlife Research* **10**, 319–327.

Green R (1956) Notes on the inland breeding of the red-capped dotterel. *Emu* **56**, 140–143.

Gremillet D, Tuschy I and Kierspel M (1998) Body temperature and insulation in diving great cormorants and European shags. *Functional Ecology* **12**, 386–394.

Gwynne AJ (1932) Notes on the red-capped dotterel and the period of incubation of its eggs. *Emu* **32**, 117–118.

Hall R (1903) Notes on a collection of bird-skins from the Fitzroy River, north-western Australia. III. *Emu* **3**, 40–43.

Halse SA and Jaensch RP (1989) Breeding seasons of waterbirds in south-western Australia: the importance of rainfall. *Emu* **89**, 232–249.

Halse SA, Williams MR, Jaensch RP and Lane JAK (1993) Wetland characteristics and waterbird use of wetlands in south-western Australia. *Wildlife Research* **20**, 103–125.

Hanscombe SA (1915) Observations on the nankeen night heron (*Nycticorax caledonicus*). *Emu* **15**, 132–134.

Harper (1990) Waterbird dynamics at Bool Lagoon, South Australia, 1983–87. *Australian Wildlife Research* **17**, 113–122.

Heather BD (1973) The black-fronted dotterel (*Charadrius melanops*) in the Wairarapa. *Notornis* **20**, 251–261.

Heather BD (1977) Foot-trembling by the black-fronted dotterel. *Notornis* **24**, 1–8.

Herring M (2005) 'Threatened species and farming: brolga – management of breeding wetlands in northern Victoria'. Arthur Rylah Institute for Ecological Research: Melbourne.

Higgins PJ (1999) *Handbook of Australian, New Zealand and Antarctic Birds. Vol. 4: Parrots to Dollarbird.* Oxford University Press: Melbourne.

Higgins PJ and Davies SJJF (1996) *Handbook of Australian, New Zealand and Antarctic Birds. Vol. 3: Snipe to Pigeons.* Oxford University Press: Melbourne.

Higgins PJ and Peter JM (2002) *Handbook of Australian, New Zealand and Antarctic Birds. Vol. 6: Pardalotes to Shrike-thrushes.* Oxford University Press: Melbourne.

Higgins PJ, Peter JM and Steele WK (2001) *Handbook of Australian, New Zealand and Antarctic Birds. Vol. 5: Tyrant fly-catchers to Chats.* Oxford University Press: Melbourne.

Higgins PJ, Peter JM and Cowling SJ (2006) *Handbook of Australian, New Zealand and Antarctic Birds. Vol. 7: Boatbill to Starlings.* Oxford University Press: Melbourne.

Hindwood KA (1939) Birds of the Murrumbidgee Irrigation Area, N.S.W. *Emu* **39**, 219–232.

Hindwood KA and Hoskin ES (1954) The waders of Sydney (County of Cumberland), New South Wales. *Emu* **54**, 217–255.

Hobbs JN (1956a) A flood year in the Riverina. *Emu* **56**, 349–352.

Hobbs JN (1956b) Night-heron's feeding habits. *Emu* **56**, 61.

Hobbs JN (1957a) Feeding habits of some water birds. *Emu* **57**, 216.

Hobbs JN (1957b) Notes on the pink-eared duck. *Emu* **57**, 265–268.

Hobbs JN (1958) Some notes on grebes. *Emu* **58**, 129–132.

Hobbs JN (1961) The birds of south-west New South Wales. *Emu* **61**, 21–55.

Hobbs JN (1972) Breeding of red-capped dotterel at Fletcher's Lake, Dareton, NSW. *Emu* **72**, 121–125.

Hobbs JN (1976) Birds feeding on house mice. *Emu* **76**, 219–220.

Hopkins N (1947) Birds of Townsville and district. *Emu* **47**, 334–347.

Horton W (1973) Caspian terns at Mount Isa, Queensland. *Australian Bird Bander* **11**, 51–55.

Howe FE (1909) In the Mallee. *Emu* **9**, 227–234.

Jackson SW (1918) Haunts of the letter-winged kite (*Elanus scriptus*, Gould), with an account of its breeding habits, and notes on some other species. *Emu* **18**, 160–172.

Jackson C (2004) *Cygnus atratus*. Animal Diversity Web. University of Michigan: Michigan. <animaldiversity.ummz.umich.edu>.

Jackson RW and Lyall H (1964) An account of the establishment of the Australian coot in the Rotorua district with some notes on its nesting habits. *Notornis* **11**, 82–86.

Jaensch RP, Vervest RM and Hewish MJ (1988) 'Waterbirds in nature reserves of south-western Australia 1981–1985: reserve accounts'. Royal Australian Ornithologists Union: Melbourne.

Jenkins KM and Boulton AJ (2003) Connectivity in a dryland river: short-term aquatic microinvertebrate recruitment following floodplain inundation. *Ecology* **84**, 2708–2723.

Jones AB (1980) The effects of weather on a caspian tern colony. *Notornis* **27**, 405–406.

Junk WJ, Bayley PB and Sparks RE (1989) The flood pulse concept in river-floodplain system. *Special Publication, Canadian Journal of Fisheries and Aquatic Sciences* **106**, 110–127.

Kear J (2005) *Ducks, Geese and Swans*. Oxford University Press: Oxford and New York.

King AJ, Humphries P and Lake PS (2003) Fish recruitment on floodplains: the roles of patterns of flooding and life history characteristics. *Canadian Journal of Fisheries and Aquatic Sciences* **60**, 773–786.

Kingsford RT (1986) Reproductive biology and habitat use of the maned duck *Chenonetta jubata* (Latham). PhD thesis. University of Sydney.

Kingsford RT (1989) Food of the maned duck *Chenonetta jubata* during the breeding season. *Emu* **89**, 119–124.

Kingsford RT and Auld KM (2005) Waterbird breeding and environmental flow management in the Macquarie Marshes, arid Australia. *River Research and Applications* **21**, 187–200.

Kingsford RT and Johnsons W (1998) Impact of water diversions on colonially-nesting waterbirds in the Macquarie Marshes of arid Australia. *Colonial Waterbirds* **21**, 159–170.

Kingsford RT and Norman FI (2002) Australian waterbirds: products of the continent's ecology. *Emu* **102**, 47–69.

Kingsford RT, Wong PS, Braithwaite LW and Maher MT (1999) Waterbird abundance in eastern Australia, 1983–92. *Wildlife Research* **26**, 351–366.

Kushlan J (1993) Colonial waterbirds as bioindicators of environmental change. *Colonial Waterbirds* **16**, 223–251.

Lavery HJ (1986) Breeding seasons of birds in north-eastern Australia. First Supplement, 1967–74. *Emu* **86**, 111–113.

Le Souef D (1902) Descriptions of birds' eggs from the Port Darwin district, northern Australia. II. *Emu* **2**, 139–159.

Leslie DJ (2001) Effect of river management on colonially-nesting waterbirds in the Barmah-Millewa forest, south-eastern Australia. *Regulated Rivers: Research and Management* **17**, 21–36.

Lord EAR (1951) Banded plover, *Zonifer tricolor. Emu* **51**, 79–80.

Lowe VT (1957) Glossy ibis notes. *Emu* **57**, 347–348.

Lowe VT (1965) Notes on the musk-duck. *Emu* **65**, 279–290.

Lowe KW (1982) Feeding behaviour and diet of royal spoonbills *Platalea regia* in Westernport Bay, Victoria. *Emu* **82**, 163–168.

Lowe KW (1983a) Egg size, clutch size and breeding success of the glossy ibis *Plegadis falcinellus. Emu* **83**, 31–34.

Lowe KW (1983b) Feeding behaviour and diet of the white-faced heron *Ardea novaehollandiae* in Westernport Bay, Victoria. *Corella* **7**, 101–108.
Lowe KW (1989) Notes on the breeding of the Pacific heron *Ardea pacifica* near Balranald, New South Wales. *Corella* **13**, 8–89.
Loyn RH (1978) A survey of birds in Westernport Bay, Victoria, 1973–74. *Emu* **78**, 11–19.
Lyle GW (1979) Australian little grebe on Lake Okareka. *Notornis* **20**, 279–280.
Macdonald R (1968) The Australian coot established on Virginia Lake, Wanganui. *Notornis* **15**, 234–237.
MacGillvray W (1922) The nesting of the Australian pelican (*Pelicanus conspicillatus*). *Emu* **22**, 162–174.
MacGillvray W (1925) Birds of the Capricorn Islands. *Emu* **25**, 229–238.
Maclean GL (1977) Comparative notes on black-fronted and red-kneed dotterels. *Emu* **77**, 199–207.
Maddock M (1986) Fledging success of egrets in dry and wet seasons. *Corella* **10**, 101–107.
Maddock M and Baxter GS (1991) Breeding success of egrets related to rainfall: a six-year Australian study. *Colonial Waterbirds* **14**, 133–139.
Maher M (1991) Waterbirds back o'Bourke: an inland perspective on the conservation of Australian waterbirds. PhD thesis. University of New England.
Maher M and Carpenter SM (1984) Benthic studies of waterfowl breeding habitat in south-western New South Wales. II. Chironomid populations. *Australian Journal of Marine and Freshwater Research* **35**, 97–110.
Marchant S (1988) Nesting behaviour of the Pacific heron *Ardea pacifica. Australian Birds* **21**, 61–65.
Marchant S and Higgins PJ (1990) *Handbook of Australian, New Zealand and Antarctic Birds. Vol. 1: Ratites to Ducks*. Oxford University Press: Melbourne.
Marchant S and Higgins PJ (1993) *Handbook of Australian, New Zealand and Antarctic Birds. Vol. 2: Raptors to Lapwings*. Oxford University Press: Melbourne.
Martin PR, Thomson BG and Witts SJ (1979) Niche separation in three species of waterbirds. *Corella* **3**, 1–6.
Martin JM, French K and Major RE (2007) The pest status of Australian white ibis (*Threskiornis molucca*) in urban situations and the effectiveness of egg-oil in reproductive control. *Wildlife Research* **34**, 319–324.
Masters JR and Milhinch AL (1974) Birds of the shire of Northam, about 100 km east of Perth. *Emu* **74**, 228–244.
Mathews GM (1910) On the birds of north-west Australia. III. *Emu* **10**, 103–110.
McCracken KG, Afton AD and Paton DC (2000) Nests and eggs of musk ducks (*Biziura lobata*) at Murray Lagoon, Cape Gantheaume Conservation Park, Kangaroo Island, South Australia. *South Australian Ornithologist* **33**, 65–70.
McEvey AR (1965) The birds of the Rutherglen district. *Emu* **65**, 1–56.
McEvey AR and Middleton WG (1968) Birds and vegetation between Perth and Adelaide (results of the Harold Hall Australian Expedition, No. 12). *Emu* **68**, 161–212.
McGarvie AM and Templeton MT (1974) Additions to the birds of King Island, Bass Strait. *Emu* **74**, 91–96.
McGill AR (1943) The red-kneed dotterel in coastal south-eastern Australia. *Emu* **43**, 225–228.
McGilp JN (1922) Birds of Lake Frome District, South Australia. *Emu* **22**, 237–243.
McKenzie HR (1967) Foods of the pukeko. *Notornis* **14**, 41–42.
McKeown KC (1943) The food of cormorants and other fish-eating birds. *Emu* **43**, 259–269.

McKilligan NG (1975) Breeding and movements of the straw-necked ibis in Australia. *Emu* **75**, 199–212.

McKilligan N (2005) *Herons, Egrets and Bitterns: Their Biology and Conservation in Australia.* CSIRO Publishing: Melbourne.

McLean JC (1901) *Porphyrio melanonotus* in New Zealand. *Emu* **1**, 52–53.

Meine CD and Archibald GW (1996) 'The cranes: status survey and conservation action plan'. IUCN: Gland, Switzerland and Cambridge, UK.

Middleton AJ (1987) A Caspian tern feeding on the ground. *Notornis* **34**, 116.

Miller B (1979) Ecology of the little black cormorant, *Phalacrocorax sulcirostris*, and little pied cormorant, *P. melanoleucos*, in inland New South Wales. I. Food and feeding habits. *Wildlife Research* **6**, 79–95.

Miller B (1980) Ecology of the little black cormorant, *Phalacrocorax sulcirostris*, and little pied cormorant, *P. melanoleucos*, in inland New South Wales. II. Proximate control of reproduction. *Wildlife Research* **7**, 85–101.

Mills JA (1973) The influence of age and pair-bond on the breeding biology of the red-billed gull *Larus novaehollandiae scopulinus*. *Journal of Animal Ecology* **42**, 147–162.

Moore PJ (1984) Foraging and social behaviour of the white-faced heron at Pauatahanui Inlet. *Notornis* **31**, 285–299.

Morris AK, McGill AR and Holmes G (1981) 'Handlist of the birds of New South Wales'. New South Wales Field Ornithologists Club: Sydney.

Morse FC (1922) Birds of the Moree District. *Emu* **22**, 24–36.

Murray MD and Carrick R (1964) Seasonal movements and habitats of the silver gull, *Larus novaehollandiae* Stephens, in south-eastern Australia. *CSIRO Wildlife Research* **9**, 160–188.

Napier JR (1973) Crested grebe and coot nesting in Tasmania. *Australian Bird Watcher* **5**, 7.

Nelson JB (2005) *Pelicans, Cormorants, and Their Relatives: The Pelecaniformes.* Oxford University Press: Oxford, UK.

Nicholls B (1942) A trip to Mungeranie, Central Australia. *Emu* **42**, 45–59.

Norman FI (1974) Notes on the breeding of the pied cormorant near Werribee, Victoria, in 1971, 1972 and 1973. *Emu* **74**, 223–227.

Norman FI (1982) Eggs, egg-laying and incubation in the chestnut teal. *Emu* **82**, 195–198.

Norman FI (1983) Grey teal, chestnut teal and Pacific black duck at a saline habitat in Victoria. *Emu* **83**, 262–271.

Norman FI and Brown RS (1988) Aspects of the distribution and abundance of chestnut teal in south-eastern Australia. *Emu* **88**, 70–80.

Norman FI and Hurley VG (1984) Gonad measurements and other parameters from chestnut teal *Anas castanea* collected in the Gippsland Lakes region, Victoria. *Emu* **84**, 52–55.

Norman FI and Mumford L (1982) Food of the chestnut teal, *Anas castanea*, in the Gippsland Lakes region of Victoria. *Wildlife Research* **9**, 151–155.

Norman FI and Mumford L (1985) Studies on the purple swamphen, *Porphyrio porphyrio*, in Victoria. *Wildlife Research* **12**, 263–278.

Norman FI, Thomson LW and Hamilton JG (1979) Use of habitat and diurnal activity of Pacific black duck, chestnut teal and grey teal at Serendip, Victoria. *Emu* **79**, 54–62.

Norton SPW (1922) Bird notes from Boree (New England Plateau). *Emu* **22**, 39–44.

O'Donnell CFJ (1982) Food and feeding behaviour of the southern crested grebe on the Ashburton lakes. *Notornis* **29**, 151–156.

Oiseaux (2009a) Little egret – *Egretta garzetta*. Ecopains d'abord, France. http://www.oiseaux.net/birds/little.egret.html.

Oiseaux (2009b) Whiskered tern – *Childonias hybridus*. Ecopains d'abord, France. http://www.oiseaux.net/birds/whiskered.tern.html.

Owen KL and Sell MG (1985) The birds of Waimea Inlet. *Notornis* **32**, 271–309.

Pennycook CS (1949) Caspian tern colony. *Notornis* **3**, 129.

Pierce RJ (1980) Seasonal and long-term changes in bird numbers at Lake Wainono. *Notornis* **27**, 21–44.

Pierce RJ (1983) The Charadriiformes of a high-country river valley. *Notornis* **30**, 169–185.

Pierce RJ (1984) Breeding success of isolated pairs of Caspian terns in Canterbury. *Notornis* **31**, 185–190.

Powlesland RG and Robertson HA (1987) Changes in gull numbers over 25 years and notes on other birds of the Otaki-Ohau coast. *Notornis* **34**, 327–388.

Pullen RC (1967) Nesting success of a Pied Stilt colony. *Notornis* **14**, 76–79.

Recher HR and Holmes RT (1982) 'The foraging behaviour of herons and egrets on the Magela Creek flood plain, Northern Territory'. Technical memorandum 4, Supervising Scientist for the Alligator Rivers Region. Australian Government Publishing Service: Canberra.

Recher HF, Holmes RT, Davis WE Jr and Morton S (1983) Foraging behaviour of Australian herons. *Colonial Waterbirds* **6**, 1–10.

Riggert TL (1977) The biology of the mountain duck on Rottnest Island, Western Australia. *Wildlife Monographs* **52**, 3–67.

Roshier DA, Robertson AI, Kingsford RT and Green DG (2001a) Continental-scale interactions with temporary resources may explain the paradox of large populations of desert waterbirds in Australia. *Landscape Ecology* **16**, 547–556.

Roshier DA, Whetton PH, Allan RJ and Robertson AI (2001b) Distribution and persistence of temporary wetland habitats in arid Australia in relation to climate. *Austral Ecology* **26**, 371–384.

Roshier DA, Robertson AI and Kingsford RT (2002) Responses of waterbirds to flooding in an arid region of Australia and implications for conservation. *Biological Conservation* **106**, 399–411.

Ross D (1978) Birds on reed mats. *Canberra Bird Notes* **4**, 22.

Rydzewski W (1978) The longevity of ringed birds. *Ring* **96–97**, 218–262.

Sansom O, Bell BD, Andrews T and Wilson RA (1954) Visitation of glossy ibis. *Notornis* **6**, 18–19.

Schodde R (1982) Origin, adaptation and evolution of birds in arid Australia. In *Evolution of the Flora and Fauna of Arid Australia*. (Eds WR Barker and PJM Greenslade) pp. 191–224. Peacock: South Australia.

Schweigman P (1999) Breeding age of royal spoonbills in New Zealand. *Notornis* **46**, 502.

Scott A (1997) 'Relationships between waterbird ecology and river flows in the Murray-Darling Basin'. CSIRO Land and Water: Canberra.

Sedgwick EH and Morrison PC (1948) Observations on the lower Murchison RAOU Camp, September 1948. *Emu* **48**, 212–242.

Serventy DL (1938) Notes on some fish-eating birds other than cormorants. *Emu* **38**, 510–512.

Serventy DL (1947) The birds of the Swan River District, Western Australia. *Emu* **47**, 241–286.

Serventy DL (1977) Seabird Islands No. 49: Fisher Island, Tasmania. *Corella* **1**, 60–32.

Serventy DL and Marshall AJ (1957) Breeding periodicity in Western Australian birds: with an account of unseasonal nestings in 1953 and 1955. *Emu* **57**, 99–126.

Sharland MSR (1942) Black-fronted dotterel in Tasmania. *Emu* **42**, 52–53.

Sharland M (1957) Egrets at Ulmarra, NSW. *Emu* **57**, 295–301.

Sibson RB, Medway DG, Kennedy-Smuts C, Drew J and Grant G (1972) The spread of the black-fronted dotterel. *Notornis* **19**, 83–85.

Small MM (1960) The coot colony at Lake Hayes. *Notornis* **9**, 61.

Smith GC (1991) 'The silver gull: problems, biology and management for the Sydney-Wollongong Region'. NSW National Parks and Wildlife Service: Sydney.

Smith GC and Carlile N (1993) Food and feeding ecology of breeding silver gulls (*Larus novaehollandiae*) in urban Australia. *Colonial Waterbirds* **16**, 9–16.

Smith ACM and Munro U (2008) Cannibalism in the Australian pelican (*Pelecanus conspicillatus*) and Australian white ibis (*Threskiornis molucca*). *Waterbirds* **31**, 632–635.

Smith A and O'Connor L (1955) Breeding of marsh terns on Coode Island. *Emu* **55**, 255–256.

Staav R and Fransson T (2008) 'EURING list of longevity records for European birds'. European Union for Bird Ringing. <www.euring.org>.

Stidolph RHD (1931) Birds of the Wairarapa Plains. *Emu* **31**, 136–139.

Stidolph RHD (1938) Birds of the Wairarapa Lake District. *Emu* **38**, 344–355.

Stokes AF (1949) Stilts nesting at Ardmore, 1947–48 season. *Notornis* **3**, 107–108.

Stone AC (1912) Birds of Lake Boga, Victoria. *Emu* **12**, 112–122.

Storer RW (1987) Morphology and relationships of the hoary-headed grebe and the New Zealand dabchick. *Emu* **87**, 150–157.

Storr GM (1964a) The avifauna of Rottnest Island, Western Australia. I. Marine birds. *Emu* **64**, 48–60.

Storr GM (1964b) The avifauna of Rottnest Island, Western Australia. II. Lake and littoral birds. *Emu* **64**, 105–113.

Tarr HE (1960) Nesting of the Caspian tern. *Australian Bird Watcher* **1**, 78–81.

Thomas DG (1968) Waders of Hobart. *Emu* **68**, 99–125.

Thomas DG (1969) Breeding biology of the Australian spur-winged plover. *Emu* **69**, 81–102.

Tingay A, Tingay S and Goodsell J (1977) Report of a management program for black swans in south-western Australia. *Emu* **77**, 185–187.

van Tets GF, D'Andria AH and Slater E (1967) Nesting distribution and nomenclature of Australasian vanelline plovers. *Emu* **67**, 84–93.

van Tets GF, Vestjens WJM and Slater EC (1969) Orange runway lighting as a method for reducing bird strike damage to aircraft. *CSIRO Wildlife Research* **14**, 129–151.

Vestjens WJM (1975a) Breeding behaviour of the darter at Lake Cowal, NSW. *Emu* **75**, 121–131.

Vestjens WJM (1975b) Feeding behaviour of spoonbills at Lake Cowal, NSW. *Emu* **75**, 132–136.

Vestjens WJM (1977a) Breeding behaviour and ecology of the Australian pelican, *Pelecanus conspicillatus*, in New South Wales. *Wildlife Research* **4**, 37–58.

Vestjens WJM (1977b) 'Status, habitats and food of vertebrates at Lake Cowal'. CSIRO Wildlife Research Technical Memorandum 1: 21–87, Canberra.

Veynet L (2006) Brolga (*Grus rubicunda*) husbandry and handraising. In *Proceedings of National Wildlife Rehabilitation Conference 2006*. Darwin NT.

Walker KF, Sheldon F and Puckridge JT (1995) An ecological perspective on dryland river ecosystems. *Regulated Rivers: Research and Management* **11**, 85–104.

Watson IM (1955) Some species seen at the Laverton Saltworks, Victoria, 1950–1953, with notes on seasonal changes. *Emu* **55**, 224–248.

Wheeler R (1955) Charadriiformes at the Laverton Saltworks, Victoria, 1950–1953. *Emu* **55**, 279–295.

Wheeler R (1957) Altona survey group. Report No. 4. *Emu* **57**, 185–190.

Wheeler JR (1960) Blue-billed ducks at Lake Wendouree, Ballarat. *Emu* **60**, 281–284.

Wheeler WR and Watson I (1963) The silver gull, *Larus novaehollandiae* Stephens. *Emu* **63**, 99–173.

Whitlock FL (1920) Notes on Dirk Hartog Island and Peron Peninsula, Shark Bay, Western Australia. *Emu* **20**, 168–186.

Whittell HM (1932) The birds of the Bridgetown district, south-west Australia. *Emu* **32**, 182–189.

Woodall PF (1985) Waterbird populations in the Brisbane Region, 1972–83, and correlates with rainfall and water heights. *Wildlife Research* **12**, 495–506.

Chapter 4

Fish

Timothy J Ralph, Jennifer A Spencer and Thomas S Rayner

Introduction

Natural patterns of flow and floodplain inundation are regarded as key drivers of freshwater fish recruitment (Poff *et al.* 1997). Flooding, floodplain connection and periods of low flow may all be critical for freshwater fish within the Murray-Darling Basin (Humphries *et al.* 1999; King *et al.* 2003, 2009; King 2004; Balcombe *et al.* 2005; Closs *et al.* 2005; Koehn and Harrington 2006). However, significant knowledge gaps exist for most species, despite the fact that fish recruitment can act as an indicator of ecosystem productivity at all trophic levels (Kushlan 1993). It is timely that consideration be given to the roles of flooding and drying for the fish species found in lowland-dryland rivers and floodplain wetlands in the Murray-Darling Basin, since flows in these systems are increasingly threatened by the impacts of water resource developments and changes to their hydrology due to increased climate variability and climate change (CSIRO and BOM 2007; CSIRO 2008).

This chapter synthesises current knowledge of the water requirements of key fish taxa in the Murray-Darling Basin. Consideration is given to the life-cycle, habitat and dietary requirements of each species, and linkages are identified between the flow characteristics of these habitats, food availability and conditions required for successful recruitment. Although there are additional factors which may influence fish community structure and function, including competition, predation and habitat modification, the emphasis of this chapter is on the effects of flow regime and flood and drought perturbations. Of the 46 native and endemic fish species known to occur in the Murray-Darling Basin, 17 native species that occur predominantly in lowland rivers are profiled. These native fish have been divided into four groups according to their preferred water and habitat requirements:

- low-flow and wetland opportunists;
- main channel generalists and wetland opportunists;
- main channel specialists;
- flood spawners.

This grouping is based partly on a classification developed for the Murray Flow Assessment Tool (MFAT) (Young *et al.* 2003) as part of the Murray-Darling Basin Commission's *The Living Murray* initiative, which split native fish into seven groups using available information and expert opinion.

- Low-flow specialists, e.g. rainbowfish, carp gudgeons – these spawn and recruit only during low flow, in channel or floodplain habitats.

- Wetland specialists, e.g. Australian smelt, bony bream, carp gudgeons, Southern pygmy perch, hardyheads – these spawn and recruit in floodplain wetlands (opportunistically, when flooding occurs) and in lakes, anabranches and billabongs during in-channel flows.
- Main channel generalists, e.g. Australian smelt, bony bream, flathead gudgeons – these spawn and recruit during high or low flows in the main channel.
- Main channel specialists, e.g. Murray cod, trout cod – these spawn and recruit during high or low flows in the main channel. Woody debris is an important part of their habitat.
- Flood spawners, e.g. golden perch, silver perch – these spawn and recruit following flow rises. Major spawning occurs during periods of floodplain inundation.
- Macquarie perch – these require clean gravel substrate. Floodplain inundation is not required, but spawning is probably enhanced by rising flows.
- Freshwater catfish – these spawn in coarse sediment beds, usually sand or gravel, during any flow conditions.

Bony bream and Australian smelt span both the wetland and main channel generalist groups, so the two groups have been combined for the purpose of this chapter. Similarly, several species fit both the low-flow and wetland groups (e.g. carp gudgeons) and these groups have also been combined. Macquarie perch and freshwater catfish were placed in separate groups in the MFAT classification, however, they have specialised habitat and flow requirements as do many of the main channel specialists so those groups have been combined for simplicity.

This chapter also provide summaries for four of the most common alien fish species found in the Murray-Darling Basin (of the 11 alien species previously recorded; Lintermans 2007), and their likely impacts on native fish. The species profiles are based largely on the following sources: Merrick and Schmida (1984), McDowall (1996), Schiller and Harris (2001), Allen *et al.* (2003), Treadwell and Hardwick (2003) and Lintermans (2007). Other sources are also listed.

Low-flow and wetland opportunists

Olive perchlet: *Ambassis agassizii*

The olive perchlet is a small oval fish with very large eyes (Figure 4.1). It has a forked tail with olive to semi-transparent scales with a dark edge and clear fins. It is usually less than 5 cm in

Figure 4.1: Olive perchlet, *Ambassis agassizii*. Photograph: Gunther Schmida.

length, but has been recorded up to 7.5 cm (Allen *et al.* 2003). Historically the olive perchlet was found throughout the Murray-Darling Basin, but it is now largely restricted to the northern Basin. Recent records are limited to the Border River, Condamine-Balonne and the northern Darling River catchments in New South Wales, and the Nebine and Warrego catchments in Queensland (Lintermans 2007). The western Murray-Darling Basin population of the olive perchlet is listed as an endangered population in the *Fisheries Management Act 1994* (NSW). It is a protected species in South Australia (*Fisheries Act 1982*) and is extinct in Victoria (*Flora and Fauna Guarantee Act 1988*).

Reproduction and life-cycle

Much of the information available for the olive perchlet is based on studies of the closely-related *A. nigripinnis* in coastal streams of New South Wales and Queensland (Milton and Arthington 1985). Males and females mature at one year old and usually live for two years, although females can live for up to three years. The olive perchlet is thought to spawn in response to rising water temperatures (22–23°C) rather than rising water levels, typically from October to December. It has a relatively low fecundity (200–700 eggs) and eggs are small (0.7 mm diameter) adhering to vegetation and rocks. Larvae (3 mm in length) can hatch within five to seven days at 22°C and are free-swimming after nine days. Spawning is thought to occur in pre-flood periods to maximise survival of larvae and eggs before higher flows distribute them downstream (Milton and Arthington 1985). It is not known whether the olive perchlet undertakes spawning migrations in the Murray-Darling Basin (Lintermans 2007).

Habitat and trophic position

The olive perchlet prefers the vegetated edges of lakes, creeks, swamps, wetlands and rivers, typically forming schools during the day among aquatic vegetation (Allen *et al.* 2003). It is carnivorous, mostly consuming microcrustaceans (copepods and cladocera), aquatic insects and terrestrial insects, but occasionally small fish. It is most active during daylight hours (Lintermans 2007).

Water requirements

Given its short reproductive life-span (two to three years) (Milton and Arthington 1985), the olive perchlet is susceptible to increased predation pressure and short-term fluctuations in habitat, water quality and flow regime. In 2007, a large but highly localised population was discovered near Lake Brewster, in the Lachlan River system. The size of the population was thought to be related to an abundance of young-of-year fish which recruited during low-flow conditions that provided sheltered, shallow and well-vegetated habitats for spawning (McNeil *et al.* 2008). Management using environmental flows could be achieved at the scale of the Lake Brewster system, but would be difficult in a larger and more complex system, such as the Darling River. McNeil *et al.* (2008) recommended careful ongoing monitoring, research and management of the Lake Brewster population. Key knowledge gaps for this species include the water-quality tolerances (particularly temperature and salinity) of eggs, larvae and adults (Treadwell and Hardwick 2003).

Un-specked hardyhead: *Craterocephalus stercusmuscarum fulvus*

Un-specked hardyheads are small fish, up to 10 cm long but commonly only 5–6 cm long, with a small mouth, a slender body of silver grey (or varying silver gold) and a dark-edged silvery stripe from snout to tail (Figure 4.2). These fish have a fairly widespread distribution in rivers of the Murray-Darling Basin and are particularly abundant in the northern regions. Un-specked hardyheads are regarded as wetland opportunists since they tend to spawn and recruit

Figure 4.2: Un-specked hardyhead, *Craterocephalus stercusmuscarum fulvus*. Photograph: Gunther Schmida.

in anabranches, billabongs and floodplain wetlands during their life-cycle, although they will also spawn in the river channel (Young *et al.* 2003).

Reproduction and life-cycle

Spawning is known to occur between October and February, with a peak in spring when water temperatures exceed 24°C (Lintermans 2007). The spawning period of the un-specked hardyhead is not very flexible and is known to overlap with the flood period in some rivers, although there is a low likelihood of the use of the floodplain during floods (King *et al.* 2003). Fecundity is low, with typically 20–107 large oocytes and up to 2200 smaller oocytes present. Until recently, little was known about its early development, although five ontogenetic stages have now been identified for fish 0.5–1.3 cm long, with associated changes in body morphology (Close *et al.* 2005). Upstream movements through fishways have been recorded on the Murray and Murrumbidgee rivers, although migration and spawning patterns have not been recorded. The modes of larvae and juvenile movement are unknown.

Habitat and trophic position

The un-specked hardyhead is typically found around the margins of slow-flowing rivers, backwaters and billabongs, usually in shallow vegetated areas with sandy or muddy substrates (Allen *et al.* 2003). Its local distribution and abundance is patchy, but it may congregate in areas where streams flow into still water. It is unlikely to use the floodplain for spawning and recruitment (Treadwell and Hardwick 2003). The diet of the un-specked hardyhead consists mainly of small insects such as mosquito larvae, and small crustaceans (Lintermans 2007). The fish is likely to be prey for birds and larger carnivorous fish species.

Water requirements

The un-specked hardyhead can withstand variable water temperatures (9.3–36°C) and salinities (<43 000 mg L^{-1}) in a variety of floodplain and in-channel habitats. Key knowledge gaps include the habitat preferences and water-quality requirements of larvae and juveniles, direct use of the floodplain during the life-cycle (likely to be opportunistic), and impacts of flooding on water quality and dietary requirements (Treadwell and Hardwick 2003).

Murray hardyhead: *Craterocephalus fluviatilis*

The Murray hardyhead is a small slender fish with large silvery eyes and two small dorsal fins (Figure 4.3). It has a forked tail and the upper surface of the body is silver to golden. It has a silvery black stripe along its lateral line. It can be distinguished from the closely related un-specked hardyhead (*C. stercusmuscarum fulvus*) by an irregular scale pattern on the back of the head and shoulders and by dark pigment around the edge of round scales, as opposed to the regular diamond-shaped scales of the un-specked hardyhead (Ellis 2005b). The Murray hardyhead can reach 7.6 cm but is more commonly 4–5 cm (Lintermans 2007).

The Murray hardyhead occurs around the edges of lakes, backwaters and wetlands only in the lowland areas of the southern Murray-Darling Basin. It was formerly endemic to the lowland floodplains of the Murray and Murrumbidgee rivers (Ebner *et al.* 2003). The species has suffered a significant decline in distribution and abundance; it is now extinct in New South Wales and only isolated populations remain along the lower Murray River in Victoria and South Australia (Ellis 2005b). Consequently, the species is listed as endangered under the IUCN criteria and as vulnerable nationally under the *Environment Protection and Biodiversity Conservation Act 1999.* A recovery plan has been developed (Backhouse *et al.* 2008).

Reproduction and life-cycle

There is limited information on the Murray hardyhead as it was formerly often confused with other hardyhead species, including the un-specked hardyhead, Darling river hardyhead (*C. amniculus*) and Lake Eyre hardyhead (*C. eyresii*) (Ebner *et al.* 2003). Recent studies have indicated that the Murray hardyhead is a batch-spawner with a prolonged breeding season (September–February). Females lay adhesive eggs in submerged vegetation several times throughout the breeding season (Ellis 2005a). It is a short-lived species (one to two years), with first breeding generally occurring around 12 months (Ellis 2006). The stimulus for spawning is unknown, but may be triggered by increasing water temperature and day length (Ellis 2005b).

Habitat and trophic position

The Murray hardyhead prefers clear, slow-moving or still waters with sand or silt substrates. The adult and juvenile are thought to use different habitats, with the juvenile preferring well-vegetated fringes of waterbodies and the adult more commonly seen in open deep water

Figure 4.3: Murray hardyhead, *Craterocephalus fluviatilis*. Photograph: Gunther Schmida.

habitats with submerged macrophytes (Ellis 2005b). It is tolerant of high salinities (sea water, ~35 000 mg L^{-1} and above) (Wedderburn *et al.* 2008) but early life-stages, particularly eggs and larvae, may be sensitive to high levels of salinity (Ellis 2005a). Although its diet has not been fully investigated, the Murray hardyhead is thought to consume mostly zooplankton (Ellis 2005a), some aquatic insects and algae (Lintermans 2007).

Water requirements

There is limited information on the breeding biology, recruitment or spawning requirements of the Murray hardyhead. Populations of Murray hardyhead are restricted to four Victorian lakes and a series of lentic and lotic habitats along the lower Murray River (Backhouse *et al.* 2008). The latter sites, particularly Lake Alexandrina and Lake Albert, have undergone major decline during recent drought, leading to reductions in the diversity and abundance of native fish. However, the Murray hardyhead persists in the lower lakes at salinities up to 30 100 mg L^{-1} (Wedderburn and Barnes 2009). Management of habitat availability (through water levels in lakes) and salinity management through environmental water allocations are important for this species. The Murray hardyhead is able to tolerate salinities up to 85 000 mg L^{-1} in a laboratory setting (Wedderburn *et al.* 2008) and may recruit successfully up to a maximum of 7000 mg L^{-1} (Ellis 2005a). Rehabilitation of freshwater habitats is critical to the survival of the species (Wedderburn and Barnes 2009).

Carp gudgeon: *Hypseleotris* spp.

The carp gudgeon (including *Hypseleotris klunzingeri*, the western carp gudgeon) is a small fish approximately 4 cm long but up to 7 cm, with a blunt snout and a compressed body of yellow grey or green brown (Figure 4.4). It has a very widespread distribution in rivers of the Murray-Darling Basin and can be abundant. It is regarded as both a wetland and low-flow opportunist, since it tends to spawn and recruit during low flows and can utilise the main channels, floodplain wetlands and secondary channels during its life-cycle (Young *et al.* 2003).

Reproduction and life-cycle

Spawning of carp gudgeon is known to occur between October and April, initiated when water temperatures exceed 22–23°C (Treadwell and Hardwick 2003). The spawning period does not

Figure 4.4: Western carp gudgeon, *Hypseleotris klunzingeri*. Photograph: Gunther Schmida.

overlap with the flood period in some rivers, such as the Ovens River in Victoria (King *et al.* 2003). Because the eggs are usually deposited in shallow areas, rises and falls in water level can lead to egg exposure, desiccation and death (Merrick and Schmida 1984). Fecundity is moderate with up to 2000 oocytes present depending on the size of the female (Lintermans 2007). Movements through fishways have been recorded on the Murray and Murrumbidgee rivers (Lintermans 2007), although it is unclear whether such movements are for dispersal or foraging. The carp gudgeon has been known to spawn during low-flow conditions when water was not present on the floodplain, while the peak abundance of carp gudgeon coincided with a rapidly declining hydrograph and within channel flow conditions (King *et al.* 2003). The cues for spawning and the pre- and post-spawning behaviour are uncertain (Humphries *et al.* 1999).

Habitat and trophic position

The carp gudgeon prefers slow-flowing or still waters, including backwaters and billabongs, usually in vegetated areas with macrophyte beds or other aquatic plants over a range of substrates. It has been known to gather in large schools below weirs (Merrick and Schmida 1984). It has a low likelihood of using the floodplain for recruitment, although it can reside and recruit in the main channel, permanent secondary channels and billabongs (Treadwell and Hardwick 2003). For example, in the Ovens River, larvae and juveniles did not drift from the main channel onto the inundated floodplain (King *et al.* 2003). The diet of the carp gudgeon consists mainly of small aquatic invertebrates and zooplankton, with some detritus and microcrustaceans (Merrick and Schmida 1984). The fish are likely prey for larger carnivorous fish species such as golden perch (*Macquaria ambigua*), freshwater catfish (*Tandanus tandanus*) and juvenile Murray cod (*Maccullochella peelii peelii*).

Water requirements

Water levels are a key determinant of dietary variation in the carp gudgeon, as fluctuating water levels cause a decrease in the availability of higher-quality foods (e.g. chironomids) (Balcombe and Humphries 2006). In a recent study of fish movements in the Murray River and adjoining wetlands, the carp gudgeon was observed to move bidirectionally into and out of wetlands during high flows, which emphasises the importance of wetland and main-channel connections (Lyon *et al.* 2010). The temperature range of the carp gudgeon is unknown. It can handle salinities up to 38 000 mg L^{-1} and low dissolved oxygen concentrations, but there are insufficient data on whether it adapts well to poor water-quality conditions (King *et al.* 2003). Key knowledge gaps include uncertain taxonomy, water-quality tolerances, and habitat preferences of larvae and juveniles (Treadwell and Hardwick 2003).

Southern purple-spotted gudgeon: *Mogurnda adspersa*

The southern purple-spotted gudgeon is a small fish with rounded head, small mouth and rounded tail (Figure 4.5). It has two dorsal fins; the first is short and the second fin is longer and taller. Maximum length is 15.2 cm but it is more commonly 6–12 cm (Lintermans 2007). It is generally purplish brown to light blue with a white underbelly. Its sides have blackish grey blotches and scattered white and red spots, which end in large blue blotches near the tail. The fins are yellow with red spots.

The southern purple-spotted gudgeon can be found in slow-flowing rivers, creeks and billabongs (Allen *et al.* 2003). The species was formerly found throughout the Murray-Darling Basin but has undergone significant declines in the southern Basin and is presumed to be extinct in South Australia and Victoria. It is also rare in the northern Basin but there are records for the Border rivers, Moonie River, Macquarie and Condamine-Balonne catchments (Lintermans 2007). Populations of southern purple-spotted gudgeon in the Murray-Darling

Figure 4.5: Southern purple-spotted gudgeon, *Mogurnda adspersa*. Photograph: Gunther Schmida.

Basin are thought to be locally endemic and have restricted genetic diversity compared to populations in southern Queensland (Faulks *et al.* 2008). It is listed as an endangered species in New South Wales (*Fisheries Management Act 1994*), critically endangered in Victoria (*Flora and Fauna Guarantee Act 1988*) and protected in South Australia (*Fisheries Act 1982*). A captive breeding program has been initiated in New South Wales. In 2004, the southern purple-spotted gudgeon was reintroduced into Adjungbilly Creek in the Murrumbidgee catchment, but it failed to establish successfully (Gilligan 2005).

Reproduction and life-cycle

The male southern purple-spotted gudgeon performs an elaborate courtship display to initiate pairing and spawning in summer months when water temperatures exceed 20°C (Lintermans 2007). Males mature at 4.5 cm and females at 4.9 cm. Females can spawn (280–1300 eggs) several times in a spawning season, laying adhesive elongated eggs (2.0–3.8 mm length) on logs, aquatic vegetation and boulders. The eggs hatch in three to nine days. The males defend and fan the eggs until hatching (Allen *et al.* 2003; Lintermans 2007).

Habitat and trophic position

The southern purple-spotted gudgeon generally prefers deep pools in slow-moving or still waters near the headwaters of upland streams, and is often associated with vegetation and woody debris. It is carnivorous, feeding from the benthos, water column and water surface on insect larvae. It is also known to consume worms, tadpoles, small fish and some plant matter (Boxall *et al.* 2002).

Water requirements

The southern purple-spotted gudgeon is one of the few species of native fish found in the Murray-Darling Basin for which there is evidence of successful conservation using environmental flows. In 1995, a population was discovered in the Cardross lakes system, near Mildura – the first record of this species in Victoria since the 1930s. In 1996, environmental water was delivered to the lakes with the aim of maintaining water levels and water quality, and providing littoral habitat access over summer. These inflows allowed the purple-spotted gudgeon and other species to access stands of *Typha* reeds, which they used for cover. Environmental allocations of approximately 1 GL yr^{-1} have continued and the health of the native fish community has improved. The water-quality tolerances of this species are a significant knowledge gap (Treadwell and Hardwick 2003) but it is known that they do not tolerate salinities above 14 800 mg L^{-1} (James *et al.* 2003).

Murray-Darling rainbowfish: *Melanotaenia fluviatilis*

The Murray-Darling rainbowfish is a small fish, commonly around 7 cm long but up to 10 cm, with a rounded head and a deep slender silver body with greenish iridescence (coloration varies with sex; Figure 4.6). The Murray-Darling rainbowfish is fairly common and has a widespread but patchy distribution in the Murray-Darling Basin (Lintermans 2007). It is primarily regarded as a low-flow specialist since it tends to spawn and recruit during low flows in channels, but it can also use floodplain habitats (Young *et al.* 2003).

Reproduction and life-cycle

The Murray-Darling rainbowfish has a short spawning season from November to February, when water temperatures rise above 20°C, in habitats with weeds and slow-moving waters (Merrick and Schmida 1984). The spawning period of the Murray-Darling rainbowfish does not tend to overlap with the flood period and the precise cues for spawning are uncertain (Humphries *et al.* 1999). Fecundity is generally low, with only 35–333 eggs laid depending on the size of the female (Lintermans 2007). Newly hatched larvae are strong swimmers and prefer to remain near the surface, although larval development is slow over the first month of the life-cycle (Reid and Holdway 1995). Larvae and juveniles of Murray-Darling rainbowfish were not seen to drift from the main channel onto the inundated floodplain in the Ovens River (King *et al.* 2003).

Habitat and trophic position

The Murray-Darling rainbowfish prefers slow-flowing waters in lowland rivers, wetlands and billabongs, often where there is submerged vegetation and other riparian cover (Allen *et al.*

Figure 4.6: Murray-Darling rainbowfish, *Melanotaenia fluviatilis*. Photograph: Gunther Schmida.

2003). Its diet consists mainly of small aquatic and terrestrial insects, with some crustaceans and algae (Lintermans 2007). The species has been documented moving through fishways on the Murrumbidgee River, although the reasons for migration are uncertain (Lintermans 2007).

Water requirements

Knowledge gaps for this species include the movements, behaviour and water-quality tolerances of adults and juveniles and their dependence on floodplain wetlands (Treadwell and Hardwick 2003). King (2004) found a significant relationship between the Murray-Darling rainbowfish and sheltered backwater habitats in the Broken River, particularly for larvae during the night. Larvae tend to school at or near the surface, and there is a low likelihood of direct use of the floodplain for spawning and fish recruitment (King *et al.* 2003). The temperature range of the Murray-Darling rainbowfish is 18–28°C, and pH 5.3–8.5. The adult can tolerate salinities up to 30 000 mg L^{-1} while the juvenile prefers <12 000 mg L^{-1} (Treadwell and Hardwick 2003). There are insufficient data to determine whether the species can adapt well to poor water-quality conditions (e.g. low dissolved oxygen) (King *et al.* 2003).

Main channel generalists and wetland opportunists

Flat-headed gudgeon: *Philypnodon grandiceps*

Dwarf flat-headed gudgeon: *Philypnodon macrostomus*

Flat-headed gudgeons are small native fish, distinguished by compressed bodies with broad flattened heads and large mouths. Their eyes are positioned high and close together on their heads, and they have two separate dorsal fins and rounded tails. Two species inhabit the Murray-Darling Basin: *Philypnodon grandiceps* and *P. macrostomus* (Figure 4.7). The flat-headed gudgeon (*P. grandiceps*) can reach 12 cm long but is commonly 8 cm. The dorsal colour can vary from yellow green to reddish brown, typically with a series of dark blotches that fade into a lighter yellow underbelly. The dwarf flat-headed gudgeon (*P. macrostomus*) can reach a maximum length of 6.5 cm, but is more commonly 4 cm. The body and head is usually brown black with irregular blotches, and the dorsal fins have black and orange whitish bands. The flat-headed gudgeon has gill openings on the underside of its head that extend below the eye; the gill openings are narrower and do not extend to the eye in the dwarf flat-headed gudgeon. Males in both species tend to have larger mouths than females (Lintermans 2007).

The flat-headed gudgeon has a wide range in south-eastern Australia but a limited and patchy distribution in the Murray-Darling Basin, being predominantly found in the inland systems of the Murray, Lachlan and Macquarie River systems. It is regarded as a main channel generalist since it tends to spawn and recruit in the main channel during both high- and low-flow stages (Young *et al.* 2003). The dwarf flat-headed gudgeon is more common in coastal streams. In inland areas it is restricted to the southern Murray-Darling Basin, being distributed only patchily in the Macquarie, Murrumbidgee, Condamine, Murray and Boorowa River drainages, and the Lower Lakes and Mount Lofty Ranges in South Australia (Lintermans 2007).

Reproduction and life-cycle

The flat-headed gudgeon has an extended spawning season from October to April, usually when water temperatures are 15–30°C. The specific triggers for spawning are uncertain (Humphries *et al.* 1999). The spawning period of the flat-headed gudgeon overlaps with the flood period in some rivers (King *et al.* 2003), and downstream drift has been observed for larvae

Figure 4.7: Flat-headed gudgeon, *Philypnodon grandiceps* (a) and dwarf flat-headed gudgeon, *Philypnodon macrostomus* (b). Photographs: Gunther Schmida.

towards the end of the breeding season. Larvae and juveniles have been observed to drift from the main channel onto an inundated floodplain (King *et al.* 2003). Fecundity is moderate, with 500–900 eggs laid depending on the size of the female (Lintermans 2007). The frequency, cues and pre- and post-spawning behaviour of the flat-headed gudgeon are uncertain, as well as the processes of egg development, and the water-quality tolerances of eggs and larvae. Very little is known about the reproduction of the dwarf flat-headed gudgeon in the Murray-Darling Basin, although it is assumed that it is similar to that of the flat-headed gudgeon. Some information is available from fish raised in aquariums, where the dwarf flat-headed gudgeon breeds in water temperatures of 19–22°C (Lintermans 2007). Male fish guard and fan the eggs until hatching, which occur four to five days after spawning (Lintermans 2007).

Habitat and trophic position

The flat-headed gudgeon and dwarf flat-headed gudgeon both prefer slow-flowing or still waters in lowland rivers with areas of aquatic vegetation, large woody debris and a muddy substrate (Lintermans 2007). The flat-headed gudgeon has a moderate likelihood of use of floodplain for spawning and fish recruitment, although larval development is unlikely to occur on floodplain. Larvae were present in both flood and non-flood years in the Ovens River, indicating that spawning and recruitment is not reliant on floodplain inundation but that the use of temporary floodplain habitats, such as billabongs and anabranches, can occur (King *et al.* 2003). The diet consists mainly of crustaceans, small aquatic insects, tadpoles and smaller fishes (Lintermans 2007). Key knowledge gaps include the movement, habitat preference, behaviour and water-quality tolerance of larvae, as well as water-quality tolerance of the adult (Treadwell and Hardwick 2003). There is little ecological information on the dwarf flat-headed gudgeon but it is thought to be a benthic carnivore, feeding on aquatic insects, including larval stages of chironomids, mayflies and caddisflies (Lintermans 2007).

Water requirements

The flat-headed gudgeon and the dwarf flat-headed gudgeon are relatively rare species with patchy distributions in the Murray-Darling Basin, particularly in northern catchments (Lintermans 2007). As the species are infrequently encountered in fish surveys, little is known of their basic ecology or natural flow requirements. The temperature range of the flat-headed gudgeon is unknown, although it can tolerate salinities up to 40 000 mg L^{-1} and low dissolved oxygen concentrations. There are not enough data to determine whether they adapt well to poor water-quality conditions. The species are not typically targeted for environmental water delivery although they could benefit substantially, given their preference for slow-flowing areas of lowland streams or lakes and dams (Lintermans 2007).

Bony bream: *Nematalosa erebi*

The bony bream is a medium-sized fish, commonly 12–20 cm long but up to 47 cm, and weighing less than 2 kg (Figure 4.8). It has a small head and a blunt snout and is silver olive. The bony bream has a fairly widespread distribution along the lower reaches of many inland rivers of the Murray-Darling Basin, particularly in the northern and western regions. It usually occurs at elevations below 200 m. The bony bream is abundant and is regarded as a main channel generalist and wetland specialist since it tends to spawn and recruit in the main channel during both high- and low-flow stages. It also uses anabranches, billabongs and floodplain wetlands during its life-cycle (Young *et al.* 2003).

Reproduction and life-cycle

The life history of the bony bream is poorly understood, although spawning occurs annually from October to February, usually in shallow waters when temperatures are around 20°C and independently of flooding (Puckridge and Walker 1990). Individuals mature in the second or third year and reproductive output increases with size (Puckridge and Walker 1990). Fecundity is high, with hundreds of thousands of oocytes produced depending on the size of the female. Daytime upstream migrations of juveniles and adults have been recorded in the Murray and Murrumbidgee rivers, possibly related to the colonisation of new habitats by juveniles and the reproductive movements of adults (Lintermans 2007). Recruitment patterns of bony bream have been associated with high-flow events and backwater inundation in the Warrego River. Recruitment was also evident following a zero-flow period, suggesting that departures from typical flood-induced recruitment in the bony bream may reflect opportunistic spawning

Figure 4.8: Bony bream, *Nematalosa erebi*. Photograph: Gunther Schmida.

related to intermittent flooding and dry periods (Balcombe *et al.* 2006). The bony bream also showed evidence of successful spawning leading to juvenile recruitment in waterholes three months after significant summer floods in the Warrego River, which coincided with higher channel–floodplain connectivity from floods. It showed a decrease in abundance during dry spells in some waterholes (Balcombe *et al.* 2006).

Habitat and trophic position

The bony bream is found in most flowing and some standing waters, and is often recorded in turbid waters. It may form large schools in shallow reaches and backwaters. Large deep waterholes are unlikely to support high numbers of bony bream (as observed in the Warrego River) since it prefers smaller shallow waterholes with a greater amount of productive littoral zone and hence potential for aquatic food sources (Balcombe *et al.* 2006). The bony bream usually feeds during the day, and has a mainly herbivorous diet which consists of algae, detritus and aquatic plants, although some small crustaceans and insects are known to be ingested with the muddy detritus (Lintermans 2007). The bony bream tends to avoid clear cooler fast-flowing waters where food availability is more limited (Merrick and Schmida 1984). The bony bream is important prey for larger carnivorous fish such as Murray cod (*Maccullochella peelii peelii*) and golden perch (*Macquaria ambigua*), and some waterbirds.

Water requirements

Connection of the channel and floodplain generally provides important nursery areas for larval and juvenile bony bream and allows older fish to disperse and feed, increasing their chances of surviving and completing reproduction during the dry season (Humphries *et al.* 1999; King *et al.* 2003; Arthington *et al.* 2005). The bony bream has been reported in greater abundance in waterholes with more habitat heterogeneity (e.g. where in-stream benches and woody debris are present) in arid-zone systems such as Cooper Creek, as these provide additional food sources and refuges from larger predator fish (Arthington *et al.* 2005). The bony bream is known to tolerate a wide variety of temperatures (9–38°C), salinities (up to 35 000 mg L^{-1}) and pH (4.8–8.6), as well as other habitat conditions (Treadwell and Hardwick 2003). However, the species is susceptible to low temperatures and low dissolved oxygen levels. The former can cause death in cases of cold-water pollution and the latter becomes a problem when ephemeral river habitats begin to dry (Merrick and Schmida 1984; Allen *et al.* 2003). Knowledge gaps exist for the migratory behavior and water-quality tolerances of larvae, juveniles and adults (Treadwell and Hardwick 2003).

Australian smelt: *Retropinna semoni*

The Australian smelt is a small fish, up to 10 cm long but commonly 4–6 cm, with a rounded snout and a slender body of bright silver olive (sometimes almost transparent; Figure 4.9). The Australian smelt has a widespread distribution in many inland rivers of the Murray-Darling Basin, except for those in the north-western region (Llewellyn 1983; McDowall 1996; Schiller *et al.* 1997). It is abundant and is regarded as a main channel generalist and a wetland specialist, since it tends to spawn and recruit in the main channel during both high- and low-flow stages, and utilises anabranches and floodplain wetlands (Young *et al.* 2003).

Reproduction and life-cycle

The Australian smelt mostly spawns in spring when temperatures reach 11–15°C, although in the Ovens River spawning was noted to occur from September to late February (King *et al.* 2003). A study from Brisbane suggested that the Australian smelt had a preference for littoral shallows of slow-flowing pools during spawning, sites where eggs can sink and bind to the base of aquatic plants, before the larvae drift at the surface for two to three days after hatching. This may be an adaptation for dispersal during periods of rising water levels and floods (Milton and Arthington 1985). However, it was also noted that spawning during periods of low flow would probably help to ensure recruitment of eggs during the period of greatest river stability and would therefore would allow more developed juveniles, rather than larvae, to disperse in rising floodwaters (Milton and Arthington 1985; Heagney *et al.* 2009). Fecundity is low to moderate,

Figure 4.9: Australian smelt, *Retropinna semoni*. Photograph: Gunther Schmida.

with 100–1000 eggs laid, depending on the size of the female (Lintermans 2007). Larvae of Australian smelt from Ovens River indicate widespread spawning and direct use of the floodplain via drift out of the main channel. The larvae were found in both non-flood and flood years, thus spawning and recruitment were not reliant on floodplain inundation. The Australian smelt can reside and recruit in main and secondary channels (King *et al.* 2003). Upstream movements of adults and juveniles have been recorded in the Murray and Murrumbidgee rivers (Lintermans 2007). The cues for spawning are uncertain under differing flow conditions, and the water-quality tolerances of eggs and larvae remain unknown (Treadwell and Hardwick 2003).

Habitat and trophic position

The Australian smelt inhabits most reaches of inland rivers and wetlands, although it is usually found in the highest concentrations in lakes and wetlands where slow-flowing or still waters are dominant. This species tend to shoal near the surface, and around aquatic plants and woody debris (Treadwell and Hardwick 2003). Little is known about the specific habitat requirements of the Australian smelt or the limiting physical or chemical factors, although its wide distribution and high abundance suggests that it must have wide physiological tolerances (Merrick and Schmida 1984). The Australian smelt consumes a wide range of small terrestrial and aquatic insects, as well as small crustaceans (Lintermans 2007).

Water requirements

The maximum temperature limit for the Australian smelt is around 28°C and it can tolerate salinities up to 59 000 mg L^{-1}, as well as fairly low dissolved oxygen concentrations. There are insufficient data to determine whether the Australian smelt adapts well to poor water-quality conditions (Treadwell and Hardwick 2003). King (2004) found a significant relationship between smelt larvae and backwater habitats in the Broken River during the day and night, and found that habitat preferences changed from backwaters to pools as larvae developed into juveniles/adults. It is thought that larvae prefer backwaters as they provide refuge from higher water velocities and predators, and have warmer temperatures and more food than other channel habitats (Humphries *et al.* 1999; King 2004). Australian smelt larvae were present in flood and non-flood years in the Ovens River, Victoria, indicating that spawning and recruitment are not reliant on floodplain inundation and that there is use of temporary floodplain habitats, such as billabongs and anabranches (King *et al.* 2003). Hence there is a moderate likelihood of floodplain use for spawning and fish recruitment. Key knowledge gaps include the habitat preferences, behaviour and water-quality tolerance of larvae, as well as the water-quality tolerance of adults (Treadwell and Hardwick 2003).

Main channel specialists

Murray cod: *Maccullochella peelii peelii*

The Murray cod is a large and long-lived fish, commonly 55–60 cm long and 2–10 kg, but it can reach a length of 1.8 m and a weight of 113 kg (Figure 4.10). It has a short rounded snout with a large mouth and a long deep body that is cream to olive to yellow green, with speckled grey green mottles. The Murray cod has a widespread distribution in the Murray-Darling Basin. Although it was formerly highly abundant in most inland rivers, populations have declined markedly since the early 20th century (Lintermans 2007). It is nationally listed as vulnerable (*Environment Protection and Biodiversity Conservation Act 1999*). The Murray cod is regarded

Figure 4.10: Murray cod, *Maccullochella peelii peelii*. Photograph: Gunther Schmida.

as a main channel specialist since it tends to spawn and recruit during high or low flows in the main channel. Woody debris is a main component of its habitat (Young *et al.* 2003).

Reproduction and life-cycle

The Murray cod spawns from spring to summer, usually between September and December, but it may have a protracted spawning period in colder regions. Migrations of adults have been recorded; movements are related to pre- and post-spawning requirements. The Murray cod requires a minimum water temperature of 15°C. Females can deposit 10 000–90 000 eggs into depressions in the bank or woody debris (Humphries 2005). Incubation time is temperature-dependent, but it usually takes six to seven days at 22–25°C and an additional 25 days for larvae to mature (Lake 1967a, 1967b). The young disperse upstream by swimming and downstream by drifting. Free embryos may drift for up to 10 days, but more usually for five to seven days (Humphries 2005). The Murray cod does not require floods to stimulate spawning, but may have enhanced recruitment during flood years because of an increase in available food (King *et al.* 2003). In some cases, it may spawn in response to a decrease in discharge (Humphries 2005). The direct development of Murray cod means that it is able to actively search for food while retaining some yolk reserves in the larval–juvenile stages (Humphries 2005). It takes advantage of raised water temperatures (above 20°C) during spring and summer months, which allow juveniles to develop during the period of higher invertebrate production (Humphries 2005).

Habitat and trophic position

The Murray cod is found in a range of habitats from clear headwater streams to turbid lowland rivers. It prefers reaches with slow-flowing waters and high levels of cover, such as rocks, fallen trees, stumps, undercut clay banks or overhanging vegetation, often near deep pools or waterholes (Treadwell and Hardwick 2003; Lintermans 2007). The adult Murray cod is territorial, known to return to certain parts of the river following upstream spawning migrations (Lintermans 2007). It is a carnivore and its diet consists mainly of crustaceans, fish and frogs (Lintermans 2007). Although the Murray cod has been extensively studied, knowledge gaps still exist regarding the preferred habitats of larvae and juveniles, as well as general water-quality tolerances (Treadwell and Hardwick 2003).

Water requirements

The Murray cod is known to tolerate water temperatures of 10–37°C and salinity up to 15 700 mg L^{-1}. There are insufficient data to determine whether it can tolerate poor water-quality conditions (e.g. low dissolved oxygen) (Treadwell and Hardwick 2003). There is a low likelihood of use of the floodplain for fish recruitment, although juveniles have been documented drifting in both the main channel and secondary anabranches (King *et al.* 2003). Murray cod larvae are rarely found in backwater habitats that are commonly used by the larvae of other species such as the Murray-Darling rainbowfish (*Melanotaenia fluviatilis*) and the Australian smelt (*Retropinna semoni*) (King 2004). The survival and recruitment of larvae and juvenile Murray cod were thought to increase following flows in the mid Murray River, as a consequence of increased food supplies following major flooding (King *et al.* 2009).

Trout cod: *Maccullochella macquariensis*

The trout cod is a large deep-bodied fish, similar in appearance to the Murray cod (*M. peelii peelii*) except that it has a protruding upper jaw and lacks distinctive markings on its head (Figure 4.11). The trout cod can reach a maximum size of 85 cm and 16 kg but is commonly 40–50 cm and less than 5 kg (Allen *et al.* 2003). The tail is rounded and the body pattern is usually speckled blue green with black to very dark green spots across the flanks. There is often a dark stripe through the eye (Lintermans 2007).

The trout cod is known to occupy deep sections of upper and mid reaches of streams and rivers. In recent decades it undergone serious declines in range and abundance throughout the

Figure 4.11: Trout cod, *Maccullochella macquariensis*. Photograph: Gunther Schmida.

Murray-Darling Basin (Cadwallader and Gooley 1984). The trout cod was formerly widespread in the southern Murray-Darling Basin, but only three self-sustaining populations remain in the wild. The largest is along a 200 km stretch of the Murray River between Yarrawonga and Barmah. Smaller populations were translocated to the Cataract Dam on the Nepean River and to a 15 km stretch of the upper Seven Creek near Euroa, a tributary of the Goulburn River in Victoria. A long-term stocking program has reintroduced the trout cod to many rivers in the southern Basin, but successful breeding has been recorded only in the Goulburn and Cotter rivers and the Cataract Dam (TCRT 2008). The trout cod is nationally endangered (*Environment Protection and Biodiversity Conservation Act 1999*) and a recovery plan is in place (TCRT 2008).

Reproduction and life-cycle

The trout cod was formally recognised as a separate species from the Murray cod (*M. peelii peelii*) in 1972, so much of its biology is poorly known. It is believed to reach sexual maturity at three to five years of age (0.7–1.5 kg) and spawn in spring (September to November) when water temperatures are 14–22°C (Ingram and Douglas 1995). Females are oviparous and produce 1200–11 000 large adhesive eggs (2.5–3.6 mm diameter), which attach to hard substrates such as boulders and logs. Larvae hatch after five to 10 days and are 6–9 mm in length (Ingram and Rimmer 1992). The trout cod can hybridise with the Murray cod (Douglas *et al.* 1995).

Habitat and trophic position

The trout cod is known to occupy both slow- and rapid-flowing streams, the mid to upper reaches of rivers and under or adjacent to logs and boulders (Nicol *et al.* 2007; TCRT 2008). It undertakes limited movements (~60 m) and is territorial around the 'home snag', occasionally undertaking exploratory movements of 1–3 km (Koehn *et al.* 2008). The trout cod is carnivorous; its diet includes fish, yabbies, aquatic insect larvae, freshwater shrimps and prawns (Baumgartner 2007). Larvae commence feeding on zooplankton about 10 days after hatching is completed (Ingram and Rimmer 1992).

Water requirements

As a result of a range of human threats, only approximately 12 viable trout cod populations remain (Ingram *et al.* 1990; Cadwallader 2006; TCRT 2008). The health of these populations is likely to be vulnerable in all life-cycle stages, from drifting larvae to large-bodied adults (Todd *et al.* 2003; Koehn and Harrington 2006). Spawning appears to be triggered by increasing day length and water temperatures during late September to late October (Ingram and Douglas 1995), and can occur in a range of flow conditions (Koehn and Harrington 2006). The trout cod may benefit from the maintenance and restoration of environmental flows (TCRT 2008). For example, the survival and recruitment of larvae and juvenile trout cod was thought to increase following flows in the mid Murray River, as a consequence of increased food supplies following major flooding (King *et al.* 2009). Environmental watering could be used to enhance habitat availability for adult fish, particularly snags in low-velocity channels (Nicol *et al.* 2007). Key knowledge gaps for this species include the water-quality tolerances of eggs, larvae and adults (Treadwell and Hardwick 2003).

Macquarie perch: *Macquaria australasica*

The Macquarie perch is a medium-sized, laterally compressed fish of blackish or silvery grey, although sometimes it is mottled (Figure 4.12). It has a large rounded tail and the fins are coloured the same as its body. It can be recognised by its concave head profile and large white eye. The Macquarie perch can reach 46 cm in length and 3.5 kg but is more commonly 20–25 cm and less than 1.5 kg (Allen *et al.* 2003).

Figure 4.12: Macquarie perch, *Macquaria australasica*. Photograph: Gunther Schmida.

The Macquarie perch inhabits clear cool waters of rivers, lakes and reservoirs, preferring slow-flowing and deep rocky pools (Allen *et al.* 2003). Two forms of Macquarie perch are recognised: the western rivers (Murray-Darling Basin) and the eastern (coastal) rivers (Shoalhaven and Hawkesbury-Nepean systems) forms (Lintermans 2007). They are yet to be described as separate species. The western inland population has suffered a severe contraction in its distribution in the southern Basin (Cadwallader 1981). It is now confined to middle to upper reaches of the Mitta Mitta River above Darmouth Dam, the Goulburn, Murrumbidgee and Lachlan river systems and a translocated population in the Yarra River (Ingram *et al.* 2000). The Macquarie perch is listed as threatened in all states and territories where it occurs and is endangered nationally (*Environment Protection and Biodiversity Conservation Act 1999*). It is difficult to breed in captivity and stocking programs have had limited success (Ingram *et al.* 2000).

Reproduction and life-cycle

Males can mature at two years (20 cm) and females at three years old (30 cm) (Appleford *et al.* 1998). They are thought to continue spawning until 10 years of age (Cadwallader and Rogan 1977). Upstream spawning migrations are triggered by increases in water temperature, usually in spring months (October–December), when water temperatures reach 16.5°C (Cadwallader and Rogan 1977). Females spawn (50 000–100 000 eggs) immediately upstream of a fast-flowing section of water, which allows fertilised eggs to drift downstream and adhere to gravel and boulders (Cadwallader and Rogan 1977). Hatching of larvae (4.5–6.5 mm) commences five to 11 days after fertilisation, at water temperatures of 18–20°C (Ingram *et al.* 2000).

Habitat and trophic position

The Macquarie perch can be found in riffle and boulder habitats in upland streams, in larger deeper rivers with sand and clay substrates and submerged timber, and in irrigation storages (Ingram *et al.* 2000). It is mainly a benthic feeder, ingesting food by a sucking action. Its diet largely comprises aquatic insects, particularly juvenile and adult stages of flies and mosquitoes

(Order Diptera), caddisflies (Order Trichoptera) and mayflies (Order Ephemeroptera) (Cadwallader and Eden 1979). Adult and sub-adult Macquarie perch generally have small home ranges during the day, being most active during crepuscular and nocturnal hours (Lintermans 2007).

Water requirements

The Macquarie perch is primarily found in slow-flowing pools with boulder substrates and under most conditions does not exhibit specialised flow requirements. However, reproductive success is dependent on springtime flows that provide access to spawning habitats, plus antecedent conditions that provide food resources suitable for pre-spawning gonad development (i.e. high in unsaturated fatty acids; Sheikh-Eldin *et al.* 1999; Gray *et al.* 2008). Typically, upstream movement of Macquarie perch aged five to 10 years and spawning is triggered when water temperatures reach a critical 16.5°C (Cadwallader and Rogan 1977). Adhesive demersal eggs are shed into the water above riffles, coming to rest in interstitial spaces of rocks, gravel and sand (Lintermans 2007). Although the adult Macquarie perch has a salinity tolerance of more than 30 000 mg L^{-1}, eggs do not survive exposure to salinity of 4000 mg L^{-1} (O'Brien and Ryan 1997, cited in Nielsen *et al.* 2003). Flow regulation, habitat modification and in-stream structures can interrupt spawning by altering spawning cues, blocking fish movements and causing sedimentation of spawning substrates.

Freshwater catfish: *Tandanus tandanus*

The freshwater catfish is a medium-sized fish, up to 90 cm long and 6.8 kg in weight, but usually 20–60 cm long and less than 2 kg (Figure 4.13). It has a large head with a flattened snout and a ventral mouth with thick fleshy lips and barbels. It has continuous rear dorsal fins (eel-tail) along a laterally compressed body that is olive green to brown. It does not have scales. Individuals less than 30 cm long are usually mottled. The freshwater catfish has a wide distribution in the inland rivers of the Murray-Darling Basin but populations have recently declined (Lintermans 2007) despite it being one of the best studied and understood Australian native

Figure 4.13: Freshwater catfish, *Tandanus tandanus*. Photograph: Gunther Schmida.

freshwater fishes (Davis 1977a, 1977b, 1977c, 1977d). It is listed as a protected species in South Australia and as a vulnerable species in Victoria (*Flora and Fauna Guarantee Act 1988*).

Reproduction and life-cycle

The freshwater catfish has a short spawning period from spring to summer, determined by an increase in water temperature to 20–24°C (Lintermans 2007). Spawning is not related to flooding, although the available photoperiod is important. Successful spawning has been associated with gonad maturation due to a temperature rise to 24°C, rather than to secondary effects of temperature on food availability (Davis 1977d). Spawning involves direct paring; 2000–26 000 eggs are deposited in nests usually composed of sand or gravel, although the freshwater catfish also breeds in areas with muddy substrates. There is little evidence of pre- or post-spawning migration in the freshwater catfish, with the adult making limited and somewhat random movements in its local environment (Treadwell and Hardwick 2003). This suggests that local perturbations of habitat may pose the most significant risk to freshwater catfish populations.

Habitat and trophic position

The freshwater catfish is usually found near the bottoms of lakes and slow-flowing rivers, although it can spawn in channels with a sandy bed during most flow conditions (Treadwell and Hardwick 2003). Inundated secondary channels provide extra habitat for the freshwater catfish during floods, and the connection of the channels and floodplain provides increased physical habitat complexity (Balcombe *et al.* 2006). Undercut root trusses, fallen logs and smaller woody material provide greater habitat heterogeneity in waterholes and are important in terms of greater protection from predators and enhanced food supplies. The diet of the freshwater catfish consists of benthic molluscs, small crustaceans, insects, snails and small fish. The juvenile catfish relies on small fish (e.g. gudgeons) for prey to a much greater extent than the adult, which prefers shrimp and crayfish (Davis 1977b). Terrestrial invertebrate species (e.g. Coleoptera, Isoptera) have been documented in the diet of the freshwater catfish after flooding, presumably due to their increased availability following disturbance and removal from the floodplain following inundation (Davis 1977b).

Water requirements

Large deep waterholes provide important refuges for the adult freshwater catfish in dryland rivers (e.g. the Warrego River), which are important during extended periods of drought since small populations of adult fish can provide a good basis for juvenile recruitment when large waterholes join smaller dry waterholes after floods (Balcombe *et al.* 2006). The freshwater catfish does not tolerate water temperatures of less than 4°C well, but has a relatively high thermal tolerance of up to 38°C (for several days at a time). It can withstand salinity up to 17 700 mg L^{-1} (Treadwell and Hardwick 2003). Knowledge gaps include the water-quality tolerances of larvae, juveniles and adults, and the movement patterns and habitat preferences of larvae (Treadwell and Hardwick 2003).

Flood spawners

Golden perch: *Macquaria ambigua*

The golden perch is a moderate to large fish, commonly 40–50 cm long and 5 kg, but it can be up to 76 cm and 23 kg (Figure 4.14). It has a deep narrow head and tapered snout, and a body of

Figure 4.14: Golden perch, *Macquaria ambigua*. Photograph: Gunther Schmida.

olive green to bronze yellow (coloration depends on water colour). The golden perch is a long-lived species and it has a very widespread distribution in the Murray-Darling Basin (Lintermans 2007). It was once abundant but is now listed as vulnerable in Victoria (*Flora and Fauna Guarantee Act 1988*). The golden perch is regarded as a flood spawner since it tends to spawn and recruit following flow rises, and major spawning occurs when floodplains become inundated (Young *et al.* 2003).

Reproduction and life-cycle

The golden perch generally spawns during floods over an extended period from spring to summer and when water temperatures exceed 20°C (Lintermans 2007). The spawning period is flexible, responding to rises in water temperature, rising water levels and photoperiod (Treadwell and Hardwick 2003). Long-distance upstream migrations of adult golden perch have been recorded in some rivers, primarily associated with the spawning season, although some individuals have been seen to migrate downstream to spawn (Reynolds 1983). Juvenile golden perch also migrate frequently, often making up the bulk of the moving population (Treadwell and Hardwick 2003). Otherwise the post-spawning behaviour of the golden perch involves dispersal downstream during low flows. The adult tends to occupy a home range of around 100 m for weeks or months before relocating to a new site, where a new home range is established (Lintermans 2007). Fecundity is high and a female of around 2.5 kg can produce up to 500 000 eggs, which will drift downstream and disperse through main channel habitats (Lintermans 2007). The golden perch requires 33–34 hours to hatch eggs (at 20–25°C). Young are 18–20 days old at the end of larval development (Lake 1967a, 1967b).

Habitat and trophic position

The golden perch prefers warm, turbid and slow-flowing waters in lowland rivers, including backwaters, billabongs and anabranches, and is fairly well adapted to variable flow conditions (Treadwell and Hardwick 2003). Large deep waterholes in dryland rivers provide important refuges for adult golden perch during dry periods (Balcombe *et al.* 2006); this helps golden perch populations, since small numbers of adult fish can provide good sources for juvenile recruitment when large waterholes join with smaller dry waterholes after flooding. Undercut root trusses, fallen logs and smaller woody material also provide greater habitat heterogeneity and are important habitat for the golden perch. Nursery habitats for the golden perch are known to occur in the main channel (Mallen-Cooper and Stuart 2003). The golden perch is a carnivore, usually feeding during the day and night on crustaceans, aquatic insects and molluscs. Small fishes may also form a part of its diet (Lintermans 2007). Terrestrial and aquatic insects make up a large part of the diet of adult fish after floodplain inundation (Ye 2004).

Water requirements

Recruitment patterns of golden perch have been associated with high-flow events and backwater inundation in the Warrego River but recruitment was also evident following a period of zero flow, suggesting that departures from typical flood-induced recruitment in the golden perch may reflect opportunistic spawning related to intermittent flooding and dry periods (Balcombe *et al.* 2006). Recruitment periods also occurred three to nine months after significant summer floods in the Warrego River, coinciding with periods of higher channel–floodplain connectivity (Balcombe *et al.* 2006). However, in a study from the Ovens River, floods and spawning of golden perch did not overlap because necessary rises in water temperature did not coincide with the floods, suggesting a reliance on optimal environmental conditions rather than the timing of flooding alone (King *et al.* 2003). In the Murray River, strong golden perch recruitment was associated with fluctuating in-channel flows. Weak recruitment was associated with overbank flooding and floodplain inundation, although these results may have been skewed by the study sampling regime and the fish probably spawned during both periods (Mallen-Cooper and Stuart 2003). The golden perch is known to delay spawning until there are appropriate environmental conditions (Humphries and Lake 2000). The most recent evidence suggests that major flooding enhances spawning activity (King *et al.* 2009).

The golden perch is known to tolerate temperatures of 4–37°C and salinities up to 33 000 mg L^{-1}, but it does not tolerate poor water-quality conditions (e.g. low dissolved oxygen) (King *et al.* 2003). Larvae have been documented actively avoiding habitats with poor water quality (Gehrke 1991). Larvae tend to avoid areas of the floodplain and ponds with low dissolved oxygen concentrations. Overall, water quality appears to have a greater effect on the distribution of larvae than does food availability in some inundated floodplain environments (Gehrke 1991). Knowledge gaps include the habitat preferences of larvae and the specifics of adult migration patterns (Treadwell and Hardwick 2003).

Silver perch: *Bidyanus bidyanus*

The silver perch is a moderate to large fish, commonly 30–40 cm long and weighing 0.75–1.5 kg, but it can weigh 8 kg (Figure 4.15). It has a small head, a beak-like snout and is silver grey in colour, which can vary with the level of water turbidity. The silver perch has a widespread but declining distribution in rivers of the Murray-Darling Basin (Lintermans 2007). The species is listed as vulnerable in New South Wales (*Fisheries Management Act 1994*), endangered in the ACT (*Nature Conservation Act 1980*) and critically endangered in Victoria (*Flora and Fauna Guarantee Act 1988*), and is a protected species in South Australia (*Fisheries Act 1982*). The silver perch is typically regarded as a flood spawner since it tends to spawn and recruit following rises in flow; major spawning occurs when floodplains become inundated (Young *et al.* 2003).

Figure 4.15: Silver perch, *Bidyanus bidyanus*. Photograph: Gunther Schmida.

Reproduction and life-cycle

Spawning occurs from November to January, after the silver perch migrates upstream to areas behind the peak of a flood, or coinciding with a slight rise in water levels and water temperatures of 23–30°C (Allen *et al.* 2003; Lintermans 2007). Pre-spawning activity occurs at the surface. Spawning probably occurs from the late afternoon to the early evening (Lintermans 2007), usually where there is water flowing over a gravel or similar coarse substrate. The silver perch has been reported to spawn in backwaters of low-gradient channels (Treadwell and Hardwick 2003). Fecundity is high, with up to 300 000 eggs from an individual female. Dams and weirs pose barriers to migration into spawning areas; such barriers, plus alterations in the hydrological and thermal regimes, pose significant threats to the success of spawning and the recruitment of juveniles into existing populations (Treadwell and Hardwick 2003; Lintermans 2007).

Habitat and trophic position

The silver perch occurs in a variety of water conditions from turbid to standing and slow-flowing waters but it tends to school in faster-flowing waters, such as in and around small rapids or weirs. It is found in the main channel and large anabranches of lowland rivers, and is occasionally found in small anabranches and floodplain lakes during major floods (Treadwell and Hardwick 2003). The silver perch seems to prefer waters with nearby cover from littoral vegetation and large woody debris. It has a varied diet which consists of small aquatic insects, molluscs, crustaceans and aquatic plant material (Lintermans 2007). In regulated systems, there may be low likelihood of floodplain use for recruitment, as observed in the Ovens River, Victoria (King *et al.* 2003). Nursery habitats for silver perch can occur in the main channel. River regulation and the maintenance of low flows is thought to have inhibited cues for spawning and migration in the silver perch (Mallen-Cooper and Stuart 2003).

Water requirements

The silver perch has a wide thermal tolerance, surviving temperatures of 4–37°C, but it generally does not adapt well to poor water-quality conditions (e.g. high salinity and low dissolved oxygen concentrations) (King *et al.* 2003). Its spawning period can be flexible and is dependent on rising water temperatures and water levels. Larvae and juveniles probably drift from the main channel onto the inundated floodplain during floods (King *et al.* 2003). In the Murray River, the silver perch has recruited following overbank flows in some years and following rising water levels in years when floods did not exceed the channel margins (Mallen-Cooper and Stuart 2003). Hence overbank flooding is not a requirement for spawning in silver perch, since rises in water level within the main channel lead to successful spawning and recruitment. In a study from the Ovens River, floods and spawning did not seem to overlap because necessary rises in water temperature did not coincide with the floods, suggesting a reliance on optimal environmental conditions rather than flood timing alone (King *et al.* 2003). The most recent evidence suggests that major flooding enhances spawning activity in the silver perch (King *et al.* 2009). The silver perch is known to delay spawning until there are appropriate environmental conditions (Humphries and Lake 2000). Knowledge gaps include the water-quality tolerances of larvae, juvenile and adult silver perch, as well as the patterns of pre- and post-spawning migrations and behaviour (Treadwell and Hardwick 2003).

Spangled perch: *Leiopotherapon unicolor*

The spangled perch is a small to moderate-sized fish, up to 33 cm long but commonly 15 cm long and 0.5 kg (Figure 4.16). It has a rounded snout and a slender silver grey body with brownish red spots or mottles. The spangled perch has a fairly widespread distribution in the inland rivers of the Murray-Darling Basin, and is particularly common in the northern and western regions (Lintermans 2007). It is fairly abundant and is regarded as a flood spawner. Although the species can spawn and recruit in the main channel during both high- and low-flow stages, its recruitment is maximised following flooding and it uses anabranches and wetlands during flood periods (Young *et al.* 2003).

Reproduction and life-cycle

The spangled perch usually spawns in summer, from November to February, when water temperatures reach 20–22°C (Lintermans 2007). Spawning typically occurs at night in shallow backwaters or still pools where eggs can be spread over generally soft substrates. Fecundity is

Figure 4.16: Spangled perch, *Leiopotherapon unicolor*. Photograph: Gunther Schmida.

high, with 24 000–113 000 eggs per female depending on size (Lintermans 2007). Spangled perch eggs take two days to hatch at temperatures of 23–26°C, and a further 24 days to reach a post-larval stage (Llewellyn 1973). A rise in water level is not essential but can maximise recruitment. Upstream migrations of the spangled perch are associated with spawning. Downstream movements have also been documented during dispersal, as have lateral movements into flooded areas (Lintermans 2007). Juvenile spangled perch tend to school in large numbers. Key knowledge gaps include the water-quality tolerances and requirements for spawning and egg survival (Treadwell and Hardwick 2003).

Habitat and trophic position

The spangled perch is well adapted to living in diverse environments such as rivers, billabongs and waterholes in intermittent streams, and it has been observed swimming across inundated fields and floodplains (Lintermans 2007). Its diet mainly consists of small aquatic insects, crustaceans, molluscs, some small fishes and plant material (Lintermans 2007). Key knowledge gaps include the movement, habitat preference and behaviour of larvae, and habitat use by juveniles and adults (Treadwell and Hardwick 2003).

Water requirements

The spangled perch can tolerate temperatures of 5–44°C, salinity up to 35 000 mg L^{-1}, pH of 4–8.6 and fairly low dissolved oxygen concentrations (<5% saturation) (Treadwell and Hardwick 2003). However, the spangled perch cannot survive very low water temperatures in winter in the southern regions of the Murray-Darling and juveniles are more prone to mortality at environmental extremes (Gehrke *et al.* 1995). In the arid-zone river system of Cooper Creek, the spangled perch had an increased abundance in some waterholes with large total and/or effective floodplain widths (Arthington *et al.* 2005). Connection of the channel and floodplain provides important nursery areas for larval and juvenile fishes and allows older fish to disperse, feed and accumulate fat reserves to allow survival during the dry season and to complete reproduction (Humphries *et al.* 1999; King *et al.* 2003; Arthington *et al.* 2005). The spangled perch can endure dry periods in isolated waterbodies in a state of semi-aestivation by sheltering in leaf litter and mud at the base of ponds (Merrick and Schmida 1984), although laboratory tests of aestivation were inconclusive (Llewellyn 1973).

Alien species

Goldfish: *Carassius auratus*

The goldfish is an alien species similar in appearance to the European carp (*Cyprinus carpio*), but it can usually be distinguished by the lack of barbels around the mouth (Figure 4.17). Like the carp, it has a forked tail and spines at the front of the dorsal fin. It ranges in colour from olive bronze to golden (occasionally orange red), with a dark dorsal surface and silvery white underbelly. The mouth is small and does not reach back below the eye. It can reach 40 cm but is more commonly less than 20 cm in Australia (Allen *et al.* 2003).

The goldfish can be found throughout the Murray-Darling Basin but is more common in the still and slow-moving waters of lowland rivers, lakes and floodplain wetlands. Native to eastern Asia, it was introduced to Australia in the mid 1800s as an ornamental fish (Allen *et al.* 2003). It and 10 other alien fish species are recognised as key threats in New South Wales (*Fisheries Management Act 1994*). Although its impacts on Australian native fish are poorly understood, the goldfish is known to prey on the eggs, larvae and adults of native fishes in the US (Deacon *et al.* 1964) and it probably competes with native fish for food and habitat. The goldfish

Figure 4.17: Goldfish, *Carassius auratus*. Photograph: Gunther Schmida.

is often infested with parasitic copepod anchorworm (*Lernaea* sp.) and it carries the 'goldfish ulcer' disease, which can infect trout (Lintermans 2007). It has also been linked to algal blooms, as significant cyanobacteria (blue green algae) growth is stimulated after passage through goldfish intestines (Kolmakov and Gladyshev 2003).

Reproduction and life-cycle

The goldfish usually matures at 10–15 cm but can mature earlier, at 3–5 cm. It spawns in summer months when water temperatures reach 17–23°C (Lintermans 2007) but needs cold water in winter months to stimulate ova development (Froese and Pauly 2009). The goldfish is a batch spawner, laying adhesive eggs in aquatic plants at intervals of eight to 10 days (Froese and Pauly 2009). Eggs hatch within seven days (Lintermans 2007). The goldfish is known to hybridise with the carp (*Cyprinus carpio*) (Taylor and Mahon 1977).

Habitat and trophic position

This species prefers still and slow-moving waters, and is often associated with submerged or emergent aquatic plants or snags and shady areas (Morgan and Beatty 2005; Lintermans 2007). It is a benthic omnivore, feeding on organic detritus (mostly cyanobacteria), algae, plant matter, aquatic invertebrates (including nematodes, anisopteran larvae, coleopteran larvae and dipteran larvae), terrestrial insects and gambusia (*Gambusia holbrooki*) (Morgan and Beatty 2005). Its feeding behaviour is thought to increase turbidity and inhibit the growth of aquatic vegetation (Richardson *et al.* 1995). It is not known to migrate (Lintermans 2007).

Water requirements

Like many alien species, the goldfish has been able to expand its range in the Murray-Darling Basin following river regulation (Gehrke and Harris 2001). It has wide environmental tolerances and can withstand salinities up to 17 000 mg L^{-1} for limited periods (Froese and Pauly 2009). The goldfish moves into wetlands upon flooding and, with other alien fish species, can benefit from small environmental flows that target specific wetlands (e.g. Barmah Forest; King *et al.* 2007). Watering of wetlands during summer months would favour this species, as juvenile goldfish require high water temperatures for growth (Froese and Pauly 2009). Alien fish control programs usually focus on the carp, but a goldfish eradication program was set up in the Vasse River, Western Australia, because of concerns that the recently introduced goldfish might feed on fish eggs, reduce macrophytes, increase turbidity and stimulate blue green algal blooms (Morgan and Beatty 2005).

European carp: *Cyprinus carpio*

The European carp is a medium-sized alien species with a forked tail and large rounded scales (Figure 4.18). It is usually olive grey or brownish, sometimes silvery grey with a lighter underbelly. Some individuals are not fully scaled, and can have very large scales in three or four rows (mirrored form) or lack scales altogether (leathered form). The European carp has a sloping head profile with a low-set mouth and thick fleshy lips with two pairs of distinctive barbels in the corners of the mouth. Its sloping dorsal fin has a thick serrated spine on the front edge. The carp can reach a maximum size of 120 cm and 60 kg, but in the Murray-Darling Basin it is more commonly 30–40 cm and 4–5 kg (Allen *et al.* 2003; Lintermans 2007).

The European carp is the most widespread large-bodied fish in the Murray-Darling Basin and can congregate in large numbers in floodplain wetlands (Stuart and Jones 2006). It is native to central Asia and was introduced into Australia in the mid 1800s (Lintermans 2007). Although the causal links are still unclear, the carp is thought to have affected native fish species through reductions in aquatic macrophytes and increases in turbidity and channel erosion (Fletcher *et al.* 1985; King *et al.* 1997; Robertson *et al.* 1997). These impacts are exacerbated in highly regulated river systems (Gehrke and Harris 2000). The European carp carries the parasitic copepod anchorworm (*Lernaea* sp.), which has been observed on several native

Figure 4.18: European carp, *Cyprinus carpio*. Photograph: Gunther Schmida.

fish species including the Murray cod, golden perch, silver perch, Macquarie perch and freshwater catfish (Lintermans 2007).

The carp has been recognised as a significant problem in the Murray-Darling Basin and several initiatives and control programs have been set up. The Murray-Darling Basin Commission (now the Murray-Darling Basin Authority, MDBA) established the Carp Control Coordinating Group in 1999 following an agreement between state, territory and federal government agencies and it released the National Management Strategy for Carp Control in 2000 (MDBC 2000). The New South Wales government's Aquatic Biosecurity Unit has finalised a draft control plan for carp in New South Wales and the MDBA is developing a Basin Alien Fish Plan. The European carp, with 10 other alien fish known from New South Wales waterways, are recognised as a key threat in the *Fisheries Management Act 1994* (NSW).

Reproduction and life-cycle

The European carp is highly fecund and a fast breeder. It can reach very high densities in artificially watered wetlands such as Barmah-Millewa Forest (Stuart and Jones 2002, 2006). Males mature at two to three years old (30 cm) and females mature at three to four years (35 cm). In favourable conditions, each female carp can produce 80 000–1 000 000 eggs annually (Brown 1996). The abundance of common carp larvae and young-of-the-year can increase rapidly in wetlands after flooding (Stuart and Jones 2006). Spawning usually occurs in spring and summer months (October to December) when water temperatures reach 15–25°C (Stuart and Jones 2002). Eggs (0.5 mm diameter) are adhesive, attaching to logs and vegetation, and the larvae hatch in two to six days depending on water temperature (Lintermans 2007).

Habitat and trophic position

The European carp is commonly found in lowland rivers in warm, slow-moving or still waters with aquatic vegetation. Movement patterns of carp are complex, with some adult fish undertaking large upstream and downstream migrations (>200 km) and others occupying small home ranges (20 m). Most (94%) will move into adjacent floodplain habitats upon flooding (Jones and Stuart 2009). The carp is a benthic omnivore, consuming a range of prey including zooplankton, molluscs, crustaceans, freshwater insect larvae, aquatic plants and detritus (Khan 2003).

Water requirements

The carp has wide habitat tolerances and can withstand very low levels of dissolved oxygen (Lintermans 2007). It exhibits strong recruitment responses to flow events in regulated and unregulated systems. For example, King *et al.* (2003) documented an increased abundance of larval carp on the Ovens River floodplain during a natural spring–summer flood, and Rayner *et al.* (2009) reported a boom in juvenile carp following a 23 000 ML environmental flow in the Macquarie Marshes. Although main channel habitats are adequate for carp reproduction, spawning is usually concentrated in lowland floodplain habitats where warm shallow well-vegetated microhabitats are preferred for recruitment. For example, less than 1% of carp larvae in the Barmah-Millewa Forest floodplain resulted from the Murray River upstream (Stuart and Jones 2006), whereas over 98% of larvae found downstream in Torrumbarry Weir originated in floodplain habitats of the forest (Crook and Gillanders 2006). There is a real risk that small environmental flows delivered to floodplain wetlands may favour European carp over native species (Rayner *et al.* 2009). However, there are many options for the control and exclusion of the carp (fishing down, automated traps, screens and wetland drawdown) that can be applied for local populations (Roberts and Tilzey 1996).

Redfin perch: *Perca fluviatilis*

The redfin perch is a medium-sized alien fish with a relatively large head (Figure 4.19). Its body is elongated and olive grey, usually with six distinct black bands that pale to a white underbelly. There are two separate dorsal fins, and the pelvic and anal fins are bright red to orange red. The tail is slightly forked with an orange edge (Allen *et al.* 2003). The redfin perch can reach a maximum length of 60 cm (10 kg), but is more commonly 40 cm and 1–2 kg (Lintermans 2007).

The redfin perch is native to Eurasia and was introduced as an angling species to Tasmania and Victoria in the 1860s (Allen *et al.* 2003). It is now widespread in the mid and lower reaches of rivers in the southern and middle Murray-Darling Basin, but is absent from colder headwaters or warmer rivers in the north-western Basin (Lintermans 2007). The redfin perch, in addition to the other alien species described in this chapter, is recognised as a key threat in the *Fisheries Management Act 1994* (NSW). It is a voracious predator of juvenile native fish and competes with adult native fish, such as the Macquarie perch and trout cod, for food. The redfin perch is a carrier of the epizootic haematopoietic necrosis virus, which is thought to have contributed to declines in the Macquarie perch, silver perch and mountain galaxia (Langdon 1989).

Reproduction and life-cycle

The redfin perch generally matures after two to three years (15.9–22.8 cm), but males can mature at the end of their first year (10.2 cm) (Morgan *et al.* 2002). At high population densities the redfin perch will stunt its growth and mature at smaller sizes (Lintermans 2007). Gonads develop in late summer, with peak spawning in August and September when water temperatures are 11–15°C (Morgan *et al.* 2002). The redfin perch can lay thousands of eggs (2–3 mm) in gelatinous ribbons among sunken logs and aquatic vegetation (Lintermans 2007). The egg masses are unpalatable to most fish and larvae hatch within one to two weeks, forming large schools (DPI 2005).

Figure 4.19: Redfin perch, *Perca fluviatilis*. Photograph: Gunther Schmida.

Habitat and trophic position

The redfin perch prefers cool slow-flowing waters with aquatic vegetation. It is a pelagic carnivore. The juvenile fish mainly consumes small aquatic invertebrates (dragonfly nymphs); the adult targets decapods (shrimps, yabbies and other freshwater crayfish) and small fish including the western carp gudgeon, flathead gudgeon, galaxiid, gambusia and juvenile redfin perch (Lintermans 2007; Wilson *et al.* 2008). It is known to prey on trout stocks (Baxter *et al.* 1985). The redfin perch can be a solitary hunter or work in groups to herd fish or flush insects and small fish from aquatic vegetation (DPI 2005). It has been observed moving downstream with large floodwaters, and colonising new areas (Wilson *et al.* 2008).

Water requirements

The redfin perch inhabits cool slow-flowing habitats and spawns in spring when temperatures are relatively mild. However, it has broad environmental tolerances (Morgan *et al.* 2002) and can tolerate brackish water environments (Froese and Pauly 2009) and survive high water temperatures (30–31°C) for brief periods (Weatherley and Lake 1967). It is highly invasive, given appropriate conditions (Hutchison and Armstong 1993), and may exploit habitats affected by cold water pollution, where lower temperatures may limit viral infection rates (Whittington and Reddacliff 2008). At higher water temperatures, final maturation of the gonad prior to spawning is compromised, resulting in high egg mortality after fertilisation (Sanstro *et al.* 2005). Environmental water allocations could be managed to deliver water at natural ambient temperatures, to suppress reproductive success of redfin perch. Additional research is required into the effects of such strategies on native fish, such as the Macquarie perch, which are susceptible to viral infections in target river systems.

Gambusia: *Gambusia holbrooki*

The gambusia is a small-bodied alien fish featuring a rounded tail, distinctly flattened head and upturned mouth (Figure 4.20). The species exhibits strong sexual size-dimorphism; females can attain 6 cm, whereas males usually grow to about 3.5 cm (Allen *et al.* 2003). The sexes have similar coloration, being drab, pale olive greenish with a silvery blue sheen and black spots on their sides and tail (Allen *et al.* 2003).

Figure 4.20: Gambusia, *Gambusia holbrooki*. Photograph: Gunther Schmida.

The gambusia was introduced to Australia in 1925, from rivers draining into the Gulf of Mexico. The species was actively spread for control of mosquito larvae and is present in every major catchment; it is one of the most widespread and abundant species in the Murray-Darling Basin (Lintermans 2007). The gambusia is listed as noxious in New South Wales (Class 1 outside the greater Sydney area, Class 3 within Sydney) and as a key threat in the *Threatened Species Conservation Act 1995* (NSW) and *Fisheries Management Act 1994* (NSW).

Reproduction and life-cycle

Like all members of the family Poecillidae, the gambusia is a viviparous livebearer. The male anal fin is modified to form a gonopodium (penis-like structure) which is used for internal fertilisation of females. Broods average ~30 embryos, with short gestation periods (Perez-Bote and Lopez 2005) and multiple broods per year (Milton and Arthington 1983). While reproduction can occur year-round, recruitment is likely to be more successful during warmer months, particularly in temperate areas, when food availability and water temperatures are high (Rowley *et al.* 2005). High salinity has complex impacts on the breeding dynamics of the gambusia, but leads to an overall reduction in reproductive success (Alcaraz and Garcia-Berthou 2007).

Habitat and trophic position

The gambusia is found predominantly in warm shallow habitats of still and slow-flowing waterbodies (McDowall 1996). It is particularly abundant in backwater and littoral habitats with low turbidity and abundant vegetative cover (i.e. macrophytes or dense filamentous algae). The species does well in anthropogenically disturbed habitats such as urban lakes, but is sensitive to droughts (Rayner *et al.* 2009) and floods (Chapman and Warburton 2006). It is an aggressive carnivore which swims at or near the surface consuming mainly planktonic and surface invertebrates, including copepods, cladocerans, rotifers, ostracods, chironomids and other dipterans (Arthington and Marshall 1999; Mieiro *et al.* 2001). The gambusia has been shown to have negative effects on native wildlife through predation on tadpoles, eggs and fish fry (NPWS 2003). It is not known to migrate.

Water requirements

The gambusia is tolerant of high salinity (NPWS 2003), water temperatures of 0.5–38°C (Swanson and Cech 1996; Clarke *et al.* 2000) and low dissolved oxygen concentrations (Lloyd 1984). The r-selected life-history (ability to mature and reproduce quickly) of gambusia is an important factor in its success in the Murray-Darling Basin. Fast maturation, large brood size and large size of fry at birth mean that the gambusia can quickly colonise and dominate newly inundated habitats, particularly shallow floodplain wetlands. Rayner *et al.* (2009) observed strong recruitment and range expansion of gambusia in the Macquarie Marshes following a 23 000 ML managed flow event. Before the flow event the species comprised <2% of the total catch and was found at <5% of study sites; following the flow it comprised 37% of the total catch and was found at >60% of study sites. Flows which improve hydrological connectivity and increase the availability of shallow off-channel habitats with dense aquatic vegetation are likely to benefit the gambusia, particularly if such flows occur during summer months (Rayner *et al.* 2009).

Summary of water requirements

Native fish species have specific water requirements that determine their distribution and recruitment success. The native fish profiled in this chapter were split into four functional groups: low-flow and wetland opportunists, main channel generalists and wetland opportunists, main channel specialists and flood spawners (Tables 4.1–4.4). Although their optimal flooding and flow conditions are not well understood, as a general rule the timing and duration of flows can influence the length of the breeding season for fish species, flow magnitude can determine mobility of species and flow frequency can influence species richness (Puckridge *et al.* 1998; Balcombe *et al.* 2006).

Seasonality of flow is an important cue for spawning in native fish. In the presence of suitable habitat, cues for spawning in native fish are often seasonal increases in water temperature, day length and flow. In unregulated systems this normally coincides with spring and summer months, and with abundant food supplies. Therefore, most species will spawn during floods if the timing is right but many species are also able to recruit during periods of low flow (Humphries *et al.* 1999; Balcombe *et al.* 2006).

Mobility is important for the dispersal of juvenile fish and spawning migrations of some species, such as the golden perch, which can migrate large distances (>1000 km) upstream (Reynolds 1983; Humphries *et al.* 1999). However, in-channel barriers, including dams and weirs, can cause high mortality in larval fish (Baumgartner *et al.* 2006) and prevent the passage of adult fish (Baumgartner 2007).

Large-scale upstream migrations and spawning of flood spawners, such as golden and silver perch, are thought to be triggered by a rise in water level at the onset of major flooding (Reynolds 1983; King *et al.* 2009). These species have buoyant eggs that are dispersed downstream with floodwaters (Reynolds 1983; Humphries *et al.* 1999). In contrast, wetland opportunists such as gudgeons and hardyheads are relatively sedentary and have small home ranges (1–2 km) (Young 2001; Lintermans 2007).

For species that are more dependent on water temperature as a cue for spawning, such as Australian smelt (Milton and Arthington 1985) and freshwater catfish (Davis 1977d), river regulation may have a greater impact on the post-spawning recruitment of larval fish than on spawning itself (Humphries and Lake 2000). River regulation can alter the timing of flows; therefore, large flows that are released after spawning can displace eggs and damage or reduce the growth of larval fish in the main channel (Humphries *et al.* 1999; Heagney *et al.* 2009).

Fecundity varies between species, but is particularly high in bony bream and spangled perch and the alien species gambusia, that all mature at an early age (Puckridge and Walker 1990; Lintermans 2007). The gambusia, for example, can produce 50 live young and up to nine batches per year (Lintermans 2007). The freshwater catfish, golden perch and Murray cod have relatively low population resilience with minimum population doubling times of 4.5–14 years (Froese and Pauly 2009). Age at maturity for these species are three to five, two to four and four to six years respectively. In contrast, the gambusia reaches maturity within two months and the spangled perch within a year (Allen *et al.* 2003; Lintermans 2007; Froese and Pauly 2009). Although the optimal duration for flooding is not well understood for most species, it should be long enough to cover the period from ovulation to metamorphosis for multiple species (Gehrke 1991, cited in Siebentritt 1999). As a result, declines in flooding duration in the Murray-Darling Basin due to water extraction and diversion may have more heavily affected native fish species with low fecundity.

Fish recruitment (the survival of young fish into the adult population) is determined by food and habitat availability as well as by predation. Most native fish species are carnivorous to some degree. Only the bony bream is a true detritivore, as it consumes mostly benthic algae (Bunn *et al.* 2003). Adult freshwater catfish, golden perch and Murray cod are the largest predators preying on freshwater shrimps, prawns, yabbies and smaller fish (Davis 1977b; Lintermans 2007). Larvae of all native fish are carnivorous and consume zooplankton and aquatic insect larvae (Young 2001; Lintermans 2007). Access to the floodplain can increase the range

Table 4.1: Habitat and water requirements of six native fish species that are generally low-flow/wetland opportunists in the Murray-Darling Basin

Common name	Scientific name	Preferred habitat	Trophic position	Movements	Primary spawning cue	Peak spawning season	Fecundity	Response to flows
Olive perchlet	*Ambassis agassizii*	Slow or still waters, associated with aquatic vegetation	Carnivore	Not known	Water temperature	Oct–Dec	Low	Distribution of eggs and/or larvae downstream
Un-specked hardyhead	*Craterocephalus stercusmuscarum fulvus*	Slow or still waters, associated with aquatic vegetation	Carnivore	Not known; some upstream movements recorded	Water temperature	Oct–Feb	Low	Not known; may use floodplain wetlands opportunistically
Murray hardyhead (V)	*Craterocephalus fluviatilis*	Slow or still waters, associated with aquatic vegetation	Omnivore	Not known	Water temperature and day length	Sept–Feb	Low to moderate	Not known; wetlands may provide spawning habitat
Carp gudgeons	*Hypseleotris* spp.	Slow or still waters, associated with aquatic vegetation	Mid-water/ benthic carnivores	Not known; some movements through fishways	Water temperature	Oct–Apr	Moderate	Not known; but flooding can increase food availability
Southern purple-spotted gudgeon	*Mogurnda adspersa*	Slow or still deep waters, associated with aquatic vegetation	Carnivore	Not known	Water temperature	Dec–Feb	Low to moderate	Not known; but flows important for maintaining vegetation for spawning habitat
Murray-Darling rainbow-fish	*Melanotaenia fluviatilis*	Slow or still waters, associated with aquatic vegetation	Pelagic omnivore	Not known; some movements through fishways	Water temperature	Nov–Feb	Low	Not known

Species are grouped in this flow/habitat group based on the MFAT classification (Young *et al.* 2003).
V = vulnerable, E = endangered (*Environment Protection and Biodiversity Conservation Act 1999*).

Table 4.2: Habitat and water requirements of four native fish species that are generally main channel generalist/wetland opportunists in the Murray-Darling Basin

Common name	Scientific name	Preferred habitat	Trophic position	Movements	Primary spawning cue	Peak spawning season	Fecundity	Response to flows
Flat-headed gudgeon and dwarf flat-headed gudgeon	*Philypnodon grandiceps* and *Philypnodon macrostomus*	Slow or still waters, associated with aquatic vegetation	Carnivores	Not known	Water temperature and maybe rising flows	Oct–Apr	Moderate	Flows may initiate spawning; flows important for downstream drift of larvae; use floodplain wetlands opportunistically
Bony bream	*Nematalosa erebi*	Slow or still shallow turbid waters, with aquatic vegetation	Algal detritivore	Upstream movements of juveniles and adults have been recorded	Water temperature and spawning enhanced during high flows	Oct–Feb	High	Rising flows enhance spawning; floodplain habitats important for juveniles
Australian smelt	*Retropinna semoni*	Slow or still waters, associated with aquatic vegetation and woody debris	Pelagic carnivore	Some upstream movements (not associated with spawning)	Water temperature	Sept–Feb	Low to moderate	Opportunistic use of floodplain wetlands by larvae and adults

Species are grouped in this flow/habitat group based on the MFAT classification (Young *et al.* 2003).

Table 4.3: Habitat and water requirements of four native fish species that are generally main channel specialists in the Murray-Darling Basin

Common name	Scientific name	Preferred habitat	Trophic position	Movements	Primary spawning cue	Peak spawning season	Fecundity	Response to flows
Murray cod (V)	*Maccullochella peelii peelii*	Slow-flowing riverine habitats with woody debris	Carnivore	Adults move during spawning but are usually territorial around home snag	Water temperature	Sept–Dec	Low	Flows important for downstream drift of eggs and larvae; enhanced recruitment during high flows linked to increased food availability
Trout cod (E)	*Maccullochella macquariensis*	Slow- and fast-flowing upper reaches of streams	Carnivore	Limited movements, usually territorial around home snag	Water temperature and day length	Sept–Nov	Moderate	Flows may be important for distribution of drifting larvae
Macquarie perch (E)	*Macquaria australasica*	Slow-flowing clear, cool and deep rivers and lakes	Benthic carnivore	Upstream spawning migrations	Water temperature	Oct–Dec	High	Flows important for providing food for pre-gonad development in adults and for downstream drift of eggs and larvae
Freshwater catfish	*Tandanus tandanus*	Slow-flowing rivers and lakes, with sand, mud and gravel substrates	Benthic opportunistic carnivore	Limited movements, no spawning migrations	Water temperature and day length	Oct–Dec	Low	Not known; floodplain inundation may be important for food availability

Species are grouped in this flow/habitat group based on the MFAT classification (Young *et al.* 2003).
V = vulnerable, E = endangered (*Environment Protection and Biodiversity Conservation Act 1999*).

Table 4.4: Habitat and water requirements of three native fish species that generally are flood spawners in the Murray-Darling Basin

Common name	Scientific name	Preferred habitat	Trophic position	Movements	Primary spawning cue	Peak spawning season	Fecundity	Response to flows
Golden perch	*Macquaria ambigua*	Slow-flowing, turbid, deep lowland riverine habitats	Carnivore	Long-distance upstream/ downstream spawning movements in adults; juveniles dispersed downstream	Rise of water levels, water temperature and day length	Oct–Dec	High	Spawning and recruitment follows peaks in flow; inundation of floodplain maximises spawning and recruitment; flows needed for larval drift downstream
Silver perch	*Bidyanus bidyanus*	Slow-flowing turbid waters and faster-flowing riverine habitats, prefers waters with littoral vegetation and woody debris	Carnivore	Long-distance upstream/ downstream spawning movements	Rise of water levels and water temper- ature	Nov–Jan	High	Major spawning occurs when floodplains are inundated; larvae and juvenile drift onto floodplain after major flooding
Spangled perch	*Leiopotherapon unicolor*	Still or slow-flowing river- ine and wetland habitats	Carnivore	Upstream spawning migrations; lateral movements onto floodplain and downstream dispersal after high flows	Water temper- ature and rise of flows	Nov–Feb	High	Rise in water levels can maximise recruitment; inundation of floodplain increases larvae/juvenile survival and adult fitness

Species are grouped in this flow/habitat group based on the MFAT classification (Young *et al.* 2003).

Table 4.5: Habitat and water requirements of four alien fish species found in the Murray-Darling Basin, and their impacts on native fish and their habitats

Common name	Scientific name	Preferred habitat	Trophic position	Movements	Primary spawning cue	Peak spawning season	Fecundity	Impacts on native fish and their habitats	Response to flows
Goldfish	*Carassius auratus*	Still and slow-moving riverine habitats, associated with aquatic vegetation	Benthic omnivore	Not known to migrate	Water temperature	Dec–Feb	Moderate	Preys on eggs, larvae and adult native fish; vector for parasites; linked to algal blooms; feeding behaviour linked to increased turbidity and reduced aquatic macrophyte cover	Needs cold water temperatures in winter months to stimulate ova development; moves into wetlands upon flooding; juvenile goldfish requires high water temperatures for growth
European carp	*Cyprinus carpio*	Still and slow-moving riverine habitats, associated with aquatic vegetation	Benthic omnivore	Some undertake upstream/downstream movements while others have small home ranges; strong cues to move into floodplain habitats after flooding	Water temperature	Oct–Dec	High	Associated with reductions in aquatic macrophytes, increases in turbidity and channel erosion; parasite vectors; probably competes with native fish and other fauna for food and habitat resources	Most adult fish move into floodplain wetlands after flooding; spawning in these habitats increases recruitment
Redfin perch	*Perca fluviatilis*	Cool, slow-flowing waters with aquatic vegetation	Pelagic carnivore	Downstream movements recorded	Water temperature	Aug–Sept	Moderate	Competes with adult native fish for food; preys on juvenile native fish; carrier of EHNV virus	Uses floodwaters to move downstream to colonise new habitats; high water temperatures inhibit gonad development
Gambusia	*Gambusia holbrooki*	Warm shallow habitats of still and slow-flowing waterbodies	Pelagic carnivore	Not known to migrate	Water temperature but spawning can occur year-round	Dec–Feb	High	Preys on eggs and larvae of native fish and frogs	Sensitive to droughts, but can survive low water quality; after flooding can breed prolifically with multiple broods

of food resources available to larval fish, compared to drier periods when permanent waterholes can offer limited food (Balcombe *et al.* 2005; Sternberg *et al.* 2008). Where species spawn independently of flows, overbank flows can indirectly benefit their recruitment through transfers of nutrients and food from the floodplain into the river channel (Humphries *et al.* 1999; King *et al.* 2009). However, poor water quality can be lethal for fish, and larval fish will actively avoid floodplain habitats if water quality is poor (Gehrke 1991).

Altered flow regimes can cause declines in water quality, including increases in salinity and build-up of agricultural chemicals. Although most adult fish can tolerate variable salinity, some larval fish and eggs (e.g. those of the Murray hardyhead and freshwater catfish) can be vulnerable to high levels of salinity. Channelisation, removal of fringing riparian vegetation, riverbank erosion by grazing livestock and high-intensity agriculture can also increase the sediment load in river systems, leading to deleterious effects on fish during their early life-stages (Mallen-Cooper 1993). For example, increased turbidity may adversely affect the eggs of the freshwater catfish, which lays large non-adhesive eggs in gravel nests (Davis 1977d). Cold water pollution from dams and storages can also have negative affects on native fish, as many species rely on water temperature as a cue for spawning (Astles *et al.* 2003). Adult Australian smelt, Murray-Darling rainbowfish, spangled perch and golden perch are also sensitive to low water temperatures (Milton and Arthington 1985; Lintermans 2007).

In addition to the water requirements described above, many native fish species have specific habitat requirements for spawning and feeding. For example, the freshwater catfish and Macquarie perch generally prefer gravel habitats for spawning, and the Murray cod and trout cod prefer deep pools with submerged logs for feeding (Allen *et al.* 2003; Lintermans 2007). Small-bodied wetland opportunists such as the olive perchlet and Murray hardyhead prefer areas of aquatic vegetation in which to hide from predators. These areas also provide safe hatching and nursery sites for eggs and larvae (Lintermans 2007).

Hydrological modification, particularly the loss of medium flow variability, has tended to favour alien fish species in some heavily regulated catchments (Gehrke and Harris 2000). Although causal mechanisms remain unclear, alien fish species are thought to have negative impacts on water quality, aquatic vegetation and native fish recruitment and survival. They are often carriers of disease (Table 4.5). Large floodplain wetlands are believed to be significant breeding hotspots for alien species, such as carp, goldfish and gambusia, which have less-specific flow requirements and wide environmental tolerances (Morgan and Beatty 2005; Stuart and Jones 2006; Rayner *et al.* 2009). Notably, the carp can withstand very low dissolved oxygen levels and the goldfish can tolerate wide ranges in salinity (Froese and Pauly 2009). The carp and goldfish are benthic feeders and sometimes re-suspend previously deposited sediments, thus increasing turbidity and inhibiting macrophyte and algal growth (Richardson *et al.* 1995; King *et al.* 1997). These species compete directly with native fish when habitat is limited. The gambusia is known to prey on the eggs and larvae of native fish and the redfin perch is a voracious predator of smaller native fish (Lintermans 2007).

In summary, native fish species utilise a variety of floodplain habitats, including channels, anabranches, billabongs and wetlands in the Murray-Darling Basin. Their assemblages have been heavily affected by habitat modification and changes to flood regimes. Reduced water availability and quality and changes to the flow regime (including the magnitude, timing, duration and frequency of flows) have had more impact on ephemeral habitats, such as floodplain wetlands, and consequently have adversely affected fish species that preferentially utilise these habitats. However, since all fish species also use main river channels and waterholes, core in-stream habitats may provide refuge during periods of low flow. While these drought refuges may enable the survival of some native fish species, particularly in ephemeral river systems in dryland regions (Balcombe *et al.* 2006), maintenance of fish populations is dependent on the availability of suitable spawning habitat and food availability. Spawning and recruitment in many species can be maximised following rising floodwaters (Tables 4.1–4.4), either through enhancing cues for spawning or, indirectly, by increasing food availability through the transfer of nutrients from the floodplain into the river channels. Therefore, changes in the frequency or magnitude of flow events and the availability of healthy floodplain habitats may alter the reproductive capacity of fish that rely on appropriately timed flows (Murray cod, Macquarie perch), on increasing flows (golden perch, silver perch) and/or on wetland habitats (olive perchlet, Murray hardyhead). Such species are likely to be the most vulnerable to changes in the flooding regime. Other species, which use floodplain habitats opportunistically, may exhibit a greater capacity to withstand the effects of altered flow regimes due to their ability to recruit and persist within in-channel habitats.

References

Alcaraz C and Garcia-Berthou E (2007) Life history variation of invasive mosquitofish (*Gambusia holbrooki*) along a salinity gradient. *Biological Conservation* **139**, 83–92.

Allen GR, Midgley SH and Allen M (2003) *Field Guide to Freshwater Fishes of Australia*. 2nd edn. CSIRO Publishing: Melbourne.

Appleford P, Anderson TA and Gooley GJ (1998) Reproductive cycle and gonadal development of Macquarie perch, *Macquaria australasica* Cuvier (Percichthyidae), in Lake Dartmouth and tributaries of the Murray-Darling Basin, Victoria, Australia. *Marine and Freshwater Research* **49**, 163–169.

Arthington AH and Marshall CJ (1999) Diet of the exotic mosquitofish, *Gambusia holbrooki*, in an Australian lake and potential for competition with indigenous fish species. *Asian Fisheries Science* **12**, 1–16.

Arthington AH, Balcombe SR, Wilson GA, Thoms MC and Marshall J (2005) Spatial and temporal variation in fish-assemblage structure in isolated waterholes during the 2001 dry season of an arid-zone floodplain river, Cooper Creek, Australia. *Marine and Freshwater Research* **56**, 25–35.

Astles KL, Winstanley RK, Harris JH and Gehrke PC (2003) Experimental study of the effects of cold water pollution on native fish. NSW Fisheries Office of Conservation: Sydney.

Backhouse G, Lyon J and Cant B (2008) National Recovery Plan for the Murray hardyhead *Craterocephalus fluviatilis*. Dept of Sustainability and Environment: Melbourne.

Balcombe SR and Humphries P (2006) Diet of the western carp gudgeon (*Hypseleotris klunzingeri* Ogilby) in an Australian floodplain lake: the role of water level stability. *Journal of Fish Biology* **68**, 1484–1493.

Balcombe SR, Bunn SE, McKenzie-Smith FJ and Davies PM (2005) Variability of fish diets between dry and flood periods in an arid zone floodplain river. *Journal of Fish Biology* **67**, 1552–1567.

Balcombe SR, Arthington AH, Foster ND, Thoms MC, Wilson GG and Bunn SE (2006) Fish assemblages of an Australian dryland river: abundance, assemblage structure and recruitment patterns in the Warrego River, Murray-Darling Basin. *Marine and Freshwater Research* **57**, 619–633.

Baumgartner LJ (2007) Diet and feeding habits of predatory fishes upstream and downstream of a low-level weir. *Journal of Fish Biology* **70**, 879–894.

Baumgartner LJ, Reynoldson N and Gilligan DM (2006) Mortality of larval Murray cod (*Maccullochella peelii peelii*) and golden perch (*Macquaria ambigua*) associated with passage through two types of low-head weirs. *Marine and Freshwater Research* **57**, 187–191.

Baxter AF, Vallis SL and Hume DJ (1985) 'The predation of recently released rainbow trout fingerling, *Salmo gairdneri*, by redfin, *Perca fluviatilis*, in Lake Burrumbeet, October–December 1983'. Arthur Rylah Institute for Environmental Research: Melbourne.

Boxall GD, Sandberg JJ and Kroon FJ (2002) Population structure, movement and habitat preferences of the purple-spotted gudgeon, *Mogurnda adspersa*. *Marine and Freshwater Research* **53**, 909–917.

Brown P (1996) 'Carp in Australia: Fish Fact Sheets 4'. NSW Fisheries, Narrandera Fisheries Centre.

Bunn SE, Davies PM and Winning M (2003) Sources of organic carbon supporting the food web of an arid zone floodplain river. *Freshwater Biology* **48**, 619–635.

Cadwallader PL (1981) Past and present distributions and translocations of Macquarie perch *Macquaria australasica* (Pisces: Percichthyidae), with particular reference to Victoria. *Proceedings of the Royal Society of Victoria* **93**, 23–30.

Cadwallader PL (2006) Distribution of native and introduced fish in the Seven Creeks River system, Victoria. *Austral Ecology* **4**, 361–385.

Cadwallader PL and Eden AK (1979) Observations on the food of Macquarie perch, *Macquaria australasica* (Pisces: Percichthyidae), in Victoria. *Australian Journal of Marine and Freshwater Research* **30**, 401–409.

Cadwallader PL and Gooley GJ (1984) Past and present distributions and translocations of Murray cod *Maccullochella peelii peelii* and trout cod *M. macquariensis* (Pisces: Percichthyidae), in Victoria. *Proceedings of the Royal Society of Victoria* **96**, 33–43.

Cadwallader PL and Rogan PL (1977) The Macquarie perch, *Macquaria australasica* (Pisces: Percichthyidae), of Lake Eildon, Victoria. *Australian Journal of Ecology* **2**, 409–418.

Chapman P and Warburton K (2006) Postflood movements and population connectivity in gambusia (*Gambusia holbrooki*). *Ecology of Freshwater Fish* **15**, 357–365.

Clarke GM, Grosse S, Matthews M, Catling PC, Baker B, Hewitt CL, Crowther D and Sadlier SR (2000) 'State of the Environment indicators for exotic environmental pest species. Consultant report to the Dept of Environment and Heritage'. CSIRO: Canberra.

Close PG, Pusey BJ and Arthington AH (2005) Larval description of the sympatric species, *Craterocephalus stercusmuscarum stercusmuscarum* (Pisces: Atherinidae) and *Mogurnda adspersa* (Pisces: Eleotridae) from tropical streams of north-east Queensland, Australia. *Journal of Fish Biology* **66**, 668–684.

Closs GP, Balcombe SR, Driver P, McNeil DG and Shirley MJ (2005) The importance of floodplain wetlands to Murray-Darling fish: What's there? What do we know? What do we need to know? Native fish and wetlands in the Murray-Darling Basin. Canberra Workshop. pp. 14–28.

Crook DA and Gillanders BM (2006) Use of otolith chemical signatures to estimate carp recruitment sources in the mid-Murray River, Australia. *River Research and Applications* **22**, 871–879.

CSIRO (2008) Water availability in the Murray-Darling Basin: a report to the Australian Government from the CSIRO Murray-Darling Basin Sustainable Yields Project. CSIRO: Canberra.

CSIRO and BOM (2007) Climate change in Australia. Canberra.

Davis TLO (1977a) Age determination and growth of the freshwater catfish, *Tandanus tandanus* Mitchell, in the Gwydir River, Australia. *Australian Journal of Marine and Freshwater Research* **28**, 119–137.

Davis TLO (1977b) Food habits of the freshwater catfish, *Tandanus tandanus* Mitchell, in the Gwydir River, Australia, and effects associated with impoundment of this river by the Copeton Dam. *Australian Journal of Marine and Freshwater Research* **28**, 455–465.

Davis TLO (1977c) Reproductive biology of the freshwater catfish, *Tandanus tandanus* Mitchell, in the Gwydir River, Australia. I. Structure of the gonads. *Australian Journal of Marine and Freshwater Research* **28**, 139–158.

Davis TLO (1977d) Reproductive biology of the freshwater catfish, *Tandanus tandanus* Mitchell, in the Gwydir River, Australia. II. Gonadal cycle and fecundity. *Australian Journal of Marine and Freshwater Research* **28**, 159–169.

Deacon JE, Hubbs C and Zahuranec BJ (1964) Some effects of introduced fishes on the native fish fauna of southern Nevada. *Copeia* **1964**, 384–388.

Douglas JW, Gooley GJ, Ingram BA, Murray ND and Brown LD (1995) Natural hybridization between Murray cod, *Maccullochella peelii peelii* (Mitchell) and trout cod, *Maccullochella macquariensis* (Cuvier) (Percichthyidae) in the Murray River, Australia. *Marine and Freshwater Research* **46**, 729–734.

DPI (2005) Redfin perch (*Perca fluviatilis*) fact sheet. NSW Government Industry and Investment, Primary Industries Fishing and Aquaculture website: http://www.dpi.nsw.gov.au/fisheries/pests-diseases/freshwater-pests/species/redfin-perch (accessed 11 January 2010).

Ebner B, Raadik T and Ivantsoff W (2003) Threatened fishes of the world: *Craterocephalus fluviatilis* McCulloch, 1913 (Atherinidae). *Environmental Biology of Fishes* **68**, 395.

Ellis I (2005a) 'Ecology and breeding seasonality of the Murray hardyhead *Craterocephalus fluviatilis* (McCulloch), Family Atherinidae, in two lakes near Mildura, Victoria'. A report prepared for the Mallee Catchment Management Authority by the Murray-Darling Freshwater Research Centre. Murray-Darling Freshwater Research Centre: Albury.

Ellis I (2005b) 'Ecology of the Murray hardyhead *Craterocephalus fluviatilis* (McCulloch) Family Atherinidae'. A literature review prepared for the Mallee Catchment Management Authority. Technical Report 03/2005. Murray-Darling Freshwater Research Centre: Albury.

Ellis I (2006) 'Age structure and dietary analysis of the Murray hardyhead *Craterocephalus fluviatilis* (McCulloch), Family Atherinidae, in two lakes near Mildura, Victoria'. A report prepared for the Mallee Catchment Management Authority by the Murray-Darling Freshwater Research Centre. Murray-Darling Freshwater Research Centre: Albury.

Faulks LK, Gilligan DM and Beheregaray LB (2008) Phylogeography of a threatened freshwater fish (*Mogurnda adspersa*) in eastern Australia: conservation implications. *Marine and Freshwater Research* **59**, 89–96.

Fletcher AR, Morison AK and Hume DJ (1985) Effects of carp, *Cyprinus carpio* L., on communities of aquatic vegetation and turbidity of waterbodies in the Lower Goulburn River Basin. *Australian Journal of Marine and Freshwater Research* **36**, 311–327.

Froese R and Pauly D (2009) FishBase. Online: www.fishbase.org version (01/2010).

Gehrke PC (1991) Avoidance of inundated floodplain habitat by larvae of golden perch (*Macquaria ambigua* Richardson): influence of water quality or food distribution? *Marine and Freshwater Research* **42**, 707–719.

Gehrke PC and Harris JH (2000) Large-scale patterns in species richness and composition of temperate riverine fish communities, south-eastern Australia. *Marine and Freshwater Research* **51**, 165–182.

Gehrke PC and Harris JH (2001) Regional-scale effects of flow regulation on lowland riverine fish communities in New South Wales, Australia. *Regulated Rivers: Research and Management* **17**, 369–391.

Gehrke PC, Brown P, Schiller CB, Moffatt DB and Bruce AM (1995) River regulation and fish communities in the Murray-Darling river system, Australia. *Regulated Rivers: Research and Management* **11**, 363–375.

Gilligan DM (2005) Fish communities of the Murrumbidgee catchment: status and trends. NSW Dept of Primary Industries: Cronulla Fisheries Centre.

Gray SC, De Silva SS, Ingram BA and Gooley GJ (2008) Effects of river impoundment on body condition and reproductive performance of the Australian native fish, Macquarie perch (*Macquaria australasica*). *Lakes and Reservoirs: Research and Management* **5**, 281–291.

Heagney E, Fowler A, Allman R and Spencer JA (2009) 'Growth and condition of juvenile Australian smelt and bony bream in the Gwydir Wetlands: an otolith-based analysis'. Final report to the Rivers and Wetlands Unit, NSW Dept of Environment and Climate Change. University of New South Wales: Sydney.

Humphries P (2005) Spawning time and early life history of Murray cod, *Maccullochella peelii* (Mitchell) in an Australian river. *Environmental Biology of Fishes* **72**, 393–407.

Humphries P and Lake PS (2000) Fish larvae and the management of regulated rivers. *Regulated Rivers: Research and Management* **16**, 421–432.

Humphries P, King AJ and Koehn JD (1999) Fish, flows and flood plains: links between freshwater fishes and their environment in the Murray-Darling River system, Australia. *Environmental Biology of Fishes* **56**, 129–151.

Hutchison MJ and Armstong PH (1993) The invasion of a south-western Australian river system by *Perca fluviatilis*: history and probable causes. *Global Ecology and Biogeography Letters* **3**, 77–89.

Ingram BA and Douglas JW (1995) Threatened fishes of the world: *Maccullochella macquariensis* (Cuvier, 1829) (Percichthyidae). *Environmental Biology of Fishes* **43**, 38.

Ingram BA and Rimmer MA (1992) Induced breeding and larval rearing of the endangered Australian freshwater fish trout cod *Maccullochella macquariensis* (Cuvier) (Percichthyidae). *Aquaculture and Fisheries Management* **24**, 7–17.

Ingram BA, Barlow CG, Burchmore JJ, Gooley GJ, Rowland SJ and Sanger AC (1990) Threatened native fishes in Australia: some case histories. *Journal of Fish Biology* **37**, 175–182.

Ingram BA, Douglas JW and Lintermans M (2000) Threatened fishes of the world: *Macquaria australasica* Cuvier, 1830 (Percichthyidae). *Environmental Biology of Fishes* **59**, 68.

James KR, Cant B and Ryan T (2003) Responses of freshwater biota to rising salinity levels and implications for saline water management: a review. *Australian Journal of Botany* **51**, 703–713.

Jones MJ and Stuart IG (2009) Lateral movement of common carp (*Cyprinus carpio* L.) in a large lowland river and floodplain. *Ecology of Freshwater Fish* **18**, 72–82.

Khan TA (2003) Dietary studies on exotic carp (*Cyprinus carpio* L.) from two lakes of western Victoria, Australia. *Aquatic Sciences* **65**, 272–286.

King AJ (2004) Ontogenetic patterns of habitat use by fishes within the main channel of an Australian floodplain river. *Journal of Fish Biology* **65**, 1582–1603.

King AJ, Robertson AI and Healey MR (1997) Experimental manipulation of the biomass of introduced carp (*Cyprinus carpio*) in billabongs. I. Impacts on water-column properties. *Marine and Freshwater Research* **48**, 435–443.

King AJ, Humphries P and Lake PS (2003) Fish recruitment on floodplains: the roles of patterns of flooding and life history characteristics. *Canadian Journal of Fisheries and Aquatic Sciences* **60**, 773–786.

King AJ, Tonkin Z and Mahoney J (2007) 'Assessing the effectiveness of environmental flows on fish recruitment in Barmah-Millewa Forest'. Arthur Rylah Institute for Environmental Research, Dept of Sustainability and Environment: Melbourne.

King AJ, Tonkin Z and Mahoney J (2009) Environmental flow enhances native fish spawning and recruitment in the Murray River, Australia. *River Research and Applications* **25**, 1205–1218.

Koehn JD and Harrington DJ (2006) Environmental conditions and timing for the spawning of Murray cod (*Maccullochella peelii peelii*) and the endangered trout cod (*M. macquariensis*) in south-eastern Australian rivers. *River Research and Applications* **22**, 327–342.

Koehn JD, Nicol SJ, McKenzie JA, Lieschke JA, Lyon JP and Pomorin K (2008) Spatial ecology of an endangered native Australian Percichthyid fish, the trout cod *Maccullochella macquariensis. Endangered Species Research* **4**, 219–225.

Kolmakov VI and Gladyshev MI (2003) Growth and potential photosynthesis of cyanobacteria are stimulated by viable gut passage in crucian carp. *Aquatic Ecology* **37**, 237–242.

Kushlan J (1993) Colonial waterbirds as bioindicators of environmental change. *Colonial Waterbirds* **16**, 223–251.

Lake JS (1967a) Rearing experiments with five species of Australian freshwater fishes. I. Inducement to spawning. *Australian Journal of Marine and Freshwater Research* **18**, 137–153.

Lake JS (1967b) Rearing experiments with five species of Australian freshwater fishes. II. Morphogenesis and ontogeny. *Australian Journal of Marine and Freshwater Research* **18**, 155–173.

Langdon JS (1989) Experimental transmission and pathogenicity of epizootic haematopoietic necrosis virus (EHNV) in redfin perch, *Perca fluviatilis* L., and 11 other teleosts. *Journal of Fish Diseases* **12**, 295–310.

Lintermans M (2007) 'Fishes of the Murray-Darling Basin: an introductory guide'. MDBC Publication No. 10/07. Murray-Darling Basin Authority: Canberra.

Llewellyn LC (1973) Spawning, development, and temperature tolerance of the spangled perch, *Madigania unicolor* (Gunther), from inland waters in Australia. *Marine and Freshwater Research* **24**, 73–94.

Llewellyn LC (1983) The distribution of fish in New South Wales. Australian Society for Limnology Special Publication No. 7. Sydney.

Lloyd L (1984) Exotic fish: useful additions or 'animal weeds'? *Fishes of Sahul* **1**, 31–42.

Lyon J, Stuart I, Ramsey D and O'Mahony D (2010) The effect of water level on lateral movements of fish between river and off-channel habitats and implications for management. *Marine and Freshwater Research* **61**, 271–278.

Mallen-Cooper M (1993) Habitat changes and declines of freshwater fish in Australia: what is the evidence and do we need more? In *Australian Society for Fish Biology Workshop on Sustaining Fisheries Through Sustaining Habitat*. (Ed. D Hancock). Australian Government Publishing Service: Canberra.

Mallen-Cooper M and Stuart IG (2003) Age, growth and non-flood recruitment of two potamodromous fishes in a large semi-arid/temperate river system. *River Research and Applications* **19**, 697–719.

McDowall RM (Ed.) (1996) *Freshwater Fishes of South-eastern Australia*. 2nd edn. Reed Books: Sydney.

McNeil D, Wilson P, Hartwell D and Pellizzari M (2008) Olive perchlet (*Ambassis agassizii*) in the Lachlan River: population status and sustainability in the Lake Brewster Region. SARDI Research Report Series No. 309. Adelaide.

MDBC (2000) 'National Management Strategy for carp control 2000–2005'. Prepared by the Carp Control Coordinating Group. Murray-Darling Basin Commission: Canberra.

Merrick JR and Schmida GE (1984) *Australian Freshwater Fishes: Biology and Management*. Murray-Darling Basin Commission: Canberra.

Mieiro CL, Cabral JA and Marques JC (2001) Predation pressure of introduced mosquitofish (*Gambusia holbrooki* Girard), on the native zooplankton community. A case study from representative habitats in the lower Mondego River Valley (Portugal). *Limnetica* **20**, 279–292.

Milton D and Arthington AH (1983) Reproductive biology of *Gambusia affinis* (Baird and Girard), *Xiphophorus helleri* (Gunther) and *X. maculatus* (Heckel) (Pisces: Poeciliidae) in south-eastern Queensland, Australia. *Journal of Fish Biology* **23**, 23–41.

Milton DA and Arthington AH (1985) Reproductive strategy and growth of the Australian smelt, *Retropinna semoni* (Weber) (Pisces: Retropinnidae), and the olive perchlet, *Ambassis nigripinnis* (De Vis) (Pisces: Ambassidae), in Brisbane, south-eastern Queensland. *Australian Journal of Marine and Freshwater Research* **36**, 329–341.

Morgan DL and Beatty SJ (2005) Fish fauna of the Vasse River and the colonisation by feral goldfish (*Carassius auratus*). Centre for Fish and Fisheries Research, Murdoch University: Perth.

Morgan DL, Hambleton SJ, Gill HS and Beatty SJ (2002) Distribution, biology and likely impacts of the introduced redfin perch (*Perca fluviatilis*) (Percidae) in Western Australia. *Marine and Freshwater Research* **53**, 1211–1221.

Nicol SJ, Barker RJ, Koehn JD and Burgman MA (2007) Structural habitat selection by the critically endangered trout cod, *Maccullochella macquariensis*, Cuvier. *Biological Conservation* **138**, 30–37.

Nielsen DL, Brock MA, Rees GN and Baldwin DS (2003) Effects of increasing salinity on freshwater ecosystems in Australia. *Australian Journal of Botany* **51**, 655–665.

NPWS (2003) 'NSW threat abatement plan. Predation by *Gambusia holbrooki* – the plague minnow'. NSW National Parks and Wildlife Service: Sydney.

O'Brien TA and Ryan TJ (1997) 'Impact of saline drainage on key Murray-Darling Basin fish species'. Freshwater Ecology Division, Dept of Resources and Environment: Melbourne.

Perez-Bote JL and Lopez MT (2005) Life-history pattern of the introduced eastern mosquitofish, *Gambusia holbrooki* (Baird & Girard, 1854), in a Mediterranean-type river: the River Guadiana (SW Iberian Peninsula). *Italian Journal of Zoology* **72**, 241–248.

Poff NL, Allan JD, Bain MB, Karr JR, Prestegaard KL, Richter B, Sparks R and Stromberg JC (1997) The natural flow regime: a new paradigm for riverine conservation and restoration. *Bioscience* **47**, 769–784.

Puckridge JT, Sheldon F, Walker KF and Boulton AJ (1998) Flow variability and the ecology of large rivers. *Marine and Freshwater Research* **49**, 55–72.

Puckridge JT and Walker KF (1990) Reproductive biology and larval development of a gizzard shad, *Nematalosa erebi* (Günther) (Dorosomatinae: Teleostei), in the River Murray, South Australia. *Australian Journal of Marine and Freshwater Research* **41**, 695–712.

Rayner TS, Jenkins KM and Kingsford RT (2009) Small environmental flows, drought and the role of refugia for freshwater fish in the Macquarie Marshes, arid Australia. *Ecohydrology* **2**, 440–453.

Reid HP and Holdway DA (1995) Early development of Australian crimson-spotted rainbowfish, *Melanotaenia fluviatilis* (Pisces: Melanotaeniidae). *Marine and Freshwater Research* **46**, 475–480.

Reynolds LF (1983) Migration patterns of five fish species in the Murray-Darling River system. *Australian Journal of Marine and Freshwater Research* **34**, 857–871.

Richardson MJ, Whoriskey FG and Roy LH (1995) Turbidity generaton and biological impacts of an exotic fish, *Carassius auratus,* introduced into shallow seasonally anoxic ponds. *Journal of Fish Biology* **47**, 576–585.

Roberts J and Tilzey R (1996) Controlling carp: exploring the options for Australia. In *Proceedings of a Workshop.* 22–24 October 1996, Albury. CSIRO Land and Water: Griffith.

Robertson AI, Healey MR and King AJ (1997) Experimental manipulations of the biomass of introduced carp (*Cyprinus carpio*) in billabongs. II. Impacts on benthic properties and processes. *Marine and Freshwater Research* **48**, 445–454.

Rowley JJL, Rayner TS and Pyke GH (2005) New records and invasive potential of the poeciliid fish *Phalloceros caudimaculatus*. *New Zealand Journal of Marine and Freshwater Research* **39**, 1013–1022.

Sanstro O, Abrahamsson I, Anderssion J and Vetemaa M (2005) Temperature and egg development in Eurasian perch. *Journal of Fish Biology* **51**, 1015–1024.

Schiller CB and Harris JH (2001) Native and alien fish. In *Rivers as Ecological Systems: The Murray-Darling Basin*. (Ed. WJ Young) pp. 229–258. CSIRO Land and Water: Canberra.

Schiller CB, Bruce AM and Gehrke PC (1997) Distribution and abundance of native fish in New South Wales rivers. In *Fish and Rivers in Stress: The NSW Rivers Survey*. (Eds JH Harris and PC Gehrke) pp. 71–102. NSW Fisheries Office and Cooperative Research Centre for Freshwater Ecology: Canberra.

Sheikh-Eldin M, De Silva SS, Anderson TA and Gooley G (1999) Comparison of fatty acid composition of muscle, liver mature oocytes, and diets of wild and captive Macquarie perch, *Macquaria australasica*, broodfish. *Aquaculture* **144**, 201–216.

Siebentritt MA (1999) Fish and the Lower Gwydir River floodplain. Wetland Care Australia: Berri, SA.

Sternberg D, Balcombe S and Marshall J (2008) Food resource variability in an Australian dryland river: evidence from the diet of two generalist native fish species. *Marine and Freshwater Research* **59**, 137–144.

Stuart I and Jones M (2002) 'Ecology and management of common carp in the Barmah-Millewa Forest'. Freshwater Ecology, Arthur Rylah Institute for Environmental Research: Melbourne.

Stuart IG and Jones M (2006) Large, regulated forest floodplain is an ideal recruitment zone for non-native common carp (*Cyprinus carpio* L.). *Marine and Freshwater Research* **57**, 333–347.

Swanson C and Cech JJ (1996) Comments on 'Adverse assessments of *Gambusia affinis*'. *Journal of American Mosquito Control Association* **12**, 163–164.

Taylor J and Mahon R (1977) Hybridization of *Cyprinus carpio* and *Carassius auratus*, the first two exotic species in the lower Laurentian Great Lakes. *Environmental Biology of Fishes* **1**, 205–208.

TCRT (2008) Trout Cod Recovery Team National Recovery Plan for the trout cod *Maccullochella macquariensis*. Melbourne.

Todd CR, Nicol SJ and Koehn JD (2003) Density dependence uncertainty in population models for the conservation management of trout cod, *Maccullochella macquariensis*. *Ecological Modelling* **171**, 359–380.

Treadwell S and Hardwick R (2003) Review of habitat associations of native fish in the Murray-Darling Basin. Murray-Darling Basin Commission Project R2105. Sinclair Knight Merz Consulting: Armidale.

Weatherley AJ and Lake JS (1967) Introduced fish species in Australian waters. In *Australian Inland Waters and Their Fauna*. (Ed. AH Weatherley) pp. 217–239. Australian National University Press: Canberra.

Wedderburn S and Barnes T (2009) Condition monitoring of threatened fish species at Lake Alexandrina and Lake Albert (2008–2009). University of Adelaide: Adelaide.

Wedderburn SD, Walker KF and Zampatti BP (2008) Salinity may cause fragmentation of hardyhead (Teleostei: Atherinidae) populations in the River Murray, Australia. *Marine and Freshwater Research* **59**, 254–258.

Whittington RJ and Reddacliff GL (2008) Influence of environmental temperature on experimental infection of redfin perch (*Perca fluviatilis*) and rainbow trout (*Oncorhynchus mykiss*) with epizootic haematopoietic necrosis virus, an Australian iridovirus. *Australian Veterinary Journal* **72**, 421–424.

Wilson P, McNeil D and Gillanders BM (2008) 'Impacts of introduced redfin perch on native flathead gudgeons in the South Para River. SARDI Aquatic Sciences Publication No. F2007/000882-1'. Report to SA Dept of Water, Land and Biodiversity Conservation. SA Research and Development Institute (Aquatic Sciences): Adelaide.

Ye Q (2004) 'Golden perch (*Macquaria ambigua*)'. Fishery assessment report to PIRSA Fisheries for the Inland Fishery Management Committee. Adelaide.

Young WJ (2001) Rivers as ecological systems: the Murray-Darling Basin. CSIRO Land and Murray-Darling Basin Commission: Canberra.

Young WJ, Scott AC, Cuddy SM and Rennie BA (2003) Murray flow assessment tool: a technical description. Client report. CSIRO Land and Water, Canberra.

Chapter 5

Frogs

Skye Wassens

Introduction

The Australian native amphibian fauna is represented by the anurans (frogs) of which there are currently 216 named species. There is also one introduced toad, the cane toad (*Bufo marinus*). Twenty-nine species of frog occur in the Murray-Darling Basin. Like most flood-dependent species, frogs respond to various aspects of the flood pulse, most importantly the timing of inundation, the length of time that water remains pooled and the temporal frequency of flood pulses. In addition, frogs are often sensitive to spatial aspects of flooding such as proximity to permanent waterbodies and drought refuges, and the spatial extent of flooding across the floodplain. Interactions with other wetland-dependent species, particularly fish, can also influence how frogs respond during a flood pulse.

This chapter describes the habitat and breeding biology of key frog species in the Murray-Darling Basin. Water requirements are described on the basis of biology and habitat preferences. It is important to note that the water and habitat requirements of the majority of inland frog species, along with the impact of altered flooding regimes, have received virtually no attention. This chapter should serve as a general guide but, as the water requirements of frogs are poorly known, it should not be used in place of careful monitoring to determine breeding activity, tadpole development rates and metamorphosis times following flooding.

While water is the overriding factor driving frog distributions through the Murray-Darling Basin, frogs also respond to aquatic and terrestrial vegetation, predator densities, food availability and water chemistry. In many instances these are also correlated with flooding regimes, and the relative importance of each can be difficult to separate. Aquatic vegetation cover and structural complexity is important for many species of frogs and their tadpoles and are important drivers of habitat occupancy patterns and recruitment success (Healey *et al.* 1997; Mac Nally *et al.* 2009; Tarr and Babbitt 2002). Aquatic vegetation provides shelter for adult frogs and acts as a substrate for the growth of biofilms and organic matter, which are important food sources for tadpoles (Gillespie 2002; Kupferberg *et al.* 1994; Mokany 2007). Predation by introduced fish, particularly the common carp (*Cyprinus carpio*), can also greatly alter recruitment outcomes following flooding (Spencer and Wassens 2009). Declining water quality can influence the response of frogs to flooding, for example, increasing salinity can exclude sensitive frog species from affected waterbodies (Smith *et al.* 2007). These components are important when describing the patterns of habitat use and responses to flooding at local scales and when comparing relationships between individual wetlands within a single catchment. At large scales, these components exhibit considerable spatial variability and it is more practical to consider and manage flooding regimes.

The water requirements of frogs within the Murray-Darling Basin can be described in terms of local rainfall, which can create temporary pools suitable for some species, and flood pulses which are often driven by rain higher in the catchment. The relative importance of local rainfall and flood pulses is likely to vary between species and over time. For example, a population of frogs which normally depends on persistent pools created by river flow and flood pulses might breed opportunistically in rain-fed pools. In these cases, local rainfall may play a significant role in maintaining local populations between flood pulses. Conversely, flood pulses might be critical to population persistence when local rainfall is limited (Karraker and Gibbs 2009). Further research is required to describe how these complementary interactions between local rainfall and flooding influence population persistence and species distributions through the Murray-Darling Basin.

In this chapter, only species that are known or likely to occur in floodplain wetlands and rivers within the Murray-Darling Basin have been included. Species that typically rely on local rainfall or that are rare in floodplain wetlands within the Murray-Darling Basin have been excluded. Also excluded are the common spadefoot toad (*Neobatrachus sudelli*), crucifix toad (*Notaden bennetti*), salmon-striped frog (*Limnodynastes salmini*), Sloane's froglet (*Crinia sloanei*) and the wrinkled toadlet (*Uperoleia rugosa*) because the water requirements of these species are very poorly documented. Stream-breeding frog species that occur in the north and south-western slopes of the Murray-Darling Basin have also been excluded; these creek systems are typically above major dams and are less affected by river regulation than are lowland areas. This chapter includes a brief description of each species, however, as the colour, size and appearance of individuals within a species is variable, reliable identification is best achieved through the use of detailed taxonomic keys such as Anstis (2002).

The response of frogs to flood pulses can be broken down into three main phases: calling and spawning, tadpole development and metamorphosis. If, when and for how long each of these phases occurs is influenced by the characteristics of the flood pulse. The majority of frog species have a defined activity period and successful recruitment can only occur when this breeding window and the flood pulse coincide. Spring and summer are the main activity periods and some species, particularly members of the *Cyclorana* genus, can only breed successfully during very warm conditions.

Assuming that the timing of a flood pulse coincides with their normal activity period, the majority of flood-dependent species commence calling during the peak of the flood pulse when wetlands begin to fill (Wassens *et al.* 2009). However, spawning activity differs between species. Some, such as the common green tree frog (*Litoria caerulea*), spawn immediately after flooding while others delay spawning until after the flood pulse has passed (Watson *et al.* 1995). There is sometimes a significant lag time between when the males commence calling and when females are ready to spawn. Lags can be caused by females needing to build up body condition prior to spawning (Whitaker 2001) or by delaying breeding until water temperatures are ideal for tadpole development (Reading 2003). Lags of up to two months between calling and spawning were recorded for Peron's tree frog (*Litoria peronii*) in the small wetlands along the Murrumbidgee River (Wassens *et al.* 2009) and for the southern bell frog (*Litoria raniformis*) in irrigation systems and seasonally flooded wetlands (Wassens 2005). After spawning, tadpole development times can range from four weeks to 12 months depending on the species and environmental conditions (Anstis 2002). There is also considerable variability in tadpole development times within species, for example, the southern bell frog tadpole can take from two and a half to 12 months to reach metamorphosis. Tadpole development times can increase with increasing water temperature and food availability (Flecker *et al.* 1999; Purrenhage and Boone 2009). As a result, development times may be longer in wetlands flooded in autumn and winter when water temperatures are lower and plant growth may be limited, than if the same wetland were flooded in spring. Winter flooding therefore involves the upper limit for tadpole development times; potential delays in spawning due to cooler temperatures must also be taken into account.

There has been little research on the water requirements of newly metamorphosed frogs, but it is likely that individuals need to remain around the natal pool for some time in order to gain body condition. In irrigation areas, newly metamorphosed southern bell frogs remain at the natal waterbody for several weeks after the adults have abandoned it in favour of permanent waterbodies (Wassens 2005). Consequently, maintaining longer hydroperiods (five to seven months) may increase overall recruitment success for many wetland-dependent species.

This chapter discusses some of these factors, but it must be remembered that research on amphibian water requirements and responses to flooding is extremely limited. The water requirement information for many species will inevitably be revised and expanded in light of future research. The species profiles include brief physical descriptions of major groups and species and their general distribution through major river systems. Species distributions are based on National Parks and Wildlife Service ATLAS records, the AMPHIBIAWEB database (www.amphibiaweb.org) and field guides, particularly Cogger (2000) and Anstis (2002). In most cases it is not possible to include specific details on the habitat requirements of each species, such as associations with specific vegetation communities or wetland types, because there have been so few studies to quantify these relationships. Instead, a general overview of wetland habitats such as the use of rain-fed, flooded or permanent waterbodies is included. In predicting water requirements, tadpole life-spans were used as an indicator of the length of time that water needs to remain pooled (hydroperiod). The activity period for each species has been used to estimate the ideal flooding times. The capacity of species to burrow and aestivate, along with the use of refuge habitats, provides a general guide of each species' tolerance of reduced flooding frequency.

Species profiles

Eastern sign-bearing froglet: *Crinia parinsignifera*

Common eastern froglet: *Crinia signifera*

The eastern sign-bearing froglet (*Crinia parinsignifera*; Figure 5.1) and the common eastern froglet (*Crinia signifera*) are common and widespread through river and wetland systems of the Murray-Darling Basin. These are very small frogs, usually less than 20 mm in length. Members of the genus *Crinia* are best identified by their rough granular belly, a lack of webbing on fingers and toes and an elongate middle toe on the hind legs (Cogger 2000; Robinson 2000). Colours and marking vary considerably between individuals, ranging from dark brown to chestnut through to light grey. They may be solid or bicolour, with or without banding on the hind legs. In some regions, the eastern sign-bearing froglet and common eastern froglet can co-occur within a single waterbody and can be difficult to distinguish visually. However, the two species have distinctive calls, which are a useful aid for identification. The eastern sign-bearing froglet has a low-drawn squelch, or 'eeeekk', which can be singular or repeated in a rapid rhythmic sequence: 'eek-eekeek-eek-eekeek-eek'. The common eastern froglet has an easily recognisable 'crick-crick-crick'.

Figure 5.1: Eastern sign-bearing froglet, *Crinia parinsignifera*. Photograph: Skye Wassens (CSU).

Distribution

The eastern sign-bearing froglet is widespread and common throughout the warmer inland regions of the Murray-Darling Basin in both northern and southern Basins, including the Darling, Gwydir, Macquarie, Namoi, Lachlan, Murrumbidgee and Murray rivers in New South Wales and the Goulburn, Avoca and Wimmera River systems in Victoria. The common eastern froglet is very common in the cooler regions of Murray-Darling Basin, including the upper and middle reaches of the Darling, Gwydir, Macquarie, Namoi, Lachlan, Murrumbidgee and Murray rivers in New South Wales and the Ovens, Goulburn and Broken rivers in Victoria.

Reproduction and life-cycle

The eastern sign-bearing froglet usually breeds in spring and summer but will also breed in autumn and winter following flooding or heavy rain. The common eastern froglet prefers cooler temperatures and generally breeds through winter, autumn and spring, but will breed at any time depending on the availability of habitat and temperature. Eggs are laid in the water, usually attached to vegetation or woody debris. Tadpoles are very small with rounded tails and, like the adults, can be difficult to distinguish from those of other members of the genus. Like all frogs, tadpole development times are variable, influenced by water temperature, changing water level and food availability. Tadpole development time (the time from egg to frog) is six weeks to three months depending on environmental conditions (Anstis 2002).

Habitat and trophic position

The common eastern froglet and the eastern sign-bearing froglet have similar habitat requirements and are highly adaptable, occurring in rain-fed depressions, ditches, semi-permanent wetlands, oxbow lagoons, creeks and rivers, farm dams, irrigation canals and urban ponds. They prefer to breed in areas containing diverse aquatic vegetation or submerged grasses (Wassens and Maher, in press). Tadpoles are generalist detritivores and herbivores, feeding on biofilms, algae and detritus.

Water requirements

The common eastern froglet and eastern sign-bearing froglet will breed successfully in a range of waterbodies and at different times of the year. As a result they appear to be less sensitive to alterations in wetland hydrology than other wetland-dependent species. Wetlands should retain pooled water for a minimum of six weeks if flooded during spring or summer, and three months if flooded in winter. There is limited information on the capacity of these species to aestivate during dry conditions and it is not clear whether they burrow or move into newly flooded wetlands from the associated river systems during flooding. During the 2009 winter watering of wetlands in the Lowbidgee floodplain, in the lower Murrumbidgee catchment of south-western New South Wales, the eastern sign-bearing froglet was common in wetlands that had been dry since 2006.

Striped burrowing frog: *Cyclorana alboguttata*

Rough frog: *Cyclorana verrucosa*

Water-holding frog: *Cyclorana platycephala*

This genus contains the cocoon-forming species which are adapted to xeric environments (Figure 5.2). All members of this genus are capable of prolonged underground aestivation during dry conditions (Tracy *et al.* 2007). The frogs survive prolonged dry periods by forming a cocoon of sloughed skin which assists in moisture retention, and by slowing their metabolic activities to reduce water and energy consumption (Booth 2006). These species are widespread through the wet-dry tropics and breed during the wet season. As a result, the usual aestivation time is five to six months (Tracy *et al.* 2007). However, they can remain in aestivation for up to five years, enabling them to survive extended droughts (Booth 2006; Kayes *et al.* 2009).

These are large frogs with a distinctive wide jaw. The striped burrowing frog (*Cyclorana alboguttata*) ranges from light brown to olive brown and green on its back, often with darker raised stripes and spots. It often has a green or yellow green stripe down its back and a dark mask from each nostril over the tympanum to the point of the shoulder. Its toes are half-webbed. The water-holding frog (*Cyclorana platycephala*) is a plain frog, usually dull grey to brown or green, occasionally with darker patches. Like the striped burrowing frog, it has a pale green stripe down the spine but it can be distinguished from the former species by its smoother skin and fully webbed toes. The rough frog (*Cyclorana verrucosa*) is similar in appearance to the striped burrowing frog but has slightly rougher, more warty skin and less webbing on its toes.

Distribution

All three species are widespread through the north of Australia and their range extends into the northern inland section of the Murray-Darling Basin. The striped burrowing frog, water-holding frog and rough frog have an overlapping range in the northern section of the Basin and have been recorded from the Darling, Gwydir, Paroo, Barwon, Macquarie, Namoi and Warrego rivers. The three species co-occur around the Macquarie Marshes (Jo Ocock, University of New South Wales, *pers. comm.*). The geographic range of the water-holding frog extends further south than that of the rough frog and striped burrowing frog; it occurs around Willandra and Billabong creeks and the lower Lachlan River.

Reproduction and life-cycle

Breeding coincides with heavy rains between spring and early autumn. The tadpoles cannot tolerate cold water and breeding by members of this genus is restricted to the warmer months.

Figure 5.2: Striped burrowing frog, *Cyclorana alboguttata* (a). Rough frog, *Cyclorana verrucosa* (b). Water-holding frog, *Cyclorana platycephala* (c). Photographs: Jodi Rowley (Australian Museum).

Tadpole life-span within this group is extremely variable: striped burrowing frog tadpoles take up to four months to reach metamorphosis while water-holding frog tadpoles can reach metamorphosis in one month under ideal conditions (Anstis 2002).

Habitat and trophic position

Members of this genus breed in a range of wetland habitats, including ephemeral creek systems, flooded wetlands and rain-fed pools. There have been no studies describing the exact habitat preferences but, given their wide geographic range, they are expected to be generalists. Interactions with fish and other predators are unknown; in general, species that utilise temporary waterbodies have limited tolerances to aquatic predators such as fish but this relationship has not been investigated for any members of this group.

Water requirements

This group of species may respond to spring and summer flooding, but may also be influenced by heavy rainfall. The relative importance of the two processes is unclear. Assuming that flooding may drive breeding in some instances, the water-holding frog and rough frog should recruit successfully in waterbodies flooded for four and six weeks depending on environmental conditions. The striped burrowing frog requires water for three to four months for successful recruitment (Anstis 2002). This group can tolerate extended dry periods and will breed in rain-fed waterbodies. As a result, the group may be more tolerant of moderate reductions in flooding frequency than other frog species are, and could increase its abundance and distribution through wetland systems that have suffered decreased flooding frequency. This group may be negatively affected by the conversion of temporary waterbodies into permanent storages and by the regulation of rivers so that the flooding extent is reduced. Although the group may be more tolerant than some other wetland species, its capacity to survive during dry periods does have upper limits. Long-term reductions in flooding frequency to less than one in five years is likely to have a serious negative impact.

Common green tree frog: *Litoria caerulea*

A very well-known tree frog due to its habit of colonising houses and watertanks, the common green tree frog (*Litoria caerulea*) is widespread through northern Australia. It is a bright emerald with conspicuous toe disks, occasionally with white spots down its side (Figure 5.3). Metamorphs are similar in appearance to the adults but have a dark band across the eye and over the tympanum. Adults call from elevated open positions or from pipes and tree hollows (Gibbons and Lindenmayer 2000), with a low repeated 'croak-croak-croak-croak' (Barker *et al.* 1995).

Distribution

The common green tree frog is widespread through northern Australia. In the Murray-Darling Basin it is relatively common in the northern sections around the Condamine, Border, Gwydir and Macquarie rivers, where it has been recorded breeding following flooding of the Macquarie Marshes and the Lachlan River. It appears to be declining in the south and is likely to be locally extinct from the Murrumbidgee River at the southern edge of its range (Wassens 2006).

Reproduction and life-cycle

Breeding is restricted to warm summer months. Tadpole development times can be as little as one and a half months if water temperatures are around 30°C (Anstis 2002). In the wet-dry tropics, the common green tree frog is often one of the first species to breed, depositing eggs soon after waterbodies have filled. Tadpoles develop rapidly and in high numbers (Watson *et*

Figure 5.3: Common green tree frog, *Litoria caerulea*. Photograph: Steven Sass (EnviroKey).

al. 1995). Over 2000 eggs can be laid in large floating clumps which break down as the tadpoles hatch, after three to four days (Anstis 2002). Tadpoles are omnivores and grow rapidly, sometimes reaching nearly 90 mm from nose to tail.

Habitat and trophic position

During the day, the common green tree frog takes refuge in tree hollows and buildings which may be some distance from water. It is active at night and forages in arboreal habitats. The species does not form calling aggregations; instead, males establish an elevated calling position after heavy rain and females move selectively to individual males (Watson *et al.* 1995). There have been no studies of the specific breeding habitats of this species in the Murray-Darling Basin.

Water requirements

The common green tree frog has a restricted activity period and will respond only to spring and summer flooding. There have been no specific studies on the water requirements and the extent to which the distribution of the common green tree frog in the Murray-Darling Basin is linked to wetland flooding, as opposed to local rainfall, is unclear. Successful recruitment is most likely if water remains pooled for at least two months to allow tadpoles to complete metamorphosis.

Broad palmed frog: *Litoria latopalmata*

The broad palmed frog (*Litoria latapalmata*) is an attractive slender frog with powerful hind legs which give it exceptional jumping ability (Figure 5.4). Although a member of the tree frog genus Hylidae, the broad palmed frog is a ground-dwelling species and rarely climbs. Its colour patterns are highly variable, ranging from solid light to dark brown or mottled light and dark brown. It usually has a thick dark band from each nostril across the eye and tympanum, which breaks into a series of scattered spots through to the armpit. The back of the thighs are usually yellow and the belly and throat are white. The broad palmed frog has a distinctive call, a

Figure 5.4: Broad palmed frog, *Litoria latopalmata*. Photograph: Skye Wassens (CSU).

high-pitched repeated 'pew-pew-pew', which sounds similar to a person hitting a wire fence with a stick.

Distribution

The full geographic distribution of the broad palmed frog is often underreported in field guides (Lemckert *et al.* 1995; Wassens 2006). Broad palmed frogs are common in the northern Basin through the middle and upper reaches of the Gwydir, McIntyre, Peel, Namoi and Lachlan rivers and their tributaries. It also occurs through scattered locations on the middle reaches of the Murrumbidgee River, especially the mid-Murrumbidgee wetlands (Jansen and Healey 2003; Wassens 2006).

Reproduction and life-cycle

Breeding occurs in spring and summer. Eggs are deposited in a foamy mass among aquatic vegetation in warm shallow water. In the creek systems around the Lachlan River, eggs and tadpoles were observed after small in-stream flows in January (Wassens and Maher, in press). Tadpole life-span is unknown but, based on the observations of metamorphs, the time taken to reach metamorphosis is likely to be two to four months (Anstis 2002). Tadpoles feed on vegetation and sediment at the bottom of warm shallow water (Anstis 2002).

Habitat and trophic position

This species occupies a range of habitats, including river red gum (*Eucalyptus camaldulensis*) forests, black box (*Eucalyptus largiflorens*) woodlands and open agricultural lands. It is restricted to areas near permanent and semi-permanent water (Anstis 2002). In creek systems around the Lachlan River, the broad palmed frog was strongly associated with temporary waterbodies containing abundant aquatic vegetation. In the mid-Murrumbidgee wetlands it occurs in vegetated oxbow lagoons (Jansen and Healey 2003; Wassens 2006). It has been recorded calling from small backwaters and pools of the Murrumbidgee River (Skye Wassens, *pers. obs.*). In the McIntyre catchment it is common in seasonally flooded oxbow lagoons (Cowood 2008).

Water requirements

The broad palmed frog is usually associated with permanent and semi-permanent waterbodies and connected seasonally flooded habitats. Given this relationship, the broad palmed frog is expected to have limited tolerance for wetland drying; the loss of permanent and semi-permanent wetlands is likely to lead to local declines. In seasonally flooded wetland systems, the broad palmed frog is most likely to respond to spring and summer flooding. Water should remain pooled for three to four months under ideal conditions to allow tadpoles to reach metamorphosis.

Peron's tree frog: *Litoria peronii*

The Peron's tree frog (*Litoria peronii*) is a very common arboreal species and is often observed calling from trees and fallen timber within and around waterbodies. It is a medium-sized frog, usually around 50 mm (Figure 5.5). It is usually grey, but can range from grey to dark brown. Emerald flecking is common, especially in younger frogs (Cogger 2000). It has distinctive crossed pupils. The inner hind-thighs are typically bright yellow with black mottling. The Peron's tree frog has distinctive toepads which aid in climbing. Its call is a low ascending cackle, which has been likened to a person laughing manically.

Distribution

The Peron's tree frog is common and widespread throughout the Murray-Darling Basin, except the Wimmera and Avoca rivers away from the Murray River.

Reproduction and life-cycle

The adult calls from elevated positions around waterbodies only during the warmer months (October to February) or when the air temperature exceeds 20°C (Wassens *et al.* 2009). Females

Figure 5.5: Peron's tree frog, *Litoria peronii*. Photograph: Sascha Healy (DECCW).

lay around 1700 eggs singly or in small groups within leaf litter, woody debris and vegetation (Anstis 2002). Tadpoles are usually observed between January and April. They hatch after four to six days and can grow to over 70 mm. Tadpoles are active swimmers and cruise near the surface of deep water, feeding on vegetation, dead insects and occasionally mosquito larvae (Anstis 2002).

Habitat and trophic position

The Peron's tree frog utilises a wide range of permanent, semi-permanent and temporary waterbodies, including dams, creeks, ponds and seasonally flooded wetlands. It prefers deeper open ponds and rarely breeds in very shallow well-vegetated waterbodies. During the day it shelters in tree hollows and under loose bark. It is relatively common around houses and other man-made structures. The distribution is closely linked to the availability of standing timber and it may be absent from cleared areas (Hazell *et al.* 2001). In the Lowbidgee floodplain, the Peron's tree frog is more abundant in river red gum forests than in black box and lignum (*Muehlenbeckia florulenta*) wetlands further from the river, probably due to the limited availability of daytime shelter sites (Wassens *et al.* 2008).

Water requirements

The Peron's tree frog occupies a range of habitats and appears to be relatively unaffected by moderate reductions in flooding frequency. Its abundance has increased in river red gum wetlands within the Lowbidgee floodplain following significant reductions in wetland flooding, probably because it is able to breed successfully in permanent waterbodies and small residual ponds. Although calling and spawning are restricted to spring and summer, tadpoles may linger within waterbodies until April. As a result, the species is most likely to breed successfully in wetlands with long hydroperiods.

Southern bell frog: *Litoria raniformis*

The southern bell frog (*Litoria raniformis*) is a handsome wetland-dependent species (Figure 5.6). The southern bell frog has undergone major declines throughout the Murray-Darling Basin and is listed as threatened under the Commonwealth *Environment Protection and Biodiversity Conservation Act 1999.* The colour and patterning of the southern bell frog is quite variable and it is sometimes confused with other species, including the common green tree frog, Peron's tree frog and members of the *Cyclorana* genus. The southern bell frog is usually bright emerald with bronze and black patches, but can range from dark brown to grey to solid green. The frog has a distinctive white, cream or yellow skinfold from behind the eye to the groin. The inside of the thigh is bright blue; extending the hind legs to check for the blue coloration is often the best way to confirm the identity of this species.

Distribution

The southern bell frog was formerly widespread through south-eastern Australia and Tasmania. It is presently restricted to the Coleambally irrigation district south of the Murrumbidgee River; there are scattered populations along the Murrumbidgee River between Hay and Balranald, with a number of populations persisting in the Lowbidgee wetland system (Wassens 2008). The southern bell frog occurs in scattered oxbow lagoons and seasonally flooded wetlands along the Murray River, especially between Mildura and the lower lakes. Small populations are likely to occur through the Wakool region (Wassens 2008) and there is a small population on the Mitta Mitta River (David Hunter, DECCW, *pers. comm*). It also occurs in seasonally flooded wetlands in the Chowilla floodplain. The southern bell frog formerly

Figure 5.6: Southern bell frog, *Litoria raniformis*. Photograph: Sascha Healy (DECCW).

occurred through the middle and lower Lachlan River and Willandra Creek, but the current status of these populations is unknown (Wassens 2008).

Reproduction and life-cycle

The southern bell frog breeds in spring and summer following flooding. Males call while floating in the water, often among aquatic vegetation. Females can produce over 2000 eggs, which are deposited in a single layer in shallow water (Anstis 2002). Tadpoles can grow very large, sometimes over 100 mm from nose to tail. Tadpoles cruise mid-water in shallow well-vegetated sections of the waterbody. They often travel in small schools, and may school with small fish (Anstis 2002). Tadpole development times can vary from over 12 months in cooler regions (Anstis 2002) to two and a half months in warmer water (Cree 1984). Although the tadpole development times can be short, recruitment tends to be highest at sites where water remains pooled throughout spring and summer (approximately six months).

Habitat and trophic position

In the Murray-Darling Basin the southern bell frog is typically associated with large seasonally flooded wetlands which contain relatively complex aquatic vegetation communities. The southern bell frog has very limited capacity to survive dry periods and must move to permanent refuge sites such as river channels, deep persistent pools or farm dams if seasonally flooded waterbodies dry. Occupancy of wetlands is often linked to their proximity to drought refuge habitats. Recent studies indicate the common carp may significantly reduce recruitment success of the southern bell frog in seasonally flooded wetlands (Spencer and Wassens 2009).

Water requirements

Of all the wetland frog species, the southern bell frog is the most sensitive to changes in wetland hydrology. It has highly specialised water requirements. It prefers seasonally flooded waterbodies with long hydroperiods (five to seven months). It has very limited tolerance for reductions in flooding frequency leading to wetland drying, and can be excluded from wetlands if flood frequency is reduced. It is active only in spring and summer and therefore does not benefit from winter flooding unless the water remains pooled through spring and summer. It can breed successfully in permanent waterbodies as long as the common carp is absent or in low numbers (Wassens *et al.* 2008). The southern bell frog is most likely to breed successfully in waterbodies which are flooded from October to March.

Desert tree frog: *Litoria rubella*

The desert tree frog (*Litoria rubella*) is widespread through dry-wet tropics, through Queensland and the northern section of the Murray-Darling Basin. It is a relatively small frog, usually around 35 mm in length (Figure 5.7). Colour varies from light grey to fawn to red brown, with some darker flecks. It has a distinctive thick dark band from each nostril across the eye and tympanum to the armpit. It has large toepads (Barker *et al.* 1995; Cogger 2000). The call is a harsh repeated buzzing 'arrrrk-arrrrk-arrrrk', which sounds a little like a seagull.

Distribution

The desert tree frog is widespread and common through the northern section of the Murray-Darling Basin, including the Condamine, Gwydir, Macquarie, Lachlan, Warrego and Paroo rivers and the northern section of the Darling River. It is also common through the wet coastal areas of northern New South Wales and Queensland.

Reproduction and life-cycle

Breeding usually occurs during summer and autumn after heavy rain, but may be opportunistic following heavy rains at any time of year. Females lay around 400 eggs in small floating

Figure 5.7: Desert tree frog, *Litoria rubella*. Photograph: Jodi Rowley (Australian Museum).

clusters (Anstis 2002). Tadpoles are omnivores feeding on vegetation, detritus and decaying animals as well eggs of other frog species (Anstis 2002; Crossland 1998). In the northern parts of its range, ornate burrowing frog (*Limnodynastes ornatus*) tadpoles can prey heavily on desert tree frog tadpoles; this can limit its distribution to sites which are free of the ornate burrowing frog (Crossland 2000). Tadpole development is rapid, taking approximately five weeks when water temperatures are over 30°C (Anstis 2002).

Habitat

The desert tree frog often occurs around human habitation, where it shelters in watertanks, sinks and other vessels. It also shelters in cracks between rocks, in tree hollows and under loose bark (Tyler 1989). The desert tree frog prefers temporary waterbodies, but has been recorded calling from around dams and urban ponds (Anstis 2002). In the Lachlan floodplain, the desert tree frog was recorded only from small temporary overflows. It was absent from nearby permanent sites (Wassens and Maher, in press).

Water requirements

The desert tree frog has relatively modest water requirements. It requires spring and summer flooding that creates temporary pools. It occurs in the Macquarie Marshes but the extent to which it is dependent on environmental flooding is unclear (Jo Ocock, University of New South Wales, *pers. comm.*). In the Lachlan River system the species benefited from small instream environmental flows which spilt over weirs to create temporary waterbodies. In most instances the desert tree frog may be able to persist on rainfall, except where rain-fed ponds and depressions have been heavily modified.

Giant banjo frog: *Limnodynastes interioris*

Eastern banjo frog: *Limnodynastes dumerili*

Northern banjo frog: *Limnodynastes terrareginae*

These species belong to a complex of frogs called the western banjo frog (*Limnodynastes dorsalis*) complex (Martin 1972; Schauble *et al.* 2000). They are large non-cocoon-forming burrowing frogs, typically with dull to bright orange on their sides, throats and bellies (Figure 5.8). This group can be easily distinguished from other frogs by their orange colour, large size and large raised tibial gland. Member species have a distinctive 'bonk-bonk-bonk' call which sounds like a banjo being plucked, hence the common name. The group contains a number of species, three of which are important in the Murray-Darling Basin: the giant banjo frog which occurs in the northern central inland regions, the eastern banjo frog which occurs through the coastal and tablelands regions and throughout the Murray River corridor, and the northern banjo frog which occurs in the northern inland section, especially around the upper Darling River and its tributaries (Martin 1972).

Distribution

The giant banjo frog occurs through the central inland regions of New South Wales and is relatively common through the middle and lower Murrumbidgee, Lachlan, Macquarie and Darling rivers. The range of the giant banjo frog and the eastern banjo frog overlap in a small section of the Murray River around Albury. The eastern banjo frog is very common throughout coastal and tablelands regions of south-eastern Australia and the upper reaches of rivers in the northern section of the Murray-Darling Basin, including the Gwydir, Namoi, Macquarie and

Figure 5.8: Giant banjo frog, *Limnodynastes interioris*. Photograph: Skye Wassens (CSU).

Lachlan rivers. In the southern Basin the eastern banjo frog is common throughout upper tributaries of the Murrumbidgee River. It is widespread along the Murray River to South Australia, as well as in the Kiewa, Ovens, Goulburn, Campaspe, Lodden, Avoca and Wimmera rivers in Victoria. The northern banjo frog has a restricted distribution in the northern Basin, occurring in scattered locations in the Gwydir, Namoi, Castlereagh and Macquarie rivers (Martin 1972).

Reproduction and life-cycle

Males call from small depressions under vegetation at the edge of wetlands, or while partly submerged under overhanging banks or among floating vegetation (Anstis 2002). The frogs can be active at any time of the year but breeding is most likely to occur during spring and summer. In the Lowbidgee wetland system, the giant banjo frog was recorded calling in July following winter flooding, and between October and January following spring and summer flooding (Spencer and Wassens 2009; Wassens *et al.* 2008). Tadpole life-spans are variable, around three months for the giant banjo frog and northern banjo frog and over six months for the eastern banjo frog in cooler areas (Anstis 2002), but possibly less in the lower sections of the Murray.

Habitat and trophic position

There have been no studies on the specific habitat requirements of this group. Breeding occurs in a range of habitats including dams, seasonally flooded wetlands and slow-moving streams. In the Lowbidgee wetland system, the giant banjo frog is associated with seasonally flooded black box and lignum wetlands and open spike rush (*Eleocharis* sp.) wetlands within river red gum forests. In the Lachlan catchment, the giant banjo frog was recorded only from seasonally flooded creek systems. The eastern banjo frog has been recorded breeding in seasonally flooded oxbow lagoons, seasonally flooded wetlands, permanent creeks, urban ponds and farm dams.

Water requirements

There has been no research on the water requirements of this group nor on how long individuals can persist between floods. Unlike the *Cyclorana* species, this group is not known to significantly slow metabolism during dry periods. Instead, it is likely to burrow further down through the soil profile as watertables decline, in order to maintain its moisture balance (Booth 2006). The frogs often emerge from their burrows and forage during heavy rain; these foraging opportunities may extend the length of time they can persist between floods. In the Lowbidgee floodplain, the giant banjo frog has been recorded calling from sites that had not been watered for three years (Wassens and Spencer, unpubl. data). Successful recruitment may require wetlands to be flooded for around six months.

Barking marsh frog: *Limnodynastes fletcheri*

The barking marsh frog (*Limnodynastes fletcheri*) is a common frog in seasonally flooded and permanent waterbodies throughout the Murray-Darling Basin. It is a medium-sized frog of up to 50 mm, usually light grey or creamy with brown to olive green splotches (Cogger 2000; Figure 5.9). It can look similar to the spotted marsh frog (*Limnodynastes tasmaniensis*), especially when young. There is usually an apricot or rusty blush on the back of the eye, visible when looking at the frog from behind. It also has a T-shaped 'hat' which runs between both eyes. Males often have well-developed muscular forearms which may play a role in combat to maintain calling territories.

Distribution

The barking marsh frog is widespread through all but the cooler slopes and driest inland sections of the Murray-Darling Basin. It occurs through the Condamine, Gwydir, Namoi,

Figure 5.9: Barking marsh frog, *Limnodynastes fletcheri*. Photograph: Skye Wassens (CSU).

Castlereagh, Macquarie, Lachlan and Murrumbidgee rivers, and the Murray River west of Albury to South Australia. It is widespread in the Barmah-Millewa wetlands on the Murray River (Ward 2004).

Reproduction and life-cycle

Very little is known of the breeding biology of the barking marsh frog. It is most active in spring and summer, but occasionally calls during warm weather at the end of winter. Eggs are laid in foamy masses attached to vegetation (Anstis 2002). Tadpoles are large (around 70 mm) and are usually bottom-dwellers. They will swim to the water surface to feed; their diet is not known (Anstis 2002). The time for tadpoles to reach metamorphosis is four to five months in the Lowbidgee floodplain when wetlands are flooded through spring and summer (Wassens *et al.* 2008).

Habitat and trophic position

There have been few published studies describing the habitat preferences of this species. The barking marsh frog is known to occupy a range of habitats including dams, rice bays, creeks and wetlands. It has a strong preference for areas with abundant emergent vegetation, such as spike rush and cumbungi (*Typha* sp.) (Jansen and Healey 2003; Wassens *et al.* 2007). It appears to be most common in seasonally flooded wetlands with long hydroperiods of approximately six months.

Water requirements

The barking marsh frog has flooding requirements similar to those of the southern bell frog, with which it often co-occurs. It prefers wetlands with longer hydroperiods and generally occurs only if there is permanent water nearby, although it preferentially breeds in seasonally flooded sites. It has limited tolerance for wetland drying and may be excluded from wetlands that are dry for extended periods if there are no permanent refuge habitats nearby.

Spotted marsh frog: *Limnodynastes tasmaniensis*

The spotted marsh frog (*Limnodynastes tasmaniensis*) is one of the most common and widespread frogs in the Murray-Darling Basin. It is a medium-sized (35–45 mm) smart-looking frog (Figure 5.10). Colour and patterning are extremely variable. There is usually a series of regular dorsal spots, which range from green to dark brown. It may have a yellow, red or white stripe down its back. The spotted marsh frog has a raised white ridge from below the eye to the point of the shoulder. The species is represented by two distinct races which have very different calls; north of the Murray River the frog has a rapid 'uuk-uuk-uuk-uuk' or 'tok-tok-tok' and south of the Murray River the frog has a single 'poc' or 'toc'.

Distribution

The spotted marsh frog is widespread through all major river systems in the Murray-Darling Basin. It is very common and abundant through the Darling, Namoi, Gwydir, Macquarie, Lachlan, Murrumbidgee, Murray, Kiewa, Ovens, Broken, Goulburn, Lodden, Avoca and Wimmera rivers and associated wetland systems.

Reproduction and life-cycle

Breeding occurs in late winter, spring and summer. Eggs are laid as a foamy mass which sits on the water surface. Hatching occurs three to four days after laying (Anstis 2002). Tadpoles take two and a half to four months to reach metamorphosis, depending on water temperature (Anstis 2002).

Figure 5.10: Spotted marsh frog, *Limnodynastes tasmaniensis*. Photograph: Skye Wassens (CSU).

Habitat and trophic position

The spotted marsh frog will readily colonise any wet freshwater area including temporary depressions, semi-permanent and permanent wetlands, creeks, dams, irrigation canals and urban ponds. It does not have specific requirements in terms of aquatic vegetation, although tadpoles are generally more abundant in areas with aquatic vegetation (Wassens *et al.* 2008). It has limited capacity to burrow. Adults congregate around permanent water during droughts and distribution is restricted to areas with some permanent water.

Water requirements

The spotted marsh frog responds best to spring and summer flooding, but may breed opportunistically at any time of year (Anstis 2002). Following winter (May–June) watering in the Lowbidgee wetlands, the spotted marsh frog did not commence calling until August. Tadpoles require water to remain pooled for at least three months.

Summary of water requirements

The majority of wetland-dependent frogs discussed in this chapter are most active in spring and summer and have a preference for wetlands with longer hydroperiods (Table 5.1). The remaining species that are known to call through winter, such as the common eastern froglet and eastern sign-bearing froglet, desert tree frog, eastern banjo frog and giant banjo frog, have flexible requirements and will also breed through spring and summer. The timing of inundation has a strong influence on which frog species are able to successfully recruit within a wetland. It can also influence tadpole development times, which are slower in cooler water (Cree 1984). Management decisions that involve winter flooding should take into account the impacts on resident frog populations and, where possible, should maintain pooled water over spring and summer to allow recruitment of spring-breeding frogs.

Table 5.1: Breeding habitat requirements of frogs in the Murray-Darling Basin

	Distribution in the northern and southern Basin		Preferred hydrology of breeding sites			Timing of breeding				Tadpole life-span (months)
Species	Northern	Southern	Short (<3 months)	Long (3–6 months)	Permanent	Spring	Summer	Autumn	Winter	
Common eastern froglet	*	*	*	*	*	C	CM	CM	C	2–4
Eastern sign-bearing froglet	*	*	*	*	*	C	CM	CM	M	2–4
Striped burrowing frog	*			*		C	CM	M		3–4
Rough frog	*		*	*		C	C	CM		2–4?
Water-holding frog	*		*			C	CM	M		1–2
Common green tree frog	*		*	*			C	M		2–3
Broad palmed frog	*	*	*	*	*	C	CM	M		2–4?
Peron's tree frog	*	*	*	*	*	C	CM	M		3–4
Southern bell frog		*		*	*	C	CM	M		3–5
Desert tree frog	*		*	*	*	CM	CM	CM	T	1–2
Eastern banjo frog	*	*		*	*	CT	CM	CM	C	5–6
Northern banjo frog	*			*	*	C	CM	CM		2–4?
Giant banjo frog		*		*	*	CM	CM	C	C	4–6
Spotted marsh frog	*	*	*	*	*	C	CM	M		3–4
Barking marsh frog	*	*		*	*	C	CM	M		3–4

Breeding habitat is based on information from Cogger (2000), Barker and Grigg (1977), Robinson (1998) and Anstis (2002).

? = estimate based on limited data.

C = calling is likely to occur.

M = metamorphs may be present.

T = tadpoles may be present.

Many of the frog species discussed prefer wetlands with long hydroperiods (five to six months). The water-holding frog, common green tree frog and desert tree frog have relatively rapid development times and can breed successfully in rain-fed pools. However, given the wide geographic range of those species, the use of rain-fed pools (Anstis 2002; Cogger 2000) may reflect patterns of habitat use in the wet-dry tropics where rainfall is higher and persistent rain-fed pools are more common. Seasonal and overbank flooding may be more important for populations that occur in the drier inland regions of the Murray-Darling Basin. The relationship between wetland hydroperiod and the survival of metamorphs and juvenile frogs has not been investigated. Extending hydroperiods beyond the tadpole development times may increase foraging opportunities for juvenile frogs, thereby increasing their fitness and ability to survive their first winter.

With the exception of *Cyclorana* species, wetland frogs have very limited tolerances for drying and are likely to be negatively affected by reductions in the frequency of inundation. However, there are often complex interactions between recruitment success in a given year, adult mortality rates, the proximity and quality of more persistent waterbodies and local climatic conditions. These variables all influence the length of time that populations persist between floods, making it difficult to place definitive upper and lower limits on the required flooding frequency. For example, if recruitment success is poor then flooding may be required every year in order to maintain a viable population, while biannual flooding may be sufficient if levels of recruitment success are high enough to exceed mortality during non-breeding years. It is important to have a clear understanding of the rates of recruitment and mortality in the target wetland system, when developing guidelines for environmental flooding.

References

Anstis M (2002) *Tadpoles of South-eastern Australia: A Guide with Keys*. Reed New Holland: Sydney.

Barker J and Grigg GC (1977) *A Field Guide to Australian Frogs*. Rigby: Adelaide.

Barker J, Grigg G and Tyler M (1995) *A Field Guide to Australian Frogs*. 2nd edn. Rigby: Sydney.

Booth DT (2006) Effect of soil type on burrowing behavior and cocoon formation in the green-striped burrowing frog, *Cyclorana alboguttata*. *Canadian Journal of Zoology* **84**, 832–838.

Cogger HA (2000) *Reptiles and Amphibians of Australia*. Reed New Holland: Sydney.

Cowood A (2008) Floodplain billabong frog assemblages: the influence of hydrological connectivity. BAppSci thesis. University of Canberra.

Cree A (1984) Breeding biology, respiration, and larval development of two introduced frogs (*Litoria raniformis* and *L. ewingi*). *New Zealand Journal of Zoology* **11**, 179–188.

Crossland MR (1998) A comparison of cane toad and native tadpoles as predators of native anuran eggs, hatchlings and larvae. *Wildlife Research* **25**, 373–381.

Crossland MR (2000) Direct and indirect effects of the introduced toad *Bufo marinus* (Anura: Bufonidae) on populations of native anuran larvae in Australia. *Ecography* **23**, 283–290.

Flecker AS, Feifarek BP and Taylor BW (1999) Ecosystem engineering by a tropical tadpole: density-dependent effects on habitat structure and larval growth rates. *Copeia* **1999**, 495–500.

Gibbons P and Lindenmayer D (2000) *Tree Hollows and Wildlife Conservation in Australia*. CSIRO Publishing: Melbourne.

Gillespie GR (2002) Impacts of sediment loads, tadpole density, and food type on the growth and development of tadpoles of the spotted tree frog *Litoria spenceri*: an in-stream experiment. *Biological Conservation* **106**, 141–150.

Hazell D, Cunnningham R, Lindenmayer D, Mackey B and Osborne W (2001) Use of farm dams as frog habitat in an Australian agricultural landscape: factors affecting species richness and distribution. *Biological Conservation* **102**, 155–169.

Healey M, Thompson D and Robertson A (1997) Amphibian communities associated with billabong habitats on the Murrumbidgee floodplain, Australia. *Australian Journal of Ecology* **22**, 270–278.

Jansen A and Healey M (2003) Frog communities and wetland condition: relationships with grazing by domestic livestock along an Australian floodplain river. *Biological Conservation* **109**, 207–219.

Karraker NE and Gibbs JP (2009) Amphibian production in forested landscapes in relation to wetland hydroperiod: a case study of vernal pools and beaver ponds. *Biological Conservation* **142**, 2293–2302.

Kayes SM, Cramp RL, Hudson NJ and Franklin CE (2009) Surviving the drought: burrowing frogs save energy by increasing mitochondrial coupling. *Journal of Experimental Biology* **212**, 2248–2253.

Kupferberg SJ, Marks JC and Power ME (1994) Effects of variation in natural algal and detrital diets on larval anuran (*Hyla regilla*) life-history traits. *Copeia* **1994**, 446–457.

Lemckert F, Law B, Anderson J and Chidel M (1995) A further range extension of the broad-palmed frog *Litoria latopalmata*. *Herpetofauna* **25**, 12–13.

Mac Nally R, Horrocks G, Lada H, Lake PS, Thomson JR and Taylor AC (2009) Distribution of anuran amphibians in massively altered landscapes in south-eastern Australia: effects of climate change in an aridifying region. *Global Ecology and Biogeography* **18**, 575–585.

Martin AA (1972) Studies in Australian amphibia. III. The *Limnodynastes dorslis* complex (Anura: Leptodactylidae). *Australian Journal of Zoology* **20**, 165–211.

Mokany A (2007) Impact of tadpoles and mosquito larvae on ephemeral pond structure and processes. *Marine and Freshwater Research* **58**, 436–444.

Purrenhage JL and Boone MD (2009) Amphibian community response to variation in habitat structure and competitor density. *Herpetologica* **65**, 14–30.

Reading CJ (2003) The effects of variation in climatic temperature (1980–2001) on breeding activity and tadpole stage duration in the common toad, *Bufo bufo*. *Science of the Total Environment* **310**, 231–236.

Robinson M (1998) *A Field Guide to Frogs of Australia*. Australian Museum/Reed New Holland: Sydney.

Robinson M (2000) *A Field Guide to Frogs of Australia: From Port Augusta to Fraser Island, Including Tasmania*. Reed New Holland: Sydney.

Schauble CS, Moritz C and Slade RW (2000) A molecular phylogeny for the frog genus Limnodynastes (Anura: Myobatrachidae). *Molecular Phylogenetics and Evolution* **16**, 379–391.

Smith MJ, Schreiber SG, Scroggie MP, Kohout M, Ough K, Potts J, Lennie R, Turnbull D, Jin C and Clancy T (2007) Associations between anuran tadpoles and salinity in a landscape mosaic of wetlands impacted by secondary salinisation. *Freshwater Biology* **52**, 75–84.

Spencer JA and Wassens S (2009) 'Responses of waterbirds, fish and frogs to environmental flows in the Lowbidgee wetlands in 2008–09'. Sydney.

Tarr TL and Babbitt KJ (2002) Effects of habitat complexity and predator identity on predation of *Rana clamitans* larvae. *Amphibia Reptilia* **23**, 13–20.

Tracy CR, Reynolds SJ, McArthur MC, Tracy R and Christian KA (2007) Ecology of aestivation in a cocoon-forming frog, *Cyclorana australis* (Hylidae). *Copeia* **4**, 901–912.

Tyler M (1989) *Australian Frogs: A Natural History*. Reed New Holland: Sydney.

Ward PA (2004) 'Monitoring frog response to flooding in Barmah-Millewa Forest: 2003/04. Final report prepared for the Barmah-Millewa Forum'. Murray-Darling Basin Commission: Canberra.

Wassens S (2005) The use of space by the endangered southern bell frog (*Litoria raniformis*) in the semi-arid region of New South Wales, Australia. PhD thesis. Charles Sturt University.

Wassens S (2006) Frog communities of the Murrumbidgee Irrigation Area, NSW. In *Wetlands of the Murrumbidgee River Catchment: Practical Management in an Altered Environment.* (Eds IR Taylor, PA Murray and SG Taylor) pp. 86–95. Fivebough and Tuckerbil Wetlands Trust: Leeton, NSW.

Wassens S (2008) Review of the past distribution and decline of the southern bell frog *Litoria raniformis* in New South Wales. *Australian Zoologist* **34**, 446–452.

Wassens S and Maher M (in press) River regulation influences the composition and distribution of inland frog communities. *River Research and Applications.*

Wassens S, Arnaiz OL and Watts RJ (2007) *Assessing the Diversity, Abundance and Hydrological Requirements of Frog Populations at 'Burrawang West' and 'Yarnel' Lagoons, Two Small Wetlands on Anabranch Creeks of the Mid-Lachlan River.* Dept of Environment and Climate Change: Canberra.

Wassens S, Arnaiz OL, Healy S, Watts RJ and Maguire J (2008) *Hydrological and Habitat Requirements to Maintain Viable Southern Bell Frog (Litoria raniformis) Populations on the Lowbidgee Floodplain: Phase 1.* Dept of Environment and Climate Change: Canberra.

Wassens S, Healy S and Watts R (2009) *Optimising Frog Breeding Responses to Flooding in Managed Wetlands. I. Temporal Activity Pattern and Selection of Breeding Sites.* Dept of Environment and Climate Change: Albury.

Watson GF, Davies M and Tyler MJ (1995) Observations on temporary waters in north-western Australia. *Hydrobiologia* **299**, 53–73.

Whitaker BR (2001) Reproduction. In *Amphibian Medicine and Captive Husbandry.* (Eds KM Wright and BR Whitaker) pp. 285–299. Krieger Publishing: Malabar, FLA.

Chapter 6

Crustaceans and molluscs

Hugh A Jones

Introduction

Hydrological variation plays a vital role in organising the biotic diversity within riverine ecosystems (Richter *et al.* 1997) and is at the heart of several conceptual models of river function (Thorp and Delong 1994; Vannote *et al.* 1980; Ward and Stanford 1983). In this chapter, the organising role of the flood pulse is considered for freshwater molluscs (bivalves and snails) and decapod crustaceans, and the water requirements of selected species are considered. The flood pulse is clearly of central importance in the lives of those animals that occupy floodplain wetlands, but for riverine animals flowing water also plays a key role (Young 2001). A modified view of the flood pulse concept (Junk *et al.* 1989) is adopted here. While acknowledging the importance of the flood pulse as a key driver of floodplain processes, it recognises in-channel flow variability as a significant influence in the lives of riverine organisms (Puckridge *et al.* 1998; Walker *et al.* 1995).

The lowland rivers of the Murray-Darling Basin have highly variable and unpredictable flow regimes. In this chapter, attention is given to the life-history adaptations that riverine and floodplain wetland biota have evolved to cope with hydrological variability. Snails and crustaceans (e.g. branchiopods and ostracods) are abundant in floodplain wetlands and have evolved strategies to take advantage of the 'boom and bust' cycles that are characteristic of this environment. Many of these floodplain taxa are characterised by rapid development and early reproduction, high fecundity, short life-spans and oviparity to take advantage of the often-brief period when water is present (Marcus and Weeks 1997; Weeks *et al.* 1997; Williams 1985).

Snails and crustaceans have limited powers of dispersal; consequently, they must be able to resist the dry phase between floods or to utilise refugia. Inland crabs and western yabbies take refuge in burrows beneath the wetland during the inter-flood dry-period (Greenaway 1984; Williams 1985). Many floodplain snails are capable of aestivation (McMichael and Iredale 1959; Smith and Burn 1976; Walker 1998). All branchiopods and many ostracods and copepods produce desiccation-resistant cysts or eggs capable of surviving years out of water (De Deckker 1983; Ingram *et al.* 1997; Williams 1985). Only a proportion of cysts in branchiopods are activated following flood rains; the remainder stay dormant until subsequent floods (Brendonck 1996). This is considered a bet-hedging strategy to deal with the unpredictability of the wetting-drying cycle, so that the cyst bank is not exhausted if there is premature drying of the wetland (Simovich and Hathaway 1997).

Flowing water plays a central role in the lives of potamic animals (Young 2001). Many riverine snails and crustaceans have evolved behavioural and reproductive strategies to counteract downstream displacement by river flow. Most species brood their young and the planktonic larval stage has been either abbreviated or avoided. Pulmonate snails and atyid shrimps avoid strong currents and occupy still water or slow-flowing habitats in the littoral zones of rivers. Floods disperse these animals throughout the river system while, at the same time, connecting floodplain wetlands with the river and providing important nursery areas for riverine decapods and snails (Walker *et al.* 2009). Some potamic animals, such as river mussels and Murray crayfish, are stenotopic and require well-oxygenated flowing waters. Potamic animals tend to have a narrow range of environmental tolerances and are less able to cope with the drying of their habitat than floodplain taxa. Long sections of the Darling River and several of its northern tributaries dry up during prolonged droughts and waterholes are important drought refuges in these rivers.

The species profiles in this chapter bring together current knowledge of the water requirements for a selection of molluscs and crustaceans occurring within the Murray-Darling Basin. The review is restricted to native species that are common and widespread within floodplain ecosystems and for which there were sufficient ecological data. On the basis of these criteria, the Sphaeriidae, Hydrobiidae, Bithyniidae and Glacidorbidae were excluded as few members of these molluscan families occur in floodplain rivers. Species profiles for the Crustacea are limited to the decapod crustacea that have been reasonably well studied. Branchiopods (fairy shrimps, shield shrimps, clam shrimps and water fleas), ostracods and copepods are excluded because of the paucity of ecological data for Australian species, even though these crustaceans are widespread in floodplain wetlands (Ingram *et al.* 1997; Timms 2006).

For each species, a basic morphological description is provided including a comment on its taxonomic status where appropriate. This is followed by a description of its distribution within the Murray-Darling Basin and brief descriptions of habitat, diet and feeding behaviour, reproduction and life-cycle, and water requirements. The water requirements section summarises the effects of water flow on reproduction and the different developmental stages (larval, juvenile and adult). Studies of floodplain species have examined adaptations to the drying phase but very few studies have considered the effects of flood frequency, duration or timing on the population dynamics of snails and decapod crustaceans. There are almost no studies of the water requirements of riverine snails, but several studies have examined the effects of floods and the low-flow regime on riverine decapods.

Molluscs – bivalves

River mussel: *Alathyria jacksoni*

The shells are generally brown or dull black and are large and heavy, attaining a length of 170 mm (Figure 6.1). Shells are usually elongate-oval and compressed but there is some variation in shell morphology in response to hydraulic conditions (Balla and Walker 1991). The hinge is of unionid type with erect peg-like pseudocardinal teeth located above deeply imbedded anterior muscle scars (McMichael and Hiscock 1958).

The river mussel is widespread in the Murray-Darling River system and is abundant in all major rivers except some of the Victorian tributaries of the Murray River, where it is uncommon (Walker 1981b). The species also occurs sporadically in the lower reaches of the Condamine-Balonne system (the Culgoa River), which flows intermittently, but it is more common in the upper reaches of this system (Jones 2007).

Figure 6.1: River mussel, *Alathyria jacksoni*. Photograph: Hugh Jones (DECCW).

Reproduction and life-cycle

The river mussel has a unique and highly specialised life-cycle (Kat 1984). Sexes are separate and females are ovoviviparous, brooding fertilised eggs in modified regions of the inner demibranchs called marsupia. The larvae of Australian mussels are obligate parasites of fish (Walker *et al.* 2001). The larva, known as a glochidium, possesses a pair of subtriangular valves terminated by a tooth on the apex of each valve (Walker 1981b). The river mussel releases glochidia during the warmer months, with peak release periods in spring and autumn. A range of native and non-native fish species serve as hosts for the river mussel, with the notable exception of members of the carp family (Cyprinidae). Encysted glochidia metamorphose into juvenile mussels after two to three weeks, depending on the water temperature. Mortality is extremely high in the early stages of life but once the juvenile mussel survives the first few years then it is likely to live for 20–30 years. The river mussel becomes sexually mature at an age of three to four years.

Habitat and diet

The river mussel is predominantly a suspension feeder but it may also gather food from the sediments (Vaughn *et al.* 2008). It consumes a wide range of particulate matter including zooplankton and detritus but phytoplankton and bacteria are likely to be the main components of the diet (Vaughn *et al.* 2008). The mussel filters large volumes of water and, under certain conditions, may regulate phytoplankton communities (Welker and Walz 1998). It plays an integral role in the nutrient dynamics of aquatic ecosystems, storing nutrients as mussel biomass and transferring nutrients from the water column to the sediments, thereby stimulating secondary production by other organisms (Christian *et al.* 2008; Pusch *et al.* 2001; Vaughn *et al.* 2008).

The river mussel is restricted to lotic habitats of permanent rivers and creeks. The habitat requirements are reflected by its physiology, as the river mussel does not cope well under hypoxic conditions and it has a low tolerance to desiccation (Sheldon and Walker 1989; Walker 1981b). Nevertheless, the river mussel may survive for some time in damp mud, as long as it is shaded from the sun.

Water requirements

The river mussel requires permanent flowing water. River regulation, through the proliferation of weirs along the Darling and lower Murray rivers, has reduced the availability of this habitat by transforming much of the Darling and Murray rivers into long pools (Thoms *et al.* 1996; Walker 1985). The lower lacustrine sections of weir pools support very few river mussels; the few that remain are restricted to a narrow zone along the banks of the drowned river channel (Jones 2007). In other respects, flow regulation may have benefited the species. Irrigation releases from headwater dams on the eastern tributaries of the Darling River provide steady elevated base flows and the middle reaches of these rivers rarely cease to flow during the summer months (McMahon and Finlayson 2003). In addition to providing sustained flows, these 'anti-drought' regimes have drowned lateral and point bars, creating additional habitat for the river mussel. The river mussel is most abundant where it is exposed to strong water currents, such as on the bends of rivers and in fast-flowing chutes (Walker 1981b).

In the Barwon-Darling River, the distribution and abundance of the river mussel appears to be driven by the frequency and severity of droughts (Jones 2007). The mussel is restricted to the deepest parts of the channel (the thalweg and deep pools), that remain wet during the droughts that occur every several years. A prolonged drought produced exceptionally low flows in the Darling River during 2006 (MDBC 2006), resulting in the emersion and death of thousands of river mussels. This event highlights the importance of managing the low-flow regimes in dryland rivers of the Murray-Darling Basin: unless drought refuges for the river mussel are protected, there may well be a contraction in the range of this species.

Freshwater mussel populations are dependent on the well-being of their host-fish populations since the loss of these will disrupt mussel recruitment. Hence, satisfying the flow requirements of native fish (Chapter 4) should indirectly benefit river mussel populations.

Billabong mussel: *Velesunio ambiguus*

Billabong mussel shells vary greatly in shape and reach a maximum length of 100 mm. The typical form in semi-lotic creek environments is elongate-oval (Figure 6.2) but billabong forms are often round and globose (Walker 1981b). The shell is generally light and thin and this, together with the lamellar hinge dentition which is a characteristic of the genus, helps to distinguish this species from the river mussel (*Alathyria jacksoni*) and *A. condola* in the Murray-Darling Basin (McMichael and Hiscock 1958). The beaks are smooth and unsculptured but are often eroded in old specimens. Shell colour varies from yellow brown to dark brown.

The billabong mussel is widespread across the eastern half of Australia from coastal north-eastern Queensland to the south coast of Victoria, including the inland drainages of the Lake Eyre Basin and Murray-Darling Basin (Walker 1981a). It is mostly absent from the coastal streams and rivers of New South Wales but is common in sections of the Richmond, Clarence, Macleay, Hawkesbury-Nepean and Shoalhaven rivers.

Figure 6.2: Billabong mussel, *Velesunio ambiguus*. Photograph: Hugh Jones (DECCW).

Reproduction and life-cycle

The life-cycle of the billabong mussel is similar to that of the river mussel. Like the river mussel, the billabong mussel releases glochidia throughout the warmer months. The glochidia are large (210 × 247 μm) and subtriangular, similar in appearance to the glochidia of the river mussel but slightly smaller (Walker 1981b). The glochidia parasitise a range of native fish including Australian smelt, bony bream, golden perch, silver perch, catfish, Murray cod, flat-headed gudgeon, western carp-gudgeon, Mitchellian hardyhead and congolli (Hiscock 1951; Walker 1981b). Members of the carp family do not appear to be compatible hosts. Metamorphosis may take two to three weeks to complete, depending on water temperature (Walker 1981b). Larval development is arrested at water temperatures near 12°C.

Habitat and diet

The billabong mussel is common in impoundments such as farm dams, lakes, billabongs and temporary creeks throughout the Murray-Darling Basin. Information on the diet of billabong mussels is limited but, based on the diet of the closely related *V. angasi*, it is likely to be a suspension feeder, most likely consuming detritus and phytoplankton (Humphrey 1984).

Water requirements

Though not directly dependent on floods, this species inhabits flood-dependent waterbodies on river floodplains including billabongs, backwaters and anabranches. In the tropical mussel *V. angasi*, water flow strongly influenced the distribution of mussels in billabongs and the highest population densities were always found where water currents provided optimal aerobic conditions (Humphrey and Simpson 1985). A similar distribution pattern has been observed for the billabong mussel in permanent waterholes in the Culgoa River, where it was restricted to a narrow zone along the side of the thalweg (Jones 2007).

There are no published data on the minimum duration of inundation necessary to sustain populations of billabong mussel. The billabong mussel is found in waterbodies that retain water for long periods (at least several months) and are frequently flooded (e.g. every one to two years). The species will burrow into mud when its wetland habitat dries out and is able to aestivate for at least a year out of water (McMichael and Iredale 1959; Walker *et al.* 2001). To establish and maintain a self-sustaining population in a waterbody, the billabong mussel must be able to complete its life-cycle on a host-fish and juvenile mussels must grow to a stage where they are able to aestivate during droughts.

Commencement of reproduction and larval development is dependent on water temperature; the billabong mussel does not require floods as a spawning cue (Walker 1981b). The factors affecting recruitment of juveniles have not been investigated for the billabong mussel, but a study of the closely related tropical mussel, *V. angasi*, provides some insights. In *V. angasi*, wet-season flows were identified as an important determinant of juvenile recruitment in floodplain billabongs because they prolonged the breeding season and prevented the development of anoxic conditions (Humphrey and Simpson 1985). Similarly, summer floods in rivers of the Murray-Darling Basin may enhance recruitment of *V. ambiguus* by inhibiting stratification and reducing hypoxia in billabongs and secondary channels.

Basket shell: *Corbicula australis*

These small (up to 25 mm in length) bivalves have solid triangular shell valves with a rounded ventral margin. The valves are almost equilateral with the umbos slightly displaced from the centre. Shells are tan, yellow or orange, often with dark markings, and are sculptured with conspicuous concentric ridges (Ponder *et al.* 2000; Figure 6.3). The nacre on the inside of the valves has a purple hue. The lateral teeth in *Corbicula* are transversely serrated, a feature that distinguishes them from pea shells (Brunckhorst 1998). Basket shells are found in perennial rivers and streams throughout eastern Australia (Iredale 1943).

Reproduction and life-cycle

Basket shells are simultaneous hermaphrodites and are ovoviviparous, brooding their young in the inner demibranchs, traits that are widespread among the Corbiculidae (Byrne *et al.* 2000; McMahon 1983). In coastal south-eastern Australia, gametogenesis occurs throughout the year but spawning and brooding of embryos is restricted to the warmer months when water temperatures exceed 20°C (Byrne *et al.* 2000). Several clutches per year are produced, each containing 400–3000 young. The young are released at an advanced crawling stage of development. The basket shell has a short life-span, with most dying after the first year. This combination of life-history traits is well suited to frequent habitat disturbance, as populations of basket shell are able to rapidly recover from population crashes.

Habitat and diet

The basket shell is a suspension feeder, usually found buried in the sand of shallow strongly flowing rivers and irrigation channels (Brunckhorst 1998; McMichael 1967). The numbers of these bivalves occasionally reach pest proportions, with shells fouling town water supply and irrigation pipelines (Sheldon and Walker 1993a; Smith 1978; Woolford 1984).

Water requirements

Other than a dependence on flowing water, little is known of the water requirements for this species. Temperature is one of the most important exogenous factors influencing the reproductive cycle in bivalves (Mackie 1984) and the timing of reproduction in the basket shell is no

Figure 6.3: Basket shell, *Corbicula australis*. Photograph: Hugh Jones (DECCW).

exception (Byrne *et al.* 2000). Several tributaries of the Murray and Darling rivers receive cold water releases from large dams on their headwaters (Preece 2004; Ryan *et al.* 2001). Water released from the bottom of these reservoirs depresses water temperatures below the threshold for the brooding of basket shell embryos; it is likely that the reproductive output of the basket shell is limited for some distance downstream of these large dams.

Molluscs – aquatic snails

River snail: *Notopala sublineata*

Two subspecies of river snail are recognised from the Murray-Darling drainage: Hanley's river snail (*N. sublineata hanleyi*; Figure 6.4), restricted to the Murray and Murrumbidgee rivers, and the Darling River snail (*N. sublineata sublineata*). Both subspecies are medium-sized snails (15–25 mm shell length) with a globosely turbinate shell (Cotton 1935). The shells are dextrally coiled with a small umbilicus and a large aperture which is sealed by a horny operculum (Sheldon and Walker 1993b). Shell coloration is dark green to greenish brown in the Hanley's river snail and yellowish to dark brown in the Darling River snail.

The river snail was once widespread in the Murray and Darling rivers and the lower reaches of their major tributaries. In the mid 20th century, thousands of specimens of Hanley's river snail were collected from the lower Murray River (Johnston and Beckwith 1945, 1947) but the subspecies is now extinct in the main river channel. It survives only in a few irrigation pipelines (Sheldon and Walker 1993a; Walker 1996). Less attention has been given to the Darling River snail compared with its Murray River counterpart, but its number also declined in the latter half of the 20th century. Recent surveys have failed to detect the Darling River snail and it is now considered extinct (DPI 2007).

Figure 6.4: Hanley's river snail, *Notopala sublineata hanleyi*. Photograph: Dean Gilligan (DII).

Reproduction and life-cycle

Knowledge of the life-history of the Australian viviparids is poor. Sexes are separate. The right tentacle in males is modified to form a penis. Females are ovoviviparous, brooding the eggs in the pallial oviduct and giving birth to small numbers of live young (Stanisic 1998).

Habitat and diet

The river snail lives on mud, logs and rocks along the river banks, usually in deep water (Cotton 1935; Johnston and Beckwith 1945). The structure of the river snail's radula suggests that it is suited to feeding on detritus and small periphytic algae (e.g. diatoms). Dietary studies have confirmed that the river snail is a detritivore, feeding on biofilms attached to the surfaces of submerged bark, leaf litter and macrophytes (Sheldon and Walker 1997).

Water requirements

The water requirements of this snail are unknown and it has now disappeared from its natural lotic habitat. Predation by the introduced common carp (*Cyprinus carpio*) has played a part in the demise of the river snail (Walker 1998; Walker *et al.* 2009) but its disappearance also coincided with a period of increasing flow regulation in the rivers of the Murray-Darling Basin (Walker 1996).

There is circumstantial support for the hypothesis that a combination of stable water levels in weir pools and high turbidity has changed the composition of the biofilms that the river snail feeds upon, from predominantly microbial and diatom-dominated communities to filamentous algal mats. The nutritional value of the altered biofilms may no longer be sufficient to sustain the high energetic requirements of the snail's ovoviviparous reproductive strategy (Sheldon and Walker 1997). In the rivers of the Lake Eyre Basin, where the species still occurs, and in the pipeline habitats where it still thrives, biofilms are more nutritious than the filamentous algal biofilms of the main river channel. This hypothesis does not account for the disappearance of this species from less-regulated parts of the Murray-Darling Basin.

Billabong banded snail: *Notopala suprafasciata*

These medium-sized snails, reaching 200 mm in length, are very similar in appearance to the river snail (Figure 6.5). The shells of the two species are distinguished by the presence of diffuse bands and a larger, often polished shell in the billabong banded snail (Ponder *et al.* 2000). Until recently, this species was considered to be a banded form of the river snail (Sheldon and Walker 1993b) but it is now recognised as a distinct species (Ponder *et al.* 2000).

Historically, the billabong banded snail was widespread in floodplain wetlands associated with the lower reaches of the Darling River tributaries and its intersecting streams, north from Menindee Lakes. It seems to have disappeared from much of its former range and is now restricted to the northern section of the Murray-Darling Basin (Ponder *et al.* 2000).

Reproduction and life-cycle

Like other viviparids, the billabong banded snail has separate sexes. Females give birth to live young, the embryos developing in a uterine brood pouch. The clutch size has not been reported but a cluster of 20 subadults was observed in a small pool (Hugh Jones, *pers. obs.*).

Habitat and diet

In contrast to the strictly lotic environment of the river snail, the billabong banded snail inhabits billabongs, swamps close to the river and waterholes in creeks, where it is found on mud and leaves and among aquatic plants. Little is known of this species' feeding biology but it

Figure 6.5: Billabong banded snail, *Notopala suprafasciata*. Photograph: Stephanie Clarke.

is probably a detritivore, involving a combination of grazing and suspension feeding (Ponder *et al.* 2000; Sheldon and Walker 1997).

Water requirements

The habitat of the billabong banded snail implies that it is dependent on frequent overbank flooding. Little is known of the ecology of this species but it is presumed that the adult aestivates beneath logs or burrows into the mud when the waterbodies are dry, as occurs in other Australian *Notopala* (Cotton 1935).

Sculptured snail: *Plotiopsis balonnensis*

This operculate snail is distinctive and is readily distinguished from other aquatic snails by its solid sculptured shell. Shells are elongate with a high spire (occasionally decollate), six or seven angulate whorls and strong axial and spiral sculpture (Glaubrecht *et al.* 2009; Figure 6.6). Shells grow to a size of 30 mm and are usually light brown, while the body whorl may be patterned with a series of dark brown markings. Shell morphology in this species often varies widely among populations due to the clonal nature of reproduction (Stoddart 1985).

The sculptured snail occurs throughout much of Australia and is widespread in the rivers of the Murray-Darling Basin (Glaubrecht *et al.* 2009; Smith 1978). The sculptured snail has declined in the lower Murray River and has not been collected from Victorian rivers since the late 1970s (Walker *et al.* 1992). The snail appears to have declined in other parts of the Basin in recent decades; recent surveys have collected it from only a small number of localities, mostly in the north-east of the Basin.

Reproduction and life-cycle

The sculptured snail is parthenogenetic, as are many other thiarids (Healy and Wells 1998). Females are viviparous and incubate their young in a subhaemocoelic brood pouch located in the neck region, releasing three to seven fully developed juvenile snails (Glaubrecht *et al.* 2009). There is no additional information on the life-history of the sculptured snail.

Figure 6.6: Sculptured snail, *Plotiopsis balonnensis*. Photograph: Hugh Jones (DECCW).

Habitat and diet

The ecology of the sculptured snail is known only from superficial observations (Healy and Wells 1998). The sculptured snail lives on rocks and water plants, where it grazes on detritus and biofilms. It occupies river environments in both flowing and still-water habitats (Glaubrecht *et al.* 2009; Healy and Wells 1998). Large populations often occur in lentic habitats including weir pools, backwaters and lagoons adjoining the river, as well as irrigation channels and pipelines (Sheldon and Walker 1993a; Smith 1978; Woolford 1984).

Water requirements

The sculptured snail is incapable of withstanding desiccation and can survive in dryland rivers only where permanent waterholes provide refuge from droughts (Glaubrecht *et al.* 2009). The absence of ecological data makes it difficult to explain the decline of this species in the latter part of the 20th century, which coincided with rapidly increasing water resource development and the spread of common carp throughout the Murray-Darling Basin (Kingsford 2000; Koehn 2004).

Notwithstanding the potential impact of carp predation on snail populations, a possible explanation for the decline is the stabilisation of water levels behind weirs, combined with controlled releases from water storages. This may have reduced the nutritional quality of consumed food, so that it could no longer meet the energetic requirements for growth and reproduction (Sheldon and Walker 1997). Support for this explanation is undermined by the existence of sculptured snail populations in weir pools in the northern part of its range. Management strategies that reinstate substantial water level fluctuations may be beneficial to viviparous snails.

Pond snails: *Austropeplea* spp.

Lymnaeids are distinguished from other snail families by their fragile dextrally coiled shells, an inflated body whorl and the presence of a wide shell aperture without an operculum (Smith and Stanisic 1998). Pond snails possess a large fleshy foot and their eyes are located at the bases of a pair of broad triangular tentacles (Smith *et al.* 2002; Walker 1998; Figure 6.7).

Figure 6.7: Fluke pond snail, *Austropeplea huonensis*. Photograph: Louise Puslednik.

As currently recognised, two species of native pond snails occur in the Murray-Darling Basin: the fluke pond snail *Austropeplea huonensis* and the bubble pond snail *A. lessoni*. The taxonomic status of the fluke pond snail is unclear. This morphologically variable species comprise several distinct lineages within the Murray-Darling Basin but it is unclear whether these represent distinct species or a single geographically variable species (Boray and McMichael 1961; Puslednik *et al.* 2009).

The bubble pond snail occurs north from the southern tablelands of New South Wales. The fluke pond snail is found throughout mainland south-eastern Australia including the slopes and tablelands of New South Wales, with outlying populations in the Coonamble, Pilliga and Bingara districts (Boray 1964a; Puslednik *et al.* 2009). The two species are readily separated on the basis of size and shell shape. The adult fluke pond snail is smaller (maximum shell height of 16 mm) than the bubble pond snail, which grows to 25 mm (Boray 1969; Walker 1998). The bubble pond snail, as its name suggests, has a very round globose shell with a short spire, whereas the fluke pond snail possesses a higher spire comprising three or four shell whorls (Boray and McMichael 1961; Lloyd *et al.* 2007).

Reproduction and life-cycle

Pond snails are hermaphrodites. Most reproduction occurs via cross-fertilisation but self-fertilisation is possible (Boray 1964b). Eggs are laid in crescent-shaped gelatinous masses and the young develop directly within these, emerging as miniature forms of the adult (Boray and McMichael 1961). The fluke pond snail produces 30–60 eggs per egg mass and fecundities of more than 1200 eggs per month have been observed under optimal laboratory conditions (Boray 1964b, 1969). However, reproduction is strongly density-dependent and is reliant on an ample food supply and well-oxygenated water (Boray 1964b). The rate of early development in the fluke pond snail is temperature-dependent. Egg development ceases below 10°C and snails cease mating at about 15°C. At temperatures exceeding 25°C hatching and larval development are impaired. Snails do not reproduce at water temperatures above 30°C (Boray 1964b, 1969; Lynch 1963). Limited information is available on reproduction of the bubble pond snail, but in the closely related tropical lymnaeid *A. vinosa* egg laying ceased at 12°C; 18°C was near the lower limit for larval development in that species (Blair and Finlayson 1981).

Habitat and diet

These semi-aquatic snails are found in streams, ditches and ponds with temporary and permanent springs being especially important habitats (Boray 1964a, 1969; Walker 1998). Pond snails are usually absent from the main channels of large rivers and creeks but may be abundant in the associated backwaters, billabongs and swamps (Boray 1964a; Lynch 1965). Fluke pond snails are common in irrigation channels of the Murray, Murrumbidgee and Goulburn rivers (Boray 1969). Both the fluke and bubble pond snail favour temporary ponds and are able to aestivate in mud when the ponds dry out (Blair and Finlayson 1981; Boray 1969; Hurley *et al.* 1995; Lynch 1965).

Pond snails are usually far more abundant in temporary habitats than in permanent streams and pools (Lynch 1965). Although they aestivate in mud when the ponds dry out, seasonal drought kills many snails and usually only the small individuals survive. Permanent waterbodies serve as a drought refuge, providing a source of snails to recolonise temporary habitats when suitable conditions return (Boray 1969). Pond snails are herbivores, feeding largely on periphyton, and are generally associated with aquatic vegetation in less than 1 m of water (Blair and Finlayson 1981; Hurley *et al.* 1995).

Water requirements

Floods and droughts can have a devastating impact on pond snail populations, but both species of pond snail have a high reproductive potential and populations recover rapidly following disturbance (Blair and Finlayson 1981; Boray 1964b, 1969). Reproduction ceases at high population densities but sexual activity recommences if a flood reduces the size of the snail population (Boray 1964b). Small to moderate floods also aid dispersal. Pond snails will inhabit stagnant waters but growth and reproduction are inhibited under hypoxic conditions (Boray 1964b). Flooding can improve water quality, resulting in renewed vigour in populations of pond snails. A moderate water velocity of 15–20 cm s^{-1} in streams provides optimal conditions for algal growth, reproduction and dispersal in fluke pond snails (Boray 1969). Consequently, reproductive activity is greater in flowing streams, ponds and swamps following rain, and in backwaters and impoundments following flushing by floods and freshes (Blair and Finlayson 1981; Boray 1964b).

Pouch snails: *Glyptophysa, Isidorella* and *Amerianna* spp.

Pouch snails exhibit a variety of shell forms and ornamentation. *Glyptophysa gibbosa* has an elongate-conical sinistral shell (Figure 6.8), usually without shell sculpture, and it possesses a distinct columnellar fold which distinguishes it from morphologically similar species of *Isidorella* (Smith 1996). These two genera of pouch snails are also distinguished by differing penile morphology (Smith and Stanisic 1998).

Pouch snails (Family Planorbidae) are the dominant aquatic snails in many parts of Australia and several species occur in the Murray-Darling Basin (Smith 1978; Walker 1998). Of the pouch snails in the Murray-Darling Basin, *Glyptophysa gibbosa* and *Isidorella newcombi* are

Figure 6.8: Pouch snail, *Glyptophysa gibbosa*. Photograph: Hugh Jones (DECCW).

the most common species. *G. aliciae* is an uncommon inhabitant of wetlands of the southern part of the Basin, and *Amerianna carinata* is common in the northern part of the Basin (Ponder *et al.* 2000; Smith 1978; Walker *et al.* 2009). Little is known of the ecology of most pouch snails of the Murray-Darling Basin.

Reproduction and life-cycle

Pouch snails are simultaneous hermaphrodites. They are oviparous and produce gelatinous egg masses containing many eggs, which develop directly into tiny snails within the egg mass (Smith and Stanisic 1998). The egg masses are laid on wood and water plants. In *Isidorella newcombi* egg masses are attached to plant stems or plant debris and hatch in three to seven days (Stevens 2007). The reproductive biology of most species is unknown.

Habitat and diet

Pouch snails thrive among aquatic vegetation and leaf litter and on wood in microhabitats with lentic or slow-flowing waters. The range of habitats where they are found include permanent rivers and lentic environments, such as temporary creeks, ponds, swamps and billabongs (Smith 1978). All species graze on detritus, algae and aquatic plants.

Water requirements

Pouch snails occur across a broad range of water regimes, implying that most species do not have specific water requirements. Changes to the flow regime of sluggish rivers and streams that result in increases in flow velocity are likely to reduce the amount of habitat available to these snails.

Some species of pouch snails are adapted to seasonal drought and aestivate during dry periods (Smith and Burn 1976; Stevens 2007). This trait has contributed to *I. newcombi* becoming a major pest of rice crops in southern New South Wales (Stevens 2002). Although almost half the population of mature snails may survive the interval between consecutive rice crops, few snails survive for periods longer than this, implying that reductions in the flooding frequency of wetlands is detrimental for populations of pouch snails (Stevens 2007).

Freshwater limpets: *Ferrissia* spp.

The shells of freshwater limpets are small (<2.5 mm) and patelliform with no sign of coiling (Smith and Stanisic 1998; Figure 6.9). Two species, *Ferrissia petterdi* and *F. tasmanica*, are widely distributed within the rivers of the Murray-Darling Basin. They are distinguished by the height and shape of the shell, with *F. petterdi* having a flatter profile and a more elongate shell than *F. tasmanica* (Hubendick 1967). The Australian freshwater limpets are in need of taxonomic revision (Ponder, *pers. comm.*). A recent molecular phylogeny grouped freshwater limpets as a clade within the Planorbidae (Albrecht *et al.* 2007).

Reproduction and life-cycle

Freshwater limpets are hermaphrodites with directly developing eggs. These are laid on aquatic vegetation, stones or wood, as gelatinous capsules containing a single egg (Smith and Stanisic 1998).

Habitat and diet

Freshwater limpets are found in ponds, billabongs, streams and rivers in flow conditions ranging from stagnant pools to fast-flowing streams (Smith and Stanisic 1998). They are detritivores adhering to aquatic vegetation and leaf litter, and sometimes occurring on rocks and

Figure 6.9: Freshwater limpet, *Ferrissia petterdi*. Photograph: John Gooderham and Edward Tsyrlin.

wood (Ponder *et al.* 2000). In the lower Murray River, *F. petterdi* is associated with ribbonweed (*Vallisneria spiralis*) (Walker 1998). Limpets tolerate hypoxic conditions in pools and may aestivate during drought, as is the case with other planorbids (Richardot 1977).

Water requirements

Freshwater limpets are tolerant of altered flow regimes and remain abundant in many parts of the Murray-Darling Basin, including the lower Murray River where the abundance of other snails has declined (Sheldon and Walker 1998; Walker *et al.* 1992).

Crustacea

Western yabby: *Cherax destructor*

The term 'yabby' is commonly reserved for the group of smooth-shelled crayfish lacking spines on the carapace or abdomen (Figure 6.10), a characteristic that distinguishes them from the Murray crayfish (*Euastacus armatus*) (Olszewski 1980). Shell colour is variable, ranging from dull or dark brown to olive green to blue (Ingram *et al.* 1997; Withnall 2000).

Western yabbies are distributed over most of central and eastern Australia, including the entire Murray-Darling Basin (Horwitz and Knott 1995; Merrick 1993; Sokol 1988b). Recent molecular systematic studies reveal three geographically distinct clades of western yabbies within the Murray-Darling Basin. These occur in the Wimmera River in western Victoria, a southern lineage restricted to the Murray and Murrumbidgee rivers, and a northern lineage occupying the Darling River and its tributaries (Austin *et al.* 2003; Nguyen and Austin 2005; Nguyen *et al.* 2004). An additional species, the swamp yabby (*C. rotundus*), is restricted to floodplain wetlands along the Ovens River and the middle reaches of the Murray River, including Barmah Forest (Edney *et al.* 2002).

Reproduction and life-cycle

Yabbies have separate sexes, which can be determined from the location of the gonopores. Females have a pair of oviducts located at the base of the third pereiopod, whereas the male

Figure 6.10: Western yabby, *Cherax destructor.* Photograph: Gunther Schmida.

openings are at the base of the fifth pereiopod in association with reproductive papillae (Lake and Sokol 1986). Female yabbies can attain maturity at three to four months of age (Huner and Lindqvist 1995), although six to 12 months is more typical (Mosig 1995), and they may reproduce two to three times per breeding season (Lake and Sokol 1986). Egg numbers range from fewer than 100 to a maximum of 1200 per female depending on female size, nutrition and other environmental conditions and on genetic variability in reproductive traits (Austin 1998a, 1998b; Lake and Sokol 1986). The fertilised eggs are attached to filamentous setae on the swimmerets underneath the abdomen. Egg incubation takes about three weeks at summer water temperatures of 23–24°C, and the newly hatched young remain attached to the female for a further two to three weeks until after the third moult, when they become independent (Merrick 1993). Estimates of longevity range from a minimum of three years but it is likely that yabbies will survive for several years.

Habitat and diet

The western yabby occupies a wide range of aquatic habitats from permanently flowing rivers to farm dams (Lake and Sokol 1986). It is most abundant in temporary floodplain habitats such as billabongs, swamps, intermittently flowing creeks, irrigation channels and reservoirs (Walker 1983). In these environments, the yabby is exposed to very low dissolved oxygen concentrations and a wide range of water temperatures. It is well adapted to environmental extremes, exhibiting a high tolerance to hypoxia and wide variations in water temperature and salinity (Lake and Sokol 1986; Sokol 1988a). As oxygen levels decline, the yabby responds by increasing its respiration rate, becoming less active and vigorously ventilating with its pleopods (Merrick 1993). It may also move into shallow water or onto land to breathe air (Merrick 1993; Morris and Callaghan 1998a). At very low dissolved oxygen concentrations, anaerobic metabolism becomes important and animals become hypometabolic (Morris and Callaghan 1998b). The western yabby can survive temperatures of 1–36°C, but activity, growth and reproduction cease below 15–16°C (Frost 1975; Johnson 1979; Mills 1986). The western yabby is an osmoregulator, maintaining constant blood osmotic pressure until the salinity of the environment reaches 14‰, after which blood concentrations become isosmotic with the surrounding water. Above this threshold, behaviour is progressively impaired. Mortality rapidly increases at salinities above 16‰, although animals can cope with salinities of 25‰ for a few days (Mills and Geddes 1980; Sokol 1988a).

The yabby is primarily a detritivore, consuming decaying vegetable matter. It occasionally feeds on live plant material, carrion and invertebrates, including chironomid larvae and zooplankton (Bunn and Boon 1993; Faragher 1983; Olszewski 1980; Sokol 1988a). Recent studies indicate that invertebrates may be more important in the diet than was previously thought (Meakin *et al.* 2008). The juvenile yabby reportedly feeds on plankton (Mosig 1995).

Water requirements

The western yabby occurs in permanently flowing rivers but it is most abundant in lentic floodplain habitats, many of which are temporary waters. It is well-adapted to impermanent environments, burrowing deep into the bed of the waterbody during periods of drought where it may aestivate for several years until the water returns (NSW DPI 2005; Sokol 1988a). The flood pulse benefits the western yabby by making available abundant food resources that the yabby can access immediately when it emerges from aestivation (Junk *et al.* 1989).

Water resource development has reduced the amount of habitat available to the western yabby by reducing the frequency and extent of flooding (Kingsford 2000). In other respects, river regulation has been beneficial for the yabby, which prefers lentic habitats, and it may have become more common along the littoral margins of the Darling and Murray rivers due to the formation of extensive weir pools (Walker 1986, 1992; Walker *et al.* 1992). This species also makes extensive use of irrigation channels and man-made lakes, farm dams and bore drains throughout the Murray-Darling Basin (Lake and Sokol 1986).

Breeding occurs in the warmer months, commencing in spring when the water temperature reaches about 15°C. Flooding is not a breeding stimulus (Johnson 1979). A rheotactic response to flowing water has not previously been reported for the western yabby. Upstream movements of this species were recently observed on the Bogan River during a minor flood, with many yabbies scaling a low weir (*pers. obs.*). This is similar to the upstream migrations noted in other decapods (Lee and Fielder 1979).

Murray crayfish: *Euastacus armatus*

The Murray crayfish is a large spiny crayfish that may grow to 3 kg with adults averaging 20–30 cm in total length (Merrick 1993; Figure 6.11). Large spines are present on the carpus and merus (the jointed segments below the chelae) with smaller dorso-lateral spines and tubercles on the cephalothorax and the pereiopods (Morgan 1986). The abdomen has several rows of large spines, which are tipped white. The body colour of adults is mostly khaki green but it varies from a metallic grey to rufous or brown dorsally and is lighter underneath (Gilligan *et al.* 2007; Merrick 1993). The characteristic white or cream coloration of the chelae (claws) may extend to the cephalon in some specimens.

Of the five species of *Euastacus* that occur within the Murray-Darling Basin, the Murray crayfish has the widest distribution, occurring throughout the Murray and Murrumbidgee River systems (except for the headwaters), including several Victorian tributary rivers (the Goulburn, Ovens, Mitta Mitta and Kiewa rivers) (Gilligan *et al.* 2007; Morgan 1986). The species is also found in the upper reaches of the Macquarie River near Kandos (Morgan 1986, 1997) and in the upper Lachlan River, but these occurrences are reportedly due to the translocation of individuals in the early 20th century (Gilligan *et al.* 2007). Historically, the Murray crayfish ranged from the upper reaches of these rivers to the lowland reaches of the Murray River in South Australia (Morgan 1986), but since the 1940s the species has been in decline. It

Figure 6.11: Murray crayfish, *Euastacus armatus.* Photograph: NSW Department of Industry and Investment.

may no longer exist in the Murray River downstream of Mildura (Geddes 1990; McCarthy 2005; Walker 1982). The Murray crayfish has also declined in the Australian Capital Territory section of the Murrumbidgee River and in several New South Wales rivers including the Edwards, Wakool and Niemur rivers, where it is now rare (Geddes 1990; Lintermans and Rutzou 1991).

Reproduction and life-cycle

The Murray crayfish exhibits a number of features that are typical of k-selected species. It is slow-growing, taking five to nine years to attain sexual maturity (McCormack 2008). Female fecundity is dependent on size, varying from less than 200 eggs per year to more than 1200, but high juvenile mortality ensures a low recruitment rate (Gilligan *et al.* 2007).

Mating and spawning take place in early to mid May and the eggs are incubated over winter and spring (Gilligan *et al.* 2007). Hatching occurs from early to late October but juveniles are not released until November or December. Mortality of young crayfish is high due to predation by trout cod, Murray cod and golden perch. Introduced trout and redfin perch are probably also major predators of juvenile crayfish. Individual juveniles may moult up to 10 times in the first year, but adults (>60 mm O.C.L.) moult only in mid April to late May, immediately prior to breeding.

Habitat and diet

The Murray crayfish occurs in large and small streams flowing through cleared grazing land and sclerophyll forests (Morgan 1986). Although occurring at altitudes of 700 m it is most common below 300 m in strongly flowing, well-oxygenated waters (Geddes *et al.* 1993; Gilligan *et al.* 2007). The crayfish will construct burrows in clay banks or beneath logs and boulders. Heterogeneous habitat in the form of logs and boulders is important for this species (Gilligan *et al.* 2007). The Murray crayfish is a poor disperser, does not migrate and has a small home range (Gilligan *et al.* 2007).

In contrast to the western yabby, the Murray crayfish is active throughout the cooler months from May to October, when water temperatures are below 20°C. It will tolerate water temperatures up to 27°C but is intolerant of low dissolved oxygen concentrations (Geddes *et al.* 1993). The Murray crayfish survives natural seasonal variations in water quality by seeking shelter and remaining inactive during periods of higher water temperature and declining flow.

The Murray crayfish is omnivorous. It feeds mostly on decaying vegetation but will also feed on animal remains (Merrick 1993).

Water requirements

Flowing water is an essential component of Murray crayfish habitat as it maintains the well-oxygenated environment that is essential for the species (McCarthy 2005). Although recreational fishing pressure has depressed crayfish populations in many areas (Asmus 1999; Lintermans and Rutzou 1991; McCarthy 2005), river regulation is the major factor implicated in the demise of the Murray crayfish along the Murray River downstream of Mildura (McCarthy 2005; Walker 1982). Weirs have transformed the lower river from a lotic to lentic environment and water extractions for irrigation have further reduced flows in many areas (McCarthy 2005; Walker 1985, 1992). Maintenance of flow velocities in impounded reaches is critical for the survival of crayfish populations in these areas, as it prevents stratification which leads to the development of hypoxic conditions (Geddes *et al.* 1993).

Habitat loss, resulting from high rates of sedimentation, has been proposed as an explanation for the disappearance of this species from the South Australian section of the Murray River in the early 1960s (Walker 1982). The Murray crayfish occupies the deeper pools in lowland rivers but most of these filled with sediment following the construction of weirs. The 1956 floods also deposited large amounts of sediment in the river channel, most of which has remained due to the absence of flushing flows under the highly regulated flow regime (Walker 1982).

Bottom releases from large headwater dams have reduced summer water temperatures for long distances downstream and have been identified as a threatening process for the Murray crayfish (Gilligan *et al.* 2007; Preece 2004; Ryan *et al.* 2001). Reproduction and growth are temperature-dependent; cooler water temperatures in spring and summer may prolong larval and juvenile development, with potentially detrimental effects on crayfish populations.

Maintenance of baseline flows in the main channels of rivers via controlled releases is necessary to meet breeding requirements and to provide the strong flows that the Murray crayfish requires (Gilligan *et al.* 2007; McCarthy 2005).

Common freshwater shrimp: *Paratya australiensis*

The common freshwater shrimp is a small (up to 30 mm body length) translucent shrimp with a rostrum that is long, sharp and serrated (Figure 6.12). It has well-developed eyes with a supraorbital spine, and all walking legs are characterised by a small outer branch called an exopod (Williams 1980). The first two pairs of legs are terminated in brushes.

The common freshwater shrimp is distributed in rivers and streams throughout eastern Australia from the Atherton Tablelands to Tasmania (Williams and Smith 1979). It is widespread throughout the Murray-Darling Basin, where it is one of the most abundant macroinvertebrate taxa (Boulton and Lloyd 1991; Sheldon and Walker 1998). There is substantial morphological variation within and among shrimp populations, and recent molecular evidence suggests that the taxa is composed of multiple species (Cook *et al.* 2006; Smith and Williams 1980).

Reproduction and life-cycle

Breeding is initiated in late winter to spring and continues until mid-summer in response to increasing water temperature (Hancock and Bunn 1997; Williams 1977). Life expectancy

Figure 6.12: Common freshwater shrimp, *Paratya australiensis*. Photograph: Hugh Jones (DECCW).

varies geographically, with shrimp living for one to two years (Hancock and Bunn 1997; Richardson *et al.* 2004; Williams 1977). Female shrimp breed in their second summer and most are univoltine, although occasionally two broods may be produced in a season (Richardson *et al.* 2004). Females produce 50–1000 eggs, depending on size and geographic location (Smith and Williams 1980; Walsh 1993). The eggs are brooded beneath the female's abdomen for 25 days or more, before release as a planktonic larval stage lasting 25–48 days (Walsh 1993). In the rivers of the southern Murray-Darling Basin berried females appear from October to February, larvae are mostly present from November to April, and juvenile shrimps occur from November through to June (Richardson *et al.* 2004; Walsh 1993; Williams 1977).

Habitat and diet

The common freshwater shrimp occurs in a variety of freshwater habitats including streams and rivers, lakes, farm dams and billabongs (Williams 1977). In the Murray-Darling Basin the shrimp is common in the slackwater zones of slow-flowing rivers and streams, backwaters, billabongs and anabranches where it is associated with aquatic plants and snags (Humphries *et al.* 2006; Richardson and Cook 2006; Sheldon and Walker 1998; Williams 1977). The freshwater shrimp is mainly a herbivore-detritivore, grazing on periphyton, detritus, aquatic macrophytes and cyanobacteria (Bunn and Boon 1993; Burns and Walker 2000; Piola *et al.* 2008; Reid *et al.* 2008).

Water requirements

The freshwater shrimp needs permanent water and can inhabit transient floodplain habitats only when these are connected to permanent rivers. Although it is occasionally abundant in lentic habitats such as billabongs, the freshwater shrimp is more typical of lotic habitats (Boulton and Lloyd 1991; Sheldon and Walker 1998). In rivers, the shrimp favours the still-water and slow-flowing habitats of the littoral zone, especially during its early stages of development (Humphries *et al.* 2006; Richardson and Cook 2006). Larval shrimp cannot maintain position even in slow water currents, and require the protection of slackwater habitats and billabongs to avoid being flushed downstream (Hancock and Bunn 1997). Large numbers of larvae and juvenile shrimp may be displaced by floods and strong flows (Hancock and Bunn 1997, 1999). Despite this, the freshwater shrimp is often abundant in highly regulated regimes with elevated summer flows due to irrigation releases (Richardson *et al.* 2004). Irrigation releases cause a seasonal reversal of the natural flow pattern, reducing the area of slackwater. However, both juvenile and adult shrimp will swim strongly against water currents (Hancock and Bunn 1999) which, in addition to the use of flow refugia by larvae, compensates for the downstream displacement of larvae (Richardson and Cook 2006). This species has benefited from altered flow regimes in the lower Murray River, where it dominates the macroinvertebrate community.

McCulloch's shrimp: *Caridina indistincta* spp. complex

The McCulloch's shrimp is similar in size and appearance to the common freshwater shrimp but differs in that there is no supraorbital spine above the eye, and the walking legs do not possess exopods (Williams 1980). Recent DNA studies indicate the presence of two cryptic species of *Caridina* in rivers of the Murray-Darling Basin (Page and Hughes 2007). One species is widely distributed in the rivers of the northern Murray-Darling Basin and the lower Murray River (sp. D), while a second species inhabits the Murray River and its Victorian tributaries (sp. B). The two species are distinguished by the arrangement of teeth on the rostrum (Page *et al.* 2005). The teeth on the ventral side of the rostrum extend all the way to the end in *Caridina* sp. B, but there is a patch with no teeth in *Cardinia* sp. D.

Reproduction and life-cycle

Reproduction occurs in spring and summer, with females producing 30–60 eggs which are attached to the abdominal appendages (Benzie 1982; Richardson *et al.* 2004). Larval development is abbreviated with a zoeal stage that lasts less than 24 hours, followed by several post-zoeal stages that last 20–25 days (Benzie 1982). This species probably completes its life-cycle within 12 months. In Victorian tributaries of the Murray River, berried females were present from October to February, larvae were mainly present from December to March, and juvenile shrimp were recruited into the population from December to April (Richardson *et al.* 2004).

Habitat and diet

The McCulloch's shrimp is a herbivore–detritivore occupying vegetated still waters including slackwaters in large rivers, backwaters and temporary billabongs (Richardson and Cook 2006; Sheldon and Walker 1998).

Water requirements

In the Campaspe River, the McCulloch's shrimp was absent from sections of rivers with high summer irrigation flows, probably because the flows reduce the amount of slackwater habitat, disrupting larval development and recruitment (Richardson *et al.* 2004). Both larval and adult life-stages are strongly associated with slow-flowing or still-water habitats and the species also has a short planktonic phase, which reduces downstream displacement by water currents (Richardson and Cook 2006). The timing of low flows to coincide with critical stages in the life-cycle over summer and autumn may be vital for the maintenance of shrimp populations (Richardson and Cook 2006).

Freshwater prawn: *Macrobrachium australiense*

The freshwater prawn is superficially similar to atyid shrimps except that its second pair of legs are much longer and more robust than the first pair, usually extending well beyond the body, and terminated with a set of strong pincers (Williams 1980; Figure 6.13). This is in contrast to the freshwater shrimp, which possesses brushes on the ends of the first two pairs of legs. The

Figure 6.13: Freshwater prawn, *Macrobrachium australiense*. Photograph: Neil Armstrong.

freshwater prawn is usually brown or olive grey with flecks of green and grows to a body length of 7–9 cm, excluding claws (Short 2004). This decapod occurs widely in eastern Australia and is common in all major rivers of the Murray-Darling Basin.

Reproduction and life-cycle

The life-span of the freshwater prawn is two years or more (Richardson *et al.* 2004). Reproduction occurs in spring and summer, with females producing batches of 38–197 eggs that are carried on the female pleopods (Lee and Fielder 1982c; Short 2004). Larval development is abbreviated to three stages and larvae metamorphose into postlarvae within several days (Fielder 1970). Recruitment of juveniles occurs mainly between January and March in the southern Murray-Darling Basin (Richardson *et al.* 2004).

Habitat and diet

The freshwater prawn is restricted to permanent waters and is common in the main channel of rivers, backwaters and billabongs and in perennial waterholes and impoundments, favouring snags and aquatic vegetation near the edge of the channel (Lee and Fielder 1984; Williams 1981). It is a scavenger of decaying plant and animal material (Burns and Walker 2000; Lee and Fielder 1982b).

Water requirements

The onset of breeding is related to water temperature rather than flow and there is evidence that cold water releases below dams delay the onset of breeding (Richardson *et al.* 2004). In northern populations, larval development ceases and the prawn does not spawn when water temperatures fall below about 20°C (Lee and Fielder 1981, 1982a). The temperature of flow releases from large dams is frequently much lower than this threshold and the suppression of water temperatures may extend hundreds of kilometres downstream (Preece 2004).

Juvenile and adult prawns utilise different hydraulic habitats. Larval and juvenile prawns are generally restricted to slackwater habitats of rivers whereas the adult is often found in moderately fast-flowing lotic habitats (Richardson and Cook 2006). The freshwater prawn may have only a brief planktonic phase but considerable downstream displacement of larvae and postlarvae is still possible, especially during spates or in reaches receiving irrigation releases. Slackwaters and other lentic habitats may be critical refuges for the prawn during these early life stages (Richardson *et al.* 2004). The adult prawn exhibits a strongly positive rheotactic response to flow to compensate for downstream displacement, and mass upstream migrations have been observed in this and other *Macrobrachium* species (Lee and Fielder 1979, 1984). Small weirs are not serious barriers to upstream movement and flow releases enable the prawn to overcome these barriers.

Inland crab: *Austrothelphusa transversa*

The inland crab (Figure 6.14) is a robust medium-sized crab (up to 55 mm carapace width) with a broadly oval and convex carapace, impressed with an H-shaped arrangement of grooves (Bishop 1963). The carapace ranges from olive to yellow olive, with some purplish mottling. The crab is widespread throughout inland Australia, from north-western Australia across to the Queensland coast and south to the Lake Eyre Basin and the Darling River system (Bishop 1963). It is restricted to the floodplains of the northern Murray-Darling Basin, with its southern limit near the Bogan River.

Reproduction and life-cycle

Sexes are separate with the gonopores located on the coxae of the third pereiopods in females, and on the fifth in males (Bishop 1963). The inland crab has dispensed with the planktonic larval stage of marine crabs. Instead it has direct development; the young crabs hatch from large eggs and are carried by the mother for several days.

The inland crab appears to be a seasonal breeder, despite spending the majority of its life underground (Greenaway 1984). The details of courtship and mating are unknown for this species but, in other land crabs, sperm may be stored for many months in the spermatheca of the female (Adiyodi 1988). Eggs are laid in late October to early November and are carried beneath the abdomen of the female (Greenaway 1984). Juveniles are found in December and January. Surface water does not appear to be essential for this stage of development.

Figure 6.14: Inland crab, *Austrothelphusa transversa*. Photograph: Hugh Jones (DECCW).

Habitat and diet

The limited dietary information indicates that the inland crab is omnivorous, feeding predominantly on detritus and vegetable matter but also feeding opportunistically on insects and carrion (Greenaway and MacMillen 1978).

The inland crab avoids permanent waterbodies. It inhabits floodplain wetlands including swamps, gilgais, ground tanks and temporary creeks where it spends most of its life aestivating in a burrow between floods (Davie 2002; Greenaway and Taylor 1976). The crab usually burrows 50–70 cm in heavy clay soils, sealing the mouth of the burrow with a clay plug (Greenaway and MacMillen 1978). Clay soils are critical to the survival of the inland crab during droughts. They maintain a humid microclimate in the burrow, which prevents desiccation of the crab during aestivation (Greenaway and MacMillen 1978). The inland crab is physiologically tolerant of dehydration and high haemolymph ion concentrations, has low rates of urinary and evaporative water loss, and compensates for water loss by absorbing condensation from within the burrow (Adiyodi 1988; Greenaway 1980; Greenaway and MacMillen 1978). An aestivating crab has low energy requirements, which enables it to survive long periods within the burrow (MacMillen and Greenaway 1978).

Water requirements

The inland crab occurs in flood-prone environments (MacMillen and Greenaway 1978) but the effects of frequency, duration and timing of floods on the population ecology of the species are unknown. The timing of floods may be important for successful reproduction and recruitment of juveniles into the population, especially if the species has a strictly seasonal breeding season as suggested by Greenaway (1984). This may explain why this species is restricted to the northern Murray-Darling Basin, which has a summer-dominated rainfall regime.

Summary of water requirements

The water requirements of the molluscs and decapod crustaceans discussed above are summarised in Table 6.1. Floodplain molluscs and crustaceans are typically eurytopic and are able to cope with a wide range of environmental conditions. Most snails are adapted to short interflood dry-periods, with some species being more abundant in temporary than in permanent wetlands (Lynch 1965). There is little information on how variations in flood magnitude, frequency and duration affect the population dynamics of floodplain taxa in spite of the crucial role of floods in the lives of these animals.

It is well recognised that the magnitude of floods is important in maintaining longitudinal and lateral connectivity in floodplain rivers (Walker *et al.* 1995; Ward and Stanford 1995). Reductions in flood magnitude reduce the frequency and duration of floods on the floodplain, increasing the frequency of droughts. Although droughts are a normal part of the water regime, they cause large fluctuations in snail abundance (Koch 2004) and prolonged droughts may result in local extinctions of populations. Hence, some floodplain taxa might be best viewed as meta-populations (Hanski 1998) where individual populations have a high risk of extinction and the persistence of the meta-population depends on dispersal and colonisation mediated by floods.

Lateral connectivity between the river and neighbouring floodplain habitats (e.g. anabranches and billabongs) provides primary habitats for the freshwater shrimp, the billabong mussel and several species of snail (Boulton and Lloyd 1991; Sheldon and Walker 1998). The disappearance of several species of snail from the lower Murray River may be partly due to the loss and isolation of floodplain habitats due to river regulation; however, several freshwater

Table 6.1: Water requirements for different life-stages of molluscs and decapod crustaceans found in the Murray-Darling Basin

Common name	Scientific name	Critical T_w for reproduction and development	Breeding season	Dispersal mechanisms	Hydrological conditions: Aestivation	Hydrological conditions: Larvae and juveniles	Hydrological conditions: Adults	Habitat description and physiological tolerances	Flow-related threats	Specific water requirements
Mollusca bivalvia										
River mussel	*Alathyria jacksoni*	$T_w > 12°C$	Spring–autumn	Via fish hosts	No	P	P, SF	Perennial rivers; intolerant of hypoxia and desiccation	Altered flow regime (lotic to lentic)	Maintain strong base flows; protect low flows
Billabong mussel	*Velesunio ambiguus*	$T_w > 12°C$	Spring–autumn	Via fish hosts	Yes	n.d.	TP, LS	Floodplain wetlands; eurytopic	Reductions in flood frequency	Inundation of floodplain wetlands
Basket shell	*Corbicula australis*	$T_w > 20°C$	Spring–autumn	Floods	No	n.d.	P, F	Rivers in sand and gravel (runs and riffles); environmental tolerances unknown	n.d.	Moderate–fast water velocity
Gastropoda (aquatic snails)										
River snail	*Notopala sublineata*	n.d.	n.d.	n.d.	No	n.d.	P, S	River edges; irrigation pipelines; environmental tolerances unknown	Stable water levels	n.d.
Billabong banded snail	*Notopala suprafasciata*	n.d.	Spring–summer	n.d.	Yes	T, L	T, L	Billabongs and riverine swamps; environmental tolerances unknown	Reductions in flood frequency	Seasonal summer floods
Sculptured snail	*Plotiopsis balonnensis*	n.d.	n.d.	n.d.	No	n.d.	P, LS	Rivers in slackwaters, backwaters and weir pools; environmental tolerances unknown	n.d.	n.d.
Fluke pond snail	*Austropeplea huonensis*	$15 < T_w < 30°C$	Spring–autumn	Floods, stream drift	Yes	TP, LS	TP, LS	Floodplain wetlands (irrigation channels, springs; favours temporary ponds); eurytopic	None	Moderate water velocities (15–20 cm s^{-1}) optimal for recruitment and growth
Pouch snail	*Glyptophysa gibbosa*	n.d.	n.d.	n.d.	Yes	TP, LS	T, LS	Floodplain wetlands, river littoral zones in association with water plants; eurytopic	None	None

						Hydrological conditions				
Common name	Scientific name	Critical T_w for reproduction and development	Breeding season	Dispersal mechanisms	Aestivation	Larvae and juveniles	Adults	Habitat description and physiological tolerances	Flow-related threats	Specific water requirements
Pouch snail	*Isidorella newcombi*	n.d.	Spring–summer	n.d.	Yes	TP, LS	T, L	Floodplain wetlands and irrigation bays; eurytopic	None	Favoured by seasonal spring floods
Freshwater limpet	*Ferrissia* spp.	n.d.	n.d.	n.d.	Yes	n.d.	PT, LS	Rivers and floodplain wetlands; tolerant of hypoxia	n.d.	Favoured by seasonal spring floods
Crustacea										
Western yabby	*Cherax destructor*	$T_w > 15°C$	Spring–autumn	Floods	Yes	PT, LS	PT, LS	Favours lentic floodplain habitats and temporary creeks; eurytopic; fossorial	Reductions in frequency and magnitude of floods	None
Murray crayfish	*Euastacus armatus*	n.d.	May–Oct (spawns early–mid May)	None	no	P, F	P, F	Rivers and streams mainly below 300 m altitude; deep pools; stenotopic	Man-made barriers and altered flow regime (lotic to lentic)	Maintain strong base flows during breeding season
Freshwater prawn	*Macrobrachium australiense*	$T_w > 20°C$	Spring–autumn	Floods; adults migrate upstream	No	P, LS	P, SF	Rivers, backwaters and billabongs, perennial waterholes and impoundments, favouring snags and aquatic vegetation	None	Mass upstream migrations during freshes and floods
Common freshwater shrimp	*Paratya australiensis*	n.d.	Oct–Feb	Floods; positive rheotaxis in adults and juveniles	No	P, LS	P, LSF	Aquatic plants and snags in littoral zones of slow-flowing streams and rivers, anabranches, lakes, farm dams and billabongs	Summer irrigation flows may reduce recruitment	Minimise strong base flows during summer months

					Hydrological conditions					
Common name	Scientific name	Critical T_w for reproduction and development	Breeding season	Dispersal mechanisms	Aestivation	Larvae and juveniles	Adults	Habitat description and physiological tolerances	Flow-related threats	Specific water requirements
McCull-och's shrimp	*Caridina* sp. B	n.d.	Oct–Feb	Floods and strong base flows	No	P, L	P, LS	Slackwater zones in rivers, backwaters, anabranches, billabongs	Summer irrigation flows	Provision of slackwater zones, backwaters during breeding and recruitment seasons
Inland crab	*Austrothel-phusa trans-versa*	n.d.	Nov–Jan (spawns late Oct–Nov)	Floods	Yes	T, L	T, L	Floodplain wetlands; eurytopic (very high desiccation toler-ance); fossorial	Reduction in flood fre-quency and magnitude	None

Thermal thresholds for reproduction and growth (T_w) and breeding seasons are pertinent to the timing of flooding in wetland habitats.

Critical T_w are critical threshold water temperatures below which reproductive activity and larval development cease.

T = temporary waters; P = permanent waters; L = lentic; S = slow current; F = moderate–fast current; n.d. = no data.

snail species still make occasional brief appearances in the wake of now-infrequent floods (Walker 1992).

Flood frequency is an important factor affecting floodplain snails, although the optimal frequency for most snail species is unknown. Newcomb's pouch snail, a significant pest of rice crops, thrives under a flooding regime of regular annual floods. A successful method of population control involves prolonging the dry phase to 18 months, which results in very high mortality among aestivating snails (Stevens 2007).

The duration of flooding affects the period available for growth, reproduction and recruitment. At one extreme, conversion of temporary floodplain habitats into permanent waterbodies may eliminate crustaceans that produce resting stages, by removing the exogenous cues required to break dormancy (Brendonck 1996). At the other extreme, reductions in flooding frequency and duration can be detrimental if taxa do not have time to complete their life-cycle.

The timing of the flood pulse is critical for reproduction and growth in floodplain snails, since these functions are inhibited at low water temperatures (Table 6.1). Summer floods may enhance growth and juvenile recruitment in some species by preventing the establishment of anoxic conditions in weir pools and floodplain habitats (Boray 1964b; Humphrey and Simpson 1985).

The low-flow regime affects the distribution of the potamophilous molluscs and crustaceans. The river mussel and Murray crayfish prefer microhabitats with strongly flowing currents (Gilligan *et al.* 2007; Walker 1981b) and the freshwater prawn is mainly associated with moderate water currents. In contrast, the larval and juvenile stages of atyid prawns require areas of still or slow-flowing water (Richardson and Cook 2006) and a reversal in the seasonality of the flow regime, caused by summer irrigation flows, has been detrimental to the McCulloch's shrimp (Richardson *et al.* 2004). Summer irrigation releases may have been beneficial to the river mussel and the Murray crayfish, which require well-oxygenated flowing waters. Although the evidence is equivocal, seasonal variability in the low-flow regime may be critical for viviparous river snails because of potential impacts of water level fluctuations on biofilm composition, the staple food of river snails (Sheldon and Walker 1997).

References

Adiyodi RG (1988) Reproduction and development. In *Biology of the Land Crabs.* (Eds WW Burggren and BR McMahon) pp. 139–185. Cambridge University Press: Cambridge.

Albrecht C, Kuhn K and Streit B (2007) A molecular phylogeny of Planorboidea (Gastropoda, Pulmonata): insights from enhanced taxon sampling. *Zoologica Scripta* **36**, 27–39.

Asmus M (1999) Effects of recreational fishing on populations of Murray crayfish *Euastacus armatus* in the Murrumbidgee River. Hons thesis. Charles Sturt University.

Austin CM (1998a) A comparison of clutch and brood size in the red claw, *Cherax quadricarinatus* (von Martens) and the yabby, *C. destructor* Clark (Decapoda: Parastacidae). *Aquaculture* **167**, 135–145.

Austin CM (1998b) Intraspecific variation in clutch and brood size and rate of development in the yabby, *Cherax destructor* (Decapoda: Parastacidae). *Aquaculture* **167**, 147–159.

Austin CM, Nguyen TTT, Meewan MM and Jerry DR (2003) The taxonomy and phylogeny of the '*Cherax destructor*' complex (Decapoda: Parastacidae) examined using mitochondrial 16S sequences. *Australian Journal of Zoology* **51**, 99–110.

Balla SA and Walker KF (1991) Shape variation in the Australian freshwater mussel *Alathyria jacksoni* Iredale (Bivalvia, Hyriidae). *Hydrobiologia* **220**, 89–98.

Benzie JAH (1982) The complete larval development of *Caridina mccullochi* Roux, 1926 (Decapoda, Atyidae) reared in the laboratory. *Journal of Crustacean Biology* **2**, 493–513.

Bishop JA (1963) The Australian freshwater crabs of the family Potamonidae (Crustacea: Decapoda). *Australian Journal of Marine and Freshwater Research* **14**, 218–238.

Blair D and Finlayson CM (1981) Observations on the habitat and biology of a lymnaeid snail, *Austropeplea vinosa* (Gastropoda, Pulmonata), an intermediate host for avian schistosomes in tropical Australia. *Australian Journal of Marine and Freshwater Research* **32**, 757–767.

Boray JC (1964a) Studies on the ecology of *Lymnaea tomentosa*, the intermediate host of *Fasciola hepatica*. I. History, geographical distribution and environment. *Australian Journal of Zoology* **12**, 217–230.

Boray JC (1964b) Studies on the ecology of *Lymnaea tomentosa*, the intermediate host of *Fasciola hepatica*. II. The sexual behaviour of *Lymnaea tomentosa*. *Australian Journal of Zoology* **12**, 231–237.

Boray JC (1969) Experimental fascioliasus in Australia. In *Advances in Parasitology*. (Ed. B Dawes) pp. 95–210. Academic Press: London.

Boray JC and McMichael DF (1961) The identity of the Australian Lymnaeid snail host of *Fasciola hepatica* L. and its response to environment. *Australian Journal of Marine and Freshwater Research* **12**, 150–163.

Boulton AJ and Lloyd LN (1991) Macroinvertebrate assemblages in floodplain habitats of the lower River Murray, South Australia. *Regulated Rivers: Research and Management* **6**, 183–201.

Brendonck L (1996) Diapause, quiescence, hatching requirements: what we can learn from large freshwater branchiopods (Crustacea: Branchiopoda: Anostraca, Notostraca, Conchostraca). *Hydrobiologia* **320**, 85–97.

Brunckhorst DJ (1998) Superfamily Corbiculoidea. In *Mollusca: The Southern Synthesis*. (Eds PL Beesley, GJB Ross and A Wells) pp. 352–355. CSIRO Publishing: Melbourne.

Bunn SE and Boon PI (1993) What sources of organic carbon drive food webs in billabongs? A study based on stable isotope analysis. *Oecologia* **96**, 85–94.

Burns A and Walker KF (2000) Biofilms as food for decapods (Atyidae, Palaemonidae) in the River Murray, South Australia. *Hydrobiologia* **437**, 83–90.

Byrne M, Phelps H, Church T, Adair V, Selvakumaraswamy P and Potts J (2000) Reproduction and development of the freshwater clam *Corbicula australis* in south-east Australia. *Hydrobiologia* **418**, 185–197.

Christian AD, Crump BG and Berg DJ (2008) Nutrient release and ecological stoichiometry of freshwater mussels (Mollusca:Unionidae) in 2 small, regionally distinct streams. *Journal of the North American Benthological Society* **27**, 440–450.

Cook BD, Baker AM, Page TJ, Grant SC, Fawcett JH, Hurwood DA and Hughes JM (2006) Biogeographic history of an Australian freshwater shrimp, *Paratya australiensis* (Atyidae): the role of life history transition in phylogeographic diversification. *Molecular Ecology* **15**, 1083–1093.

Cotton BC (1935) The Australian viviparous river snails. *Victorian Naturalist* **52**, 96–99.

Davie PJF (2002) Parathelphusidae. In *Zoological Catalogue of Australia. Vol. 19.3B. Malacostraca: Eucarida (part 2): Decapoda – Anomura, Brachyura* pp. 377–380. CSIRO Publishing/ABRS: Melbourne.

De Deckker P (1983) Notes on the ecology and distribution of non-marine ostracods in Australia. *Hydrobiologia* **106**, 223–234.

DPI (2007) 'Recovery Plan for the endangered river snail *Notopala sublineata*'. NSW Dept of Primary Industries: Sydney.

Edney GN, McNeill DG and Lawler SH (2002) The swamp yabby (*Cherax* sp.) of the Murray River catchment. *Victorian Naturalist* **119**, 200–204.

Faragher RA (1983) Role of the crayfish *Cherax destructor* Clark as food for trout in Lake Eucumbene, New South Wales. *Australian Journal of Marine and Freshwater Research* **34**, 407–417.

Fielder DR (1970) The larval development of *Macrobrachium australiense* Holthius, 1950 (Decapoda, Palaemonidae) reared in the laboratory. *Crustaceana* **18**, 60–74.

Frost JV (1975) Australian crayfish. In *Freshwater Crayfish – Vol. 2. Proceedings of the Second International Symposium on Freshwater Crayfish.* Baton Rouge, Louisiana, 1974. (Ed. JW Avault, Jr) pp. 87–96. International Association of Astacology: Hinterthal, Austria.

Geddes MC (1990) Crayfish. In *The Murray.* (Eds N Mackay and D Eastburn) pp. 303–307. Murray-Darling Basin Commission: Canberra.

Geddes MC, Musgrove RJ and Campbell NJH (1993) The feasibility of re-establishing the River Murray crayfish, *Euastacus armatus*, in the lower River Murray. *Freshwater Crayfish* **9**, 368–379.

Gilligan D, Rolls R, Merrick J, Lintermans M, Duncan P and Kohen J (2007) 'Scoping the knowledge requirements for Murray crayfish (*Euastacus armatus*)'. NSW Dept of Primary Industries, MDBC Project No. 05/1066.

Glaubrecht M, Brinkmann N and Pöppe J (2009) Diversity and disparity 'down under': systematics, biogeography and reproductive modes of the 'marsupial' freshwater Thiaridae (Caenogastropoda, Cerithioidea) in Australia. *Zoosystematics and Evolution* **85**, 199–275.

Greenaway P (1980) Water balance and urine production in the Australian arid zone crab *Holthuisana transversa. Journal of Experimental Biology* **87**, 237–246.

Greenaway P (1984) Survival strategies in desert crabs. In *Arid Australia.* (Eds HG Cogger and E Cameron) pp. 145–152. Australian Museum: Sydney.

Greenaway P and MacMillen RE (1978) Salt and water balance in the terrestrial phase of the inland crab *Holthuisana (Austrothelphusa) transversa* Martens (Parathelphusoidea: Sundathelphusidae). *Physiological Zoology* **51**, 217–229.

Greenaway P and Taylor HH (1976) Aerial gas exchange in Australian arid-zone crab, *Parathelphusa transversa* von Martens. *Nature* **262**, 711–713.

Hancock MA and Bunn SE (1997) Population dynamics and life history of *Paratya australiensis* Kemp, 1917 (Decapoda: Atyidae) in upland rainforest streams, south-eastern Queensland, Australia. *Marine and Freshwater Research* **48**, 361–369.

Hancock MA and Bunn SE (1999) Swimming response to water current in *Paratya australiensis* Kemp, 1917 (Decapoda, Atyidae) under laboratory conditions. *Crustaceana* **72**, 313–323.

Hanski I (1998) Metapopulation dynamics. *Nature* **396**, 41–49.

Healy JM and Wells FE (1998) Family Thiaridae. In *Mollusca: The Southern Synthesis.* (Eds PL Beesley, GJB Ross and A Wells) pp. 727–729. CSIRO Publishing: Melbourne.

Hiscock ID (1951) A note on the life history of the Australian freshwater mussel, *Hyridella australis* Lam. *Transactions of the Royal Society of South Australia* **74**, 146–148.

Horwitz P and Knott B (1995) The distribution and spread of the yabby *Cherax* complex in Australia: speculations, hypotheses and the need for research. *Freshwater Crayfish* **10**, 81–91.

Hubendick B (1967) Studies on Ancylidae: the Australian, Pacific and neotropical formgroups. *Acta Regiae Societatis scientiarum et litterarum Gothoburgensis. Zoologica* **1**, 1–52.

Humphrey CL (1984) Biology and ecology of the freshwater mussel *Velesunio angasi* (Bivalvia: Hyriidae) in the Magela Creek, Alligator Rivers region, Northern Territory. PhD thesis. University of New England.

Humphrey CL and Simpson RD (1985) 'The biology and ecology of *Velesunio angasi* (Bivalvia: Hyriidae) in the Magela Creek, Northern Territory (4 parts)'. Office of the Supervising Scientist for the Alligator Rivers region, Open File Record 38: Canberra.

Humphries P, Cook RA, Richardson AJ and Serafini LG (2006) Creating a disturbance: manipulating slackwaters in a lowland river. *River Research and Applications* **22**, 525–542.

Huner JV and Lindqvist OV (1995) Physiological adaptations of freshwater crayfishes that permit successful aquacultural enterprises. *American Zoologist* **35**, 12–19.

Hurley M, Hearnden MN and Kay BH (1995) Distribution of aquatic pulmonate snail species in contrasting habitats within the Ross River reservoir (Stage 2a), tropical northern Australia. *Marine and Freshwater Research* **46**, 1033–1038.

Ingram BA, Hawking JH and Shiel RJ (1997) *Aquatic Life in Freshwater Ponds: A Guide to the Identification and Ecology of Life in Aquaculture Ponds and Farm Dams in South-eastern Australia.* Co-operative Research Centre for Freshwater Ecology: Albury.

Iredale T (1943) A basic list of the freshwater Mollusca of Australia. *Australian Zoologist* **10**, 188–230.

Johnson HT (1979) Reproduction, development and breeding activity in the freshwater crayfish, *Cherax destructor* Clark. MSc thesis. University of Sydney.

Johnston TH and Beckwith AC (1945) Larval trematodes from Australian freshwater molluscs. Part X. *Transactions of the Royal Society of South Australia* **69**, 229–242.

Johnston TH and Beckwith AC (1947) Larval trematodes from Australian freshwater molluscs. Part XII. *Transactions of the Royal Society of South Australia* **71**, 324–333.

Jones HA (2007) The influence of hydrology on freshwater mussel (Bivalvia: Hyriidae) distributions in a semi-arid river system, the Barwon-Darling River and intersecting streams. In *Animals of Arid Australia: Out on Their Own?* (Eds CR Dickman, S Burgin and D Lunney) pp. 132–142. Royal Zoological Society: Sydney.

Junk WJ, Bayley PB and Sparks RE (1989) The flood pulse concept in river-floodplain systems. In *Proceedings of the International Large River Symposium*. (Ed. DP Dodge). *Canadian Fisheries and Aquatic Sciences, Special Publication* **106**, 110–127.

Kat PW (1984) Parasitism and the Unionacea (Bivalvia). *Biological Reviews* **59**, 189–207.

Kingsford RT (2000) Ecological impacts of dams, water diversions and river management on floodplain wetlands in Australia. *Austral Ecology* **25**, 109–127.

Koch M (2004) Observations on the reproduction and population structure of the caenogastropod, *Gabbia vertiginosa* Frauenfeld, 1862 (Rissooidea: Bithyniidae). *Molluscan Research* **24**, 65–73.

Koehn JD (2004) Carp (*Cyprinus carpio*) as a powerful invader in Australian waterways. *Freshwater Biology* **49**, 882–894.

Lake PS and Sokol A (Eds) (1986) 'Ecology of the yabby *Cherax destructor* Clark (Crustacea: Decapoda: Parastacidae) and its potential as a sentinel animal for mercury and lead pollution'. Australian Water Resources Council Technical Paper. Australian Government Publishing Service: Canberra.

Lee CL and Fielder DR (1979) A mass migration of the freshwater prawn, *Macrobrachium australiense* Holthuis, 1950 (Decapoda, Palaemonidae). *Crustaceana* **37**, 219–222.

Lee CL and Fielder DR (1981) The effect of salinity and temperature on the larval development of the freshwater prawn, *Macrobrachium australiense* Holthius, 1950 from south-eastern Queensland, Australia. *Aquaculture* **26**, 167–172.

Lee CL and Fielder DR (1982a) Induced spawning in the freshwater prawn, *Macrobrachium australiense* Holthuis, 1950 (Crustacea: Decapoda: Palaemonidae). *Aquaculture* **29**, 45–52.

Lee CL and Fielder DR (1982b) Maintenance and reproductive behaviour in the freshwater prawn *Macrobrachium australiense* Holthius (Crustacea: Decapoda: Palaemonidae). *Australian Journal of Marine and Freshwater Research* **33**, 629–646.

Lee CL and Fielder DR (1982c) Reproductive cycle and female breeding dress in the freshwater prawn *Macrobrachium australiense* Holthius, 1950 (Crustacea: Decapoda: Palaemonidae). *Proceedings of the Royal Society of Queensland* **93**, 71–77.

Lee CL and Fielder DR (1984) Swimming response to water current stimulus in the freshwater prawn *Macrobrachium australiense* Holthius 1950 (Crustacea: Decapoda: Palaemonidae). *Crustaceana* **46**, 249–256.

Lintermans M and Rutzou T (1991) 'The status, distribution and management of the Murray crayfish *Euastacus armatus* in the Australian Capital Territory'. Research report 6. ACT Parks and Conservation Service: Canberra.

Lloyd J, Boray J and Campbell N (2007) Identifying liver fluke snails. In 'Prime Fact'. pp. 1–5. NSW Dept of Primary Industries: Orange.

Lynch JJ (1963) The ecology of *Lymnaea tomentosa* Pfeiffer. PhD thesis. University of Adelaide.

Lynch JJ (1965) The ecology of *Lymnaea tomentosa* (Pfeiffer 1855) in South Australia. *Australian Journal of Zoology* **13**, 461–474.

Mackie GL (1984) Bivalves. In *The Mollusca. Vol. 7. Reproduction.* (Ed. KM Wilbur) pp. 351–418. Academic Press: London.

MacMillen RE and Greenaway P (1978) Adjustments of energy and water metabolism to drought in an Australian arid-zone crab. *Physiological Zoology* **51**, 230–240.

Marcus V and Weeks S (1997) The effects of pond duration on the life history traits of an ephemeral pond crustacean, *Eulimnadia texana*. *Hydrobiologia* **359**, 213–221.

McCarthy B (2005) 'Distribution of the Murray crayfish (*Euastacus armatus*) in the Mallee region of the River Murray. Report to the Mallee Catchment Management Authority and the Murray-Darling Freshwater Research Centre'. Technical report 2/2005. Murray-Darling Freshwater Research Centre: Mildura.

McCormack RB (2008) *The Freshwater Crayfish of New South Wales, Australia.* Australian Aquatic Biological: Karuah.

McMahon RF (1983) Ecology of an invasive pest bivalve, *Corbicula*. In *The Mollusca*. (Ed. KM Wilber) pp. 505–561. Academic Press: New York.

McMahon TA and Finlayson BL (2003) Droughts and anti-droughts: the low flow hydrology of Australian rivers. *Freshwater Biology* **48**, 1147–1160.

McMichael DF (1967) Australian freshwater mollusca and their probable evolutionary relationships: a summary of present knowledge. In *Australian Inland Waters and Their Fauna. Eleven Studies.* (Ed. AH Weatherly) pp. 123–149. ANU Press: Canberra.

McMichael DF and Hiscock ID (1958) A monograph of the freshwater mussels (Mollusca: Pelecypoda) of the Australian region. *Australian Journal of Marine and Freshwater Research* **9**, 372–527.

McMichael DF and Iredale T (1959) The land and freshwater mollusca of Australia. In *Biogeography and Ecology in Australia.* (Eds A Keast, RL Crocker and CS Christian) vol. 8, pp. 224–245. Monographiae Biologicae. Junk: Den Haag.

MDBC (2006) 'River Murray system: drought update, July 2006'. Murray-Darling Basin Commission, Ref. 06/17190: Canberra.

Meakin CA, Qin JG and Mair GC (2008) Feeding behaviour, efficiency and food preference in yabbies *Cherax destructor. Hydrobiologia* **605**, 29–35.

Merrick JR (1993) *Freshwater Crayfishes of New South Wales.* Linnean Society of New South Wales: Sydney.

Mills BJ (1986) Aquaculture of yabbies. In *Proceedings of the First Australian Freshwater Aquaculture Workshop.* (Ed. LF Reynolds) pp. 89–98. Dept of Agriculture: Narrandera.

Mills BJ and Geddes MC (1980) Salinity tolerance and osmoregulation of the Australian freshwater crayfish *Cherax destructor* Clark (Decapoda: Parastacidae). *Australian Journal of Marine and Freshwater Research* **31**, 667–676.

Morgan GJ (1986) Freshwater crayfish of the genus *Euastacus* (Decapoda: Parastacidae) from Victoria. *Memoirs of the Museum of Victoria* **47**, 1–57.

Morgan GJ (1997) Freshwater crayfish of the genus *Euastacus* Clark (Decapoda: Parastacidae) from New South Wales, with a key to all species of the genus. *Records of the Australian Museum, Supplement* **23**, 1–110.

Morris S and Callaghan J (1998a) The emersion response of the Australian yabby *Cherax destructor* to environmental hypoxia and the respiratory and metabolic responses to consequent air breathing. *Journal of Comparative Physiology B* **168**, 389–398.

Morris S and Callaghan J (1998b) Respiratory and metabolic responses of the Australian yabby *Cherax destructor* to progressive and sustained environmental hypoxia. *Journal of Comparative Physiology B: Biochemical Systemic and Environmental Physiology* **168**, 377–388.

Mosig J (1995) *Australian Yabby Farmer.* Agmedia: Melbourne.

Nguyen TTT and Austin CM (2005) Phylogeny of the Australian freshwater crayfish *Cherax destructor*-complex (Decapoda: Parastacidae) inferred from four mitochondrial gene regions. *Invertebrate Systematics* **19**, 209–216.

Nguyen TTT, Austin CM, Meewan MM, Schultz M and Jerry DR (2004) Phylogeography of the freshwater crayfish *Cherax destructor* Clark (Parastacidae) in inland Australia: historical fragmentation and recent range expansion. *Biological Journal of the Linnean Society* **83**, 539–550.

NSW DPI (2005) 'Yabby: aquaculture prospects'. NSW Dept of Primary Industries: Orange.

Olszewski P (1980) *A Salute to the Humble Yabby.* Angus and Robertson: Sydney.

Page TJ, Choy SC and Hughes JM (2005) The taxonomic feedback loop: symbiosis of morphology and molecules. *Biology Letters* **1**, 139–142.

Page TJ and Hughes JM (2007) Radically different scales of phylogeographic structuring within cryptic species of freshwater shrimp (Atyidae: Caridina). *Limnology and Oceanography* **52**, 1055–1066.

Piola RF, Suthers IM and Rissik D (2008) Carbon and nitrogen stable isotope analysis indicates freshwater shrimp *Paratya australiensis* Kemp, 1917 (Atyidae) assimilate cyanobacterial accumulations. *Hydrobiologia* **608**, 121–132.

Ponder WF, Clark SA and Dallwitz MJ (2000) *Freshwater and Estuarine Molluscs. An Interactive, Illustrated Key for New South Wales.* CSIRO Publishing: Melbourne.

Preece RM (2004) 'Cold water pollution below dams in New South Wales. A desktop assessment.' NSW Dept of Infrastructure, Planning and Natural Resources: Sydney.

Puckridge JT, Sheldon F, Walker KF and Boulton AJ (1998) Flow variability and the ecology of large rivers. *Marine and Freshwater Research* **49**, 55–72.

Pusch M, Siefert J and Walz N (2001) Filtration and respiration rates of two unionid species and their impact on the water quality of a lowland river. In *Ecology and Evolution of the Freshwater Mussels Unionoida.* (Eds G Bauer and K Wächtler) pp. 317–326. Springer-Verlag: Berlin.

Puslednik L, Ponder WF, Dowton M and Davis AR (2009) Examining the phylogeny of the Australasian Lymnaeidae (Heterobranchia: Pulmonata: Gastropoda) using mitochondrial, nuclear and morphological markers. *Molecular Phylogenetics and Evolution* **52**, 643–659.

Reid DJ, Quinn GP, Lake PS and Reich P (2008) Terrestrial detritus supports the food webs in lowland intermittent streams of south-eastern Australia: a stable isotope study. *Freshwater Biology* **53**, 2036–2050.

Richardot M (1977) Ecological factors inducing estivation in freshwater limpet *Ferissia waltieri* (Basomatophora, Ancylidae). I. Oxygen content, organic matter content and pH of the water. *Malacological Review* **10**, 159–170.

Richardson AJ and Cook RA (2006) Habitat use by caridean shrimps in lowland rivers. *Marine and Freshwater Research* **57**, 695–701.

Richardson AJ, Growns JE and Cook RA (2004) Distribution and life history of caridean shrimps in regulated lowland rivers in southern Australia. *Marine and Freshwater Research* **55**, 295–308.

Richter BD, Baumgartner JV, Wigington R and Braun DP (1997) How much water does a river need? *Freshwater Biology* **37**, 231–249.

Ryan T, Webb A, Lennie R and Lyon J (2001) 'Status of cold water releases from Victorian dams'. Dept of Natural Resources and Environment, Arthur Rylah Institute for Environmental Research: Melbourne.

Sheldon F and Walker KF (1989) Effects of hypoxia on oxygen consumption by two species of freshwater mussel (Unionacea, Hyriidae) from the River Murray. *Australian Journal of Marine and Freshwater Research* **40**, 491–499.

Sheldon F and Walker KF (1993a) Pipelines as a refuge for freshwater snails. *Regulated Rivers: Research and Management* **8**, 295–299.

Sheldon F and Walker KF (1993b) Shell variation in Australian *Notopala* (Gastropoda: Prosobranchia: Viviparidae). *Journal of the Malacological Society of Australia* **14**, 59–71.

Sheldon F and Walker KF (1997) Changes in biofilms induced by flow regulation could explain extinction of aquatic snails in the lower River Murray, Australia. *Hydrobiologia* **347**, 97–108.

Sheldon F and Walker KF (1998) Spatial distribution of littoral invertebrates in the lower Murray-Darling River system, Australia. *Marine and Freshwater Research* **49**, 171–182.

Short JW (2004) A revision of Australian river prawns, *Macrobrachium* (Crustacea: Decapoda: Palaemonidae). *Hydrobiologia* **525**, 1–100.

Simovich MA and Hathaway SA (1997) Diversified bet-hedging as a reproductive strategy of some ephemeral pool Anostracans (Branchiopoda). *Journal of Crustacean Biology* **17**, 38–44.

Smith B and Stanisic J (1998) Pulmonata. In *Mollusca: The Southern Synthesis. Fauna of Australia.* (Eds PL Beesley, GJB Ross and A Wells) pp. 1037–1125. CSIRO Publishing: Melbourne.

Smith BJ (1978) Molluscs of the Murray-Darling River system. *Proceedings of the Royal Society of Victoria* **90**, 203–209.

Smith BJ (1996) 'Identification keys to the families and genera of bivalve and gastropod molluscs found in Australian inland waters'. Identification Guide No. 6. Murray-Darling Freshwater Research Centre: Albury.

Smith BJ and Burn RF (1976) *Glyptophysa cosmeta* (Iredale, 1943) in Victoria (Lymnaeoidea: Planorbidae), with notes on aestivation. *Journal of the Malacological Society of Australia* **3**, 175–176.

Smith BJ, Reid S and Ponder WF (2002) *Superorder: Pulmonata. Slugs and Snails.* Dept of Environment, Water, Heritage and the Arts: Canberra. http://www.deh.gov.au/cgi-bin/abrs/fauna/details.pl?pstrVol=PULMONATA;pstrTaxa=1;pstrChecklistMode=1.

Smith MJ and Williams WD (1980) Infraspecific variations within the Atyidae: a study of morphological variation within a population of *Paratya australiensis* (Crustacea: Decapoda). *Marine and Freshwater Research* **31**, 397–407.

Sokol A (1988a) The Australian yabby. In *Freshwater Crayfish.* (Eds DM Holdich and RS Lowry) pp. 401–425. Croom Helm: London.

Sokol A (1988b) Morphological variation in relation to the taxonomy of the destructor group of the genus *Cherax. Invertebrate Taxonomy* **2**, 55–79.

Stanisic J (1998) Superfamily Ampullarioidea. In *Mollusca: The Southern Synthesis. Fauna of Australia.* (Eds PL Beesley, GJB Ross and A Wells) pp. 706–707. CSIRO Publishing: Melbourne.

Stevens MM (2002) *Planorbidae* and *Lymnaeidae* as pests of rice, with particular reference to *Isidorella newcombi (Adams & Angus).* In *Molluscs as Crop Pests.* (Ed. GM Barker) pp. 217–233. CABI: Wallingford, UK.

Stevens MM (2007) Rice. In *Pests of Field Crops and Pastures: Identification and Control.* (Ed. PT Bailey) pp. 279–296. CSIRO Publishing: Melbourne.

Stoddart JA (1985) Analysis of species lineages of some Australian thiarids (Thiaridae, Prosobranchia, Gastropoda) using the evolutionary species concept. *Journal of the Malacological Society of Australia* **7**, 7–16.

Thoms MC, Sheldon F, Roberts J, Harris J and Hillman TJ (1996) *Scientific Panel Assessment of Environmental Flows for the Barwon-Darling River.* NSW Dept of Land and Water Conservation: Sydney.

Thorp JH and Delong MD (1994) The riverine productivity model: an heuristic view of carbon sources and organic processing in large river ecosystems. *Oikos* **70**, 305–308.

Timms BV (2006) *Class BRANCHIOPODA Latreille, 1817.* Australian Biological Resources Study: Canberra. http://www.environment.gov.au/biodiversity/abrs/online-resources/fauna/afd/taxa/BRANCHIOPODA.

Vannote RL, Minshall GW, Cummins KW, Sedell JR and Cushing CE (1980) The river continuum concept. *Canadian Journal of Fisheries and Aquatic Sciences* **37**, 130–137.

Vaughn CC, Nichols SJ and Spooner DE (2008) Community and foodweb ecology of freshwater mussels. *Journal of the North American Benthological Society* **27**, 409–423.

Walker KF (1981a) The distribution of freshwater mussels (Mollusca: Pelecypoda) in the Australian zoogeographic region. In *Ecological Biogeography of Australia.* (Ed. A Keast) pp. 1233–1249. W. Junk: The Hague.

Walker KF (Ed.) (1981b) 'Ecology of freshwater mussels in the River Murray'. Australian Water Resources Council Technical Paper. Australian Government Publishing Service: Canberra.

Walker KF (1982) The plight of the Murray crayfish in South Australia. *Red Gum (S.A.Conservation Council)* **6**, 2–6.

Walker KF (1983) The Murray is a floodplain river. *South Australian Naturalist* **58**, 29–33.

Walker KF (1985) A review of the ecological effects of river regulation in Australia. *Hydrobiologia* **125**, 111–129.

Walker KF (1986) The Murray-Darling river system. In *The Ecology of River Systems.* (Eds BR Davies and KF Walker) pp. 631–659. W. Junk: Dordrecht.

Walker KF (1992) The River Murray, Australia: a semi-arid lowland river. In *The Rivers Handbook.* (Eds PA Calow and GE Petts) pp. 472–492. Blackwell Scientific: Oxford.

Walker KF (1996) The river snail *Notopala hanleyi*: an endangered pest. *Xanthopus (Nat. Conserv. Soc. S.A.)* **March**, 1–5.

Walker KF (1998) Molluscs of inland waters. In *Mollusca: The Southern Synthesis. Fauna of Australia.* (Eds PL Beesley, GJB Ross and A Wells) pp. 59–65. CSIRO Publishing: Melbourne.

Walker KF, Thoms MC and Sheldon F (1992) Effects of weirs on the littoral environment of the River Murray, South Australia. In *River Conservation and Management.* (Eds PJ Boon, PA Calow and GE Petts) pp. 270–293. Wiley: Chichester.

Walker KF, Sheldon F and Puckridge JT (1995) A perspective on dryland river ecosystems. *Regulated Rivers: Research and Management* **11**, 85–104.

Walker KF, Byrne M, Hickey CW and Roper DS (2001) Freshwater mussels (Hyriidae) of Australasia. In *Ecological Studies: Ecology and Evolution of the Freshwater Mussels Unionoida*. (Eds G Bauer and K Wächtler) pp. 5–31. Springer: Berlin.

Walker KF, Madden CP, Williams CR, Fricker SR, Geddes MC, Goonan PM, Kokkin MJ, McEvoy PK, Shiel RJ and Tsymbal V (2009) Freshwater invertebrates. In *Natural History of the Riverland and Murraylands*. (Ed. JT Jennings) pp. 283–305. Royal Society of South Australia: Adelaide.

Walsh CJ (1993) Larval development of *Paratya australiensis* Kemp, 1917 (Decapoda: Caridea: Atyidae), reared in the laboratory, with comparisons of fecundity and egg and larval size between estuarine and riverine environments. *Journal of Crustacean Biology* **13**, 456–480.

Ward JV and Stanford JA (1983) The intermediate-disturbance hypothesis: an explanation for biotic diversity patterns in lotic ecosystems. In *Dynamics of Lotic Ecosystems*. (Eds TD Fontaine III and SM Bartell) pp. 347–356. Butterworth: Kent.

Ward JV and Stanford JA (1995) Ecological connectivity in alluvial river ecosystems and its disruption by flow regulation. *Regulated Rivers: Research and Management* **11**, 105–119.

Weeks S, Marcus V and Alvarez S (1997) Notes on the life history of the clam shrimp, *Eulimnadia texana*. *Hydrobiologia* **359**, 191–197.

Welker M and Walz N (1998) Can mussels control the plankton in rivers? A planktological approach applying a Lagrangian sampling strategy. *Limnology and Oceanography* **43**, 753–762.

Williams WD (1977) Some aspects of the ecology of *Paratya australiensis* (Crustacea: Decapoda: Atyidae). *Australian Journal of Marine and Freshwater Research* **28**, 403–415.

Williams WD (1980) *Australian Freshwater Life. The Invertebrates of Australian Inland Waters*. Macmillan: Melbourne.

Williams WD (1981) The crustacea of Australian inland waters. In *Ecological Biogeography of Australia*. (Ed. A Keast) pp. 1103–1138. W. Junk: The Hague.

Williams WD (1985) Biotic adaptations in temporary lentic waters with a special reference to those in semi-arid and arid regions. *Hydrobiologia* **125**, 85–110.

Williams WD and Smith MJ (1979) A taxonomic revision of Australian species of *Paratya* (Crustacea: Atyidae). *Australian Journal of Marine and Freshwater Research* **30**, 815–832.

Withnall F (2000) Biology of yabbies (*Cherax destructor*). *Aquaculture Notes* No. AS0004.

Woolford T (1984) A fouling bivalve, *Corbiculina australis* (Deshayes, 1830), in the Renmark irrigation pipelines: its biology and control options for the Renmark Irrigation Trust. Hons thesis. University of Adelaide.

Young W (2001) Invertebrates large and small. In *Rivers as Ecological Systems: The Murray-Darling Basin*. (Ed. W Young) pp. 223–228. Murray-Darling Basin Commission: Canberra.

Chapter 7

Impacts of hydrological changes on floodplain wetland biota

Kerrylee Rogers and Timothy J Ralph

Introduction

Floodplain wetland biota exhibit an intrinsic relationship with water whereby flooding is the 'principal driving force responsible for the existence, productivity and interactions of the major biota' (Junk *et al.* 1989, p. 10). This book documents the current knowledge base for key biota and their habitat and water requirements in the Murray-Darling Basin. However, there are knowledge gaps in our understanding of the associations between floodplain wetland biota and water, and specific relationships remain amorphous for many genera and species. Information regarding the water requirements of biota is particularly pertinent since water became a diminishing resource in the Murray-Darling Basin during the 20th century and is projected to continue its decline in the 21st century due to greenhouse warming-associated climate change.

Although there are considerable knowledge gaps about the water requirements of floodplain wetland biota, we believe that our knowledge base has developed sufficiently to establish the broad-scale impacts of hydrological changes on biota. This process primarily relies on historic observations of the response of biota to hydrological change. There were numerous opportunities for observing the response of biota in the Murray-Darling Basin during the 20th century. The majority of water resource development in the Murray-Darling Basin occurred since the 1950s, within a landscape that has undergone interdecadal and interannual climatic shifts. Both water resource development and natural climate variability have provided scientists with opportunities to monitor the response of biota. Combining our understanding of the water requirements and responses of biota to hydrological factors (Chapters 2 to 6), with projections of hydrological change – whether related to greenhouse warming, water sharing, climate variability or other factors – enables us to establish the vulnerable nature of floodplain wetlands so that management can be appropriately directed to protecting susceptible and valuable biota.

This chapter focuses on potential impacts and the likely response of biota to hydrological conditions that may occur in the Murray-Darling Basin in the 21st century. We will briefly discuss the hydrological changes that are projected to occur and apply these projections to the water needs of biota, established from the historical response of biota to hydrology. To assist resource managers with the allocation of water, the chapter will identify vulnerable biota whose water needs may be inadequately met under current flow and management regimes.

Drivers of hydrological change

Projecting the availability of water within the Murray-Darling Basin in the 21st century is not an exact science. As is the nature of forward projections, errors may be high and cumulative due to multiple influences on the variability and availability of resources. For example, future water availability is affected not only by water management and our direct manipulation of the natural flow of water throughout the Murray-Darling Basin, but by natural climatic drivers of water availability. These are affected by climate variability and climate change associated with greenhouse warming of the earth's atmosphere.

Water resource development and management

Hydrological variability is a natural feature of the lowland-dryland rivers of the Murray-Darling Basin. Considerable effort and resources have been directed towards regulating and developing the highly variable water supplies of rivers in the Basin to provide water for domestic uses and for lucrative irrigated agriculture and industry. The primary purpose of water storage and regulating structures is to capture flow in reservoirs and to allow the re-release of water during peak supply seasons or during drier years.

Water resource development and construction of water storages began around 100 years ago in the Murray-Darling Basin, but the majority of development occurred between the mid-1950s and 1990. The degree of water resource development varies across the Basin. Large public storage dams (Dartmouth, Eildon, Hume, Burrendong, Blowering, Copeton, Wyangala and Burrinjuck) store up to 1.5 times the average annual available surface water in the Murray-Darling Basin (CSIRO 2008). Public storage in the Namoi and Wimmera regions, for example, stores up to 1.5 times the average annual available surface water. In other regions, such as around the Moonie and Paroo rivers, there has been very little large-scale water resource development.

Water resource developments and land-use changes have had significant impacts on the hydrology and ecology of many lowland-dryland rivers in the Murray-Darling Basin (Kingsford 2000a; Sheldon *et al.* 2000; Thoms and Sheldon 2000; Thoms *et al.* 2005; Frazier *et al.* 2005). River regulation by large dams and abstractions for domestic and agricultural uses have generally led to a reduction in the frequency, magnitude and duration of floods in the lower reaches of many of these systems (Jolly 1996; Reid and Brooks 2000). Floods are usually less intense and tend to recede at a faster rate (Jolly 1996). The seasonality of flows has also been affected in most cases, often related to water management for industry. There is some evidence to suggest that geomorphic changes may be directly and indirectly related to hydrological changes due to river regulation and abstractions (Thoms and Walker 1993), although few studies have specifically investigated this in the Murray-Darling Basin. Recent land-use changes (e.g. clearing for agriculture) in the catchments of many lowland-dryland rivers have led to altered runoff and sediment supply regimes throughout the Basin.

Water resource development and management has led to changes in the natural drought and flood cycles in floodplain wetlands along the lower reaches of many lowland-dryland rivers in the Murray-Darling Basin (Kingsford 2000a). Lateral connections between river channels and floodplains have been altered, leading to greater ecological fragmentation (Thoms 2003). Floodplain wetlands have been particularly affected by river regulation and abstractions in terms of their spatial extent (i.e. reduced flood coverage), ecological health and biodiversity (Frazier and Page 2006; Kingsford 2000a, 2000b; Kingsford and Thomas 2004). The development of off-river water storages, canals and artificial levee banks has altered the effective wetland area of the floodplains in many lowland-dryland rivers (Thoms 2003). Even so, severe and long natural droughts are not uncommon in many subcatchments of the Murray-Darling Basin.

A goal of environmental water management is to match the ecological requirements of species within an ecosystem to the delivery of environmental water. This is difficult due to the diversity of water requirements and responses of individual species, and their interactions. There is often a limited amount of environmental water, leading to inevitable environmental trade-offs. All these factors have the potential to alter the short-, medium- and long-term ecological components of the floodplain wetlands.

Climate variability

For the purposes of this chapter, the term 'climate variability' refers to natural variability in climatic systems that causes deviations from the prevailing climatic conditions for a significant period (years to decades). Climate variability in the Murray-Darling Basin is affected by large-scale ocean-atmosphere fluctuations in the Pacific, Indian and Southern oceans that influence regional air pressure and circulation patterns, weather and rainfall. This occurs at intraseasonal, interannual and interdecadal time-scales in eastern Australia. Climatic and hydrological cycles in south-eastern Australia are strongly influenced by the following factors.

- Indian Ocean Dipole (IOD; Saji *et al.* 1999) – positive IOD events are typified by reduced winter and spring rainfall in south-eastern Australia (Drosdowsky and Chambers 2001; Nicholls 1989). Recent research indicates that the occurrence of severe droughts in south-eastern Australia may be more strongly driven by the IOD (Ummenhofer *et al.* 2009). The intensity of droughts may be suppressed or enhanced by the different phases of ENSO, SAM and IPO (Verdon-Kidd and Kiem 2009a).
- El Niño-Southern Oscillation (ENSO; Allan 1988) – negative ENSO events (El Niño) can result in warm dry conditions over south-eastern Australia and positive ENSO events (La Niña) can result in above-average rainfall and streamflow (Kiem and Franks 2001; Verdon *et al.* 2004).
- Southern Annular Mode (SAM; Thompson and Wallace 2000) – the leading mode of variability over the southern extra-tropics. The positive phase of SAM is typified by reduced autumn rainfall in south-eastern Australia (Nicholls 2010; Verdon-Kidd and Kiem 2009b).
- Interdecadal Pacific Oscillation (IPO; Power *et al.* 1999) – appears to modulate the intensity and frequency of ENSO events whereby positive phases are associated with increased frequency of El Niño events and suppressed La Niña intensity. Negative IPO phases modulated by positive ENSO events have been associated with large floods in parts of the Murray-Darling Basin. IPO appears to explain the clustering of floods and relatively long dry spells in numerous regions in the Murray-Darling Basin over multidecadal time-scales (Ralph and Hesse 2010; Verdon and Franks 2006; Verdon *et al.* 2006, 2004).

There are other modes of variability that may affect the climate of south-eastern Australia.

- Madden-Julian Oscillation (MJO; Hendon and Liebmann 1990) – an intraseasonal mode of variability that may result in rainfall variation in northern parts of the Murray-Darling Basin (Murphy and Timbal 2008).
- Subtropical Ridge (STR; Drosdowsky 2005) – a ridge of high pressure that has an annual cycle in latitude and strength and alters the progression of mid-latitude storms over Australia.

It is relatively well accepted that climate variability, particularly droughts, in south-eastern Australia is associated with a range of phenomena such as IOD, ENSO, SAM, IPO, MJO and STR. An analysis of three protracted droughts in the 20th century in south-eastern Australia identified that the severity, spatial signature, seasonality and pattern of rainfall varied considerably between the droughts (Verdon-Kidd and Kiem 2009a). A short drought in 1982–83, which recorded the highest short-term rainfall deficiencies, occurred at a time when the IOD, ENSO, IPO and SAM were all in dry phases. Although this intense drought was short, it underscores the probability of worse droughts in the future that are characterised by all modes in their dry phases for an extended period (Verdon-Kidd and Kiem 2009a).

Climate change

Increased greenhouse gas emissions during the 20th and 21st centuries are projected to alter the global climate, influencing temperature and rainfall globally and affecting water availability and water security in the Murray-Darling Basin. Global Circulation Models (GCMs) simulate climate systems and are used to project future climate change scenarios. Decoupling the influence of natural climate variability and climate change resulting from human activity is difficult, as GCMs aim to model climate systems rather than sources of climate change. For this reason, we have adopted the climate change definition used in the Intergovernmental Panel on Climate Change (IPCC) fourth assessment report:

> Climate change refers to a change in the state of the climate that can be identified (e.g. using statistical tests) by changes in the mean and/or the variability of its properties, and that persists for an extended period, typically decades or longer. It refers to any changes in climate over time, whether due to natural variability or as a result of human activity (Bernstein *et al.* 2007, p. 30).

A recent study of climate change in Australia (CSIRO and BOM 2007) projected significant warming of the Australian climate in the 21st century. The best estimate of warming for inland areas is 1–1.2°C by 2030 and 1.8–3.4°C by 2070. While precipitation is not directly influenced by greenhouse gas emissions, increased atmospheric temperatures will alter circulation (wind) patterns and consequently affect rainfall. Due to great natural variability in rainfall over Australia, GCMs are particularly sensitive to small changes in circulation and rainfall changes can vary significantly between models. Best estimates of annual precipitation change indicate decreases of 2–5% by 2030 and –30% to +20% (+5%) in central (southern) areas by 2070, with large seasonal changes.

Projected water availability

Projecting future water availability is complex due to the influence of water regulation, management, climate variability and climate change on current water availability, the inherent natural variability in river systems and errors associated with climate and hydrological modelling. Despite this complexity, considerable research effort has been and continues to be directed towards undertaking this task. The CSIRO Murray-Darling Basin Sustainable Yields Project is a world-first comprehensive attempt at projecting future water availability at a large basin scale (CSIRO 2008). The project considers the influence of catchment development, increasing groundwater extraction, increasing use of water resources and the anticipated impacts of climate change on future water availability on a regional basis (regional reports, cited in CSIRO 2008). Due to considerable variability in precipitation projections beyond 2030, projections of water availability have been limited to 2030, with trends that are projected to continue beyond that point. Due to the combined impacts of water resource development, climate variability and climate change there is a projected basin-wide decrease of 12% of available surface water compared to historic surface water availability. Decreases in surface water availability are

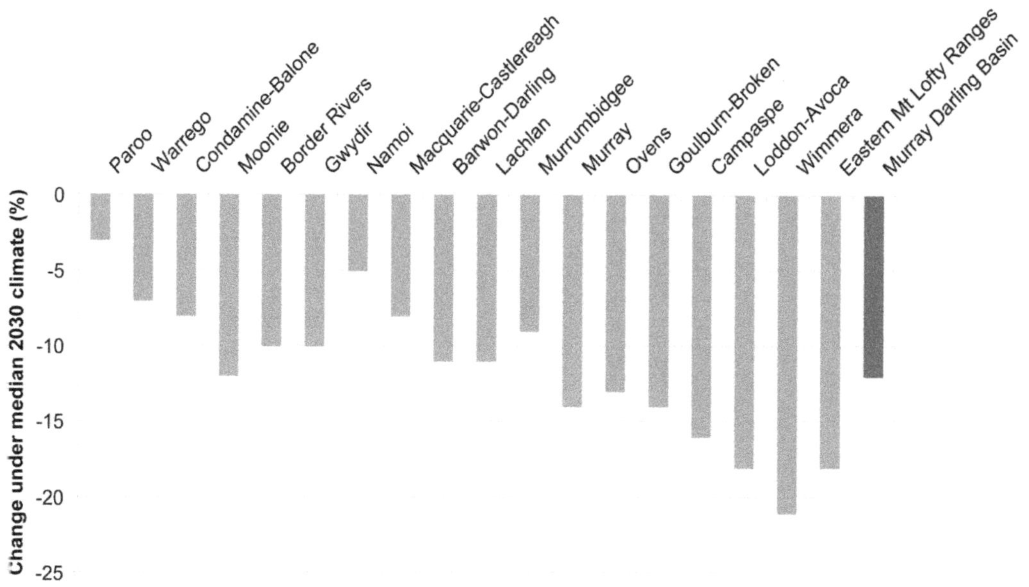

Figure 7.1: Projected change in surface water availability, from surface water availability under a historical climate without development to surface water availability under a median 2030 climate with future development in subregions of the Murray-Darling Basin. Source: CSIRO (2008) and regional reports cited within.

consistent across all regions in the Murray-Darling Basin, but the degree of change varies due to different water-sharing arrangements across the regions (Figure 7.1).

The impact of this reduced water availability is likely to be observed by all water users and will cause considerable impact to floodplain wetlands and associated biota in the Murray-Darling Basin. These impacts will largely be driven by widespread changes to the prevailing flood regimes within wetlands. The CSIRO Sustainable Yields Project identified a general trend of increasing average and maximum periods between floods (Figure 7.2) and a decreasing flood volume per year and per flood (Figure 7.3) within floodplain wetlands and other environmental indicators in the Murray-Darling Basin (Table 7.1) by 2030. Some exceptions to these trends include the Gwydir Wetlands, where little change in the average and maximum periods between floods has been projected; the Talyawalka Anabranch commence-to-flow indicator on the Barwon-Darling system where slightly more frequent, but smaller, floods are projected; and the Paroo Overflow Lakes, where little change in the average annual flood volume has been projected while the flood volume per event is projected to increase slightly due to fewer floods.

Aspects of the flood regime that were considered important for biota occupying floodplain wetlands within the Murray-Darling Basin, as detailed in Chapters 2 to 6, include flood frequency, duration, depth, timing, rate of flood rise and fall, and inter-flood dry-period. In terms of these aspects of the flood regime, the general trends of increasing average and maximum periods between floods are likely to result in a reduction in flood frequency and an increase in inter-flood dry-period for the majority of wetlands in the Murray-Darling Basin. Due to the strong relationship between flood volume, depth and duration, whereby large flood volumes are normally of greater depth across a floodplain and take longer to drain from a floodplain, the trends of decreasing flood volume per year and per event will result in a reduction in flood duration and flood depth.

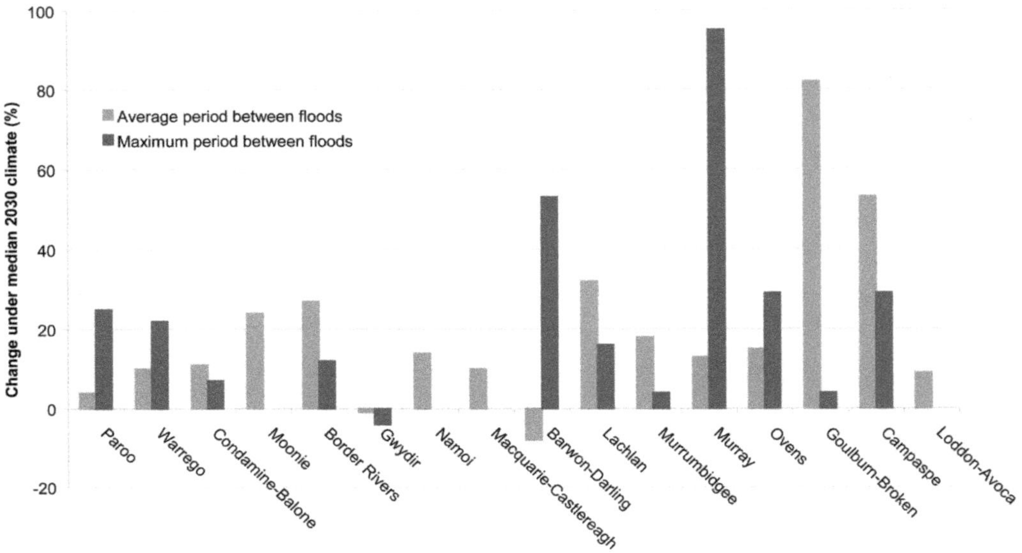

Figure 7.2: Projected change in the average and maximum periods between floods under a median 2030 climate for floodplain wetlands and other environmental indicators in subregions of the Murray-Darling Basin. Source: CSIRO (2008) and regional reports cited within.

It is difficult to infer the effect of projected hydrological changes on flood timing in the Murray-Darling Basin. Precipitation is projected to decrease in winter and spring over south-eastern Australia according to the best estimate of change by 2030 (CSIRO and BOM 2007). However, projecting the impact of these changes on flows is complex due to widespread and

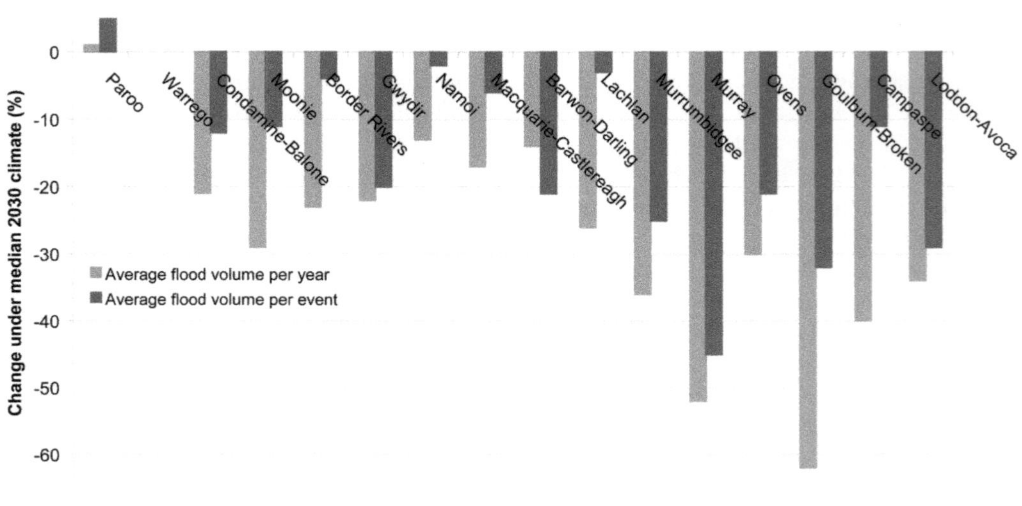

Figure 7.3: Projected change in the average flood volume per year and per event under a median 2030 climate for floodplain wetlands and other environmental indicators in subregions of the Murray-Darling Basin. Source: CSIRO (2008) and regional reports cited within.

Table 7.1: Selected environmental indicators from the Murray-Darling Basin Sustainable Yields Project used to establish a projected change in flood regimes for floodplain wetlands of the Murray-Darling Basin by 2030

Subregion	Environmental indicator	Flow indicator
Paroo	Paroo Overflow Lakes	Flows ≥170 000 ML/month at Caiwarro gauge
Warrego	Yantabulla Swamp	Flows ≥146 000 ML/month at Wyandra gauge
Condamine-Balonne	Balonne River floodplain	Flows ≥70 GL/day at St George gauge
Moonie	Floodplain wetland indicator	Flows ≥13.6 GL/day at Nindigully gauge
Border Rivers	River–floodplain connection between Goondiwindi and Mungindi	Flows ≥20 000 ML/day at Goondiwindi gauge
Gwydir	Gwydir Wetlands	Flows ≥100 GL/month at Yarraman gauge
Namoi	River–floodplain connecting flows	Flows ≥4 GL/day at Duncans Junction gauge
Macquarie-Castlereagh	Macquarie Marshes	Flows ≥200 GL June–Nov at Oxley gauge
Barwon-Darling	Talyawalka Anabranch commence-to-flow	Flows ≥30 GL/day at Wilcannia gauge
Lachlan	Great Cumbung Swamp	Flows ≥3000 ML/day 15 May–15 Nov at Booligal gauge
Murrumbidgee	Lowbidgee floodplain indicator	Flows ≥20 GL/day at Maude Weir
Murray	Barmah-Millewa Forest	Flows ≥10.6 GL/day at Yarrawonga
Ovens	Lower Ovens River floodplain	Flows ≥100 GL/week June–Nov at Mulwala gauge
Goulburn-Broken	Lower Goulburn River floodplain	Flows ≥1000 GL/month June–Nov at McCoys Bridge gauge
Campaspe	Winter–spring bankfull flows to maintain instream ecological function	Flows ≥43 GL/month June–Nov at Campaspe Siphon
Loddon-Avoca	Small winter floods to maintain riparian ecology	Flows ≥8100 ML/month July–Oct at Appin South gauge

Source: CSIRO (2008) and regional reports cited within.

large-scale water resource development. In the summer rainfall-dominated northern parts of the Murray-Darling Basin, water resource development has not altered the seasonality of flow but has dampened the amplitude of seasonal flow. In southern parts of the Basin, which are dominated by high winter rainfall, the seasonality of flow has been altered through the capture of winter flows for release in the warmer summer months. Seasonal impacts of climate change on precipitation are masked in the southern Murray-Darling Basin by reduced seasonality in flow associated with water resource development. Water resource development will result in a consistent reduction in flood variability throughout the Basin; the projected drying effects of an increase in inter-flood dry-period and reduction in flood frequency, duration and depth on floodplain wetlands will have significant implications for the survival of biota occupying floodplain wetlands of the Murray-Darling Basin.

Response of floodplain wetland biota to flood regime changes

Vegetation

There has been some discussion of the impacts of increasing and decreasing aspects of the water regime on plants (Brock and Casanova 1997; Casanova and Brock 2000; Nielsen and Brock 2009; Roberts and Marston 2000). By applying the projected changes in the flood regime

Table 7.2: Predicted plant community and functional group responses to changes in aspects of the water regime in the wet/dry ecotone

Aspects of water regime	Increase	Decrease
Frequency of flooding	Ecotone will experience wetting and drying more often May encourage amphibious plant groups	Ecotone more permanently wet or dry May favour submerged or terrestrial plants Longevity of the seed bank of amphibious plants will be important to their survival
Depth or extent of flooding	More permanent water habitat Favours competitive submerged and amphibious species	More dry habitat May favour weedy terrestrial invasion of the ecotone
Duration of flooding	Longer flooding Favours perennials and competitive submerged and amphibious species	Shorter floods Favours opportunistic short life-cycle species in all groups
Variability of flooding	More variable pattern A range of amphibious plants may cope, depending on longevity of seed bank	Less variable pattern More permanently wet or dry habitat may favour competitive submerged, amphibious and terrestrial plants
Total effect	More permanently wet May encourage competitive submerged species Species richness may decrease Survival depends on longevity of seed bank	More permanently dry May encourage weedy terrestrial species Species richness may decrease Variety of amphibious and submerged species decreases Survival depends on longevity of seed bank

Source: Brock and Casanova (1997).

throughout the Murray-Darling Basin to the water requirements of plants, it is evident that floodplain wetlands will be more permanently dry, creating conditions suitable for the encroachment of dryland species (Table 7.2).

Decreased flood frequency, depth and duration, increased inter-flood drying

Decreases in flood frequency and associated factors will affect the distribution and health of plant species and communities throughout the Murray-Darling Basin. Trees, particularly river red gum, are likely to exhibit signs of stress and a decline in reproductive ability. Since river red gum does not maintain a soil seed bank, water stress resulting from reduced flood frequency may severely limit its ability to regenerate and maintain viable populations. Black box is able to store seed within canopies and release seed when conditions are suitable; however, water stress may reduce the quantity and viability of seeds and limit the ability of black box to adapt to the altered flood regime. While both species may have some ability to regenerate in areas with ideal flood frequencies, reduced reproductive capacity may decrease the rate of regeneration and limit the capacity for communities to encroach on core wetland areas where water needs are met. Both river red gum and black box may survive under reduced flood frequencies; however, maintenance of populations will be severely limited and may result in the death of trees and loss of communities as they age. Conditions will generally be more suitable for dryland species such as poplar box (*E. populnea*), belah (*Casuarina cristata*) and wilga (*Geijera parviflora*) in northern parts of the Basin and white cypress pine (*Callitris glaucophylla*), mallee acacias (*Acacia* spp.) and *Cassinia* shrub (*Cassinia aculeate*) in southern parts of the Basin.

Unlike floodplain trees, lignum responds opportunistically to flooding by displaying lush vegetative growth and prolific seed production. Provided that flooding occurs at least every 10 years for approximately two months duration, populations may be maintained in the short term. However, reliance on one flood to maintain a lignum shrubland and/or understorey is hazardous, particularly if seed germination is limited due to other factors such as frost or low temperatures. Since seeds may be dispersed by floodwaters, there is some capacity for seed dispersal to and regeneration in core wetland areas where water needs are suitable.

Grasses, sedges, rushes, herbs and forbs occur at the ecotone between frequently wet and frequently dry environments and are regarded as amphibious species (Brock and Casanova 1997). Upper areas of the ecotone will be flooded less frequently and may no longer be regarded as part of the ecotone, resulting in shrinkage in the size of this zone. While these ecotone habitats are likely to be maintained, their contraction in size will increase competition between plants for space and resources within the ecotone. Species that can vegetatively regenerate, such as the common reed, may be able to remain within the ecotone providing the rate of regeneration is faster than the rate of ecotone contraction. In accordance with the environmental sieve model (van der Valk 1981) of succession, species that maintain a soil seed bank, such as nardoo, can respond opportunistically to floods despite increased competition for space and resources. Species reliant on seed banks may appear absent for some time, but may regenerate once water requirements for germination are suitable. Dispersal-dependent species may be severely restricted in their ability to adapt to the smaller ecotone. Flood frequency is essential for the maintenance of these populations as they are reliant on flooding for regeneration.

Provided that river flows continue within watercourses, the impact of reduced flood frequency on submerged aquatic macrophytes is limited. Populations will be sustained, but opportunities to expand into new habitats may be limited.

Decreased flood variability

A decrease in flood variability may change the flood regime from variable to more static. This will favour species that are adapted to more permanently wet or more permanently dry habitat and is likely to significantly reduce the frequency of large floods. The response of trees to a less variable water regime is difficult to predict and is likely to vary between species and sites. Provided trees can adapt to reduced flood frequency and regenerate in core wetland areas where their water requirements are met, communities may be maintained. However, it is more likely that reduced flood variability will limit the number of large floods which inundate floodplain trees, resulting in conditions more suitable for dryland trees.

Lignum, which is typically located at mid floodplain elevations, may be severely affected by a reduction in flood variability as it will limit the opportunity for lignum regeneration. An increase in the frequency of moderate floods, a likely result of reduced flood variability, may cause lignum located at lower elevations to receive flooding at higher than ideal frequencies and increase its exposure to competition from species adapted to higher flood frequencies.

Reduced flood variability will favour species suited to more static water levels or species that exhibit some plasticity towards depths of flooding. For example, although common reed prefers fluctuating water levels it can grow in static waters with depths of 20 cm above the waterline to 50 cm below. These static water conditions may become the norm for common reed within floodplain wetlands of the Murray-Darling Basin. Reduced flood variability will also favour species that require flooding for greater periods. Species such as marshwort and cumbungi will be maintained within more static water conditions. Similarly, an increase in base water flows will increase the available habitat for submerged aquatic macrophytes.

Waterbirds

As waterbirds generally rely on flooding to stimulate reproduction, flood regime changes will have a significant impact on waterbird breeding and maintenance of population numbers.

However, as waterbirds are heterotrophs and are heavily reliant on other organisms for their survival, flood regime changes are also likely to affect waterbirds indirectly through alterations to the quality and availability of habitats and consumed items.

Decreased flood frequency, depth and duration, increased inter-flood drying

A reduction in flood frequency, depth and duration and an associated increase in inter-flood drying will have marked influences on habitats utilised by waterbirds. Floodplain trees, such as river red gum and black box, have limited capacity for adapting to reduced flood frequencies. Numerous waterbirds prefer to breed in live river red gum, while others exhibit a preference for dead river red gum (Table 7.3; Higgins and Davies 1996; Leslie 2001; Marchant and Higgins 1990, 1993). The loss of vigour and death of river red gum individuals and communities may temporarily increase available breeding habitat for waterbirds that prefer to nest in dead trees, and there will be a corresponding decline in breeding habitat for waterbirds that utilise live river red gum for nesting. Similarly, the loss of vigour of floodplain trees may result in the generation of more tree hollows, thereby increasing available breeding habitat for waterbirds that nest in hollows. As lignum readily responds to flooding with lush growth, vegetative expansion and seed growth, provided that flooding occurs at least every 10 years and lasts for two months it is likely that lignum will be sustained and there will be only slight changes to this breeding habitat.

Apart from the dispersal-dependent species, many aquatic plants exhibit some adaptive capacity, such as vegetative expansion or seed banks that may enable regeneration or transgression of these habitats to core wetland areas where flood frequency requirements are suitable. If this trend occurs throughout a wetland it is feasible that foraging and/or breeding habitat for many waterbirds may be maintained, albeit at a diminished level (Higgins and Davies 1996; Marchant and Higgins 1990, 1993). This is particularly true for waterbirds that utilise reeds and rushes as breeding habitat (Table 7.3).

Flood regime changes will have significant impacts on waterbird habitats, but it is the impact on waterbird breeding that will be most alarming. Many waterbirds require flooding near or around their nest to commence breeding (Higgins and Davies 1996; Marchant and Higgins 1990, 1993). Reduced flood frequencies are likely to decrease the occurrence of suitable conditions for breeding and thus limit the breeding opportunities for these waterbirds. Waterbirds with relatively short life expectancies need to maximise breeding opportunities, to maintain population numbers. A decrease in flood frequency will reduce the number of breeding opportunities for these waterbirds and may severely affect the long-term stability of their populations.

Due to the relationship between large floods and those of long duration, reduced flood duration will affect waterbirds that require long lag times (≥6 months) to enhance their physical condition. Similarly, reduced flood duration is likely to affect waterbirds that exhibit long breeding durations (≥4 months) and may severely affect waterbirds that require long flood durations due to extremely long lag times, breeding durations or both (nine to 12 months) (Table 7.4). Some waterbirds require nests to be established in or over relatively deep water that remains for some time. Reducing flood duration and depth may cause adult waterbirds to abandon young before they have fully fledged, or may expose young to predators.

Decreased flood variability

Increased base flows and a decrease in large floods, associated with a reduction in flood variability, may affect the quality and availability of waterbird habitats. For example, floodplain trees exhibit limited capacity to adapt to fewer large floods and waterbirds that utilise these

Table 7.3: Preferred or common nesting habitat for waterbirds in the Murray-Darling Basin

Live trees[a]	Live or dead trees[a]	Tree hollows or tree hollow and ground nesting	Shrubs and/or reed nesting	Ground, floating or island nesting
Great egret	Darter	Maned duck	Blue-billed duck	Australasian shoveler
Cattle egret	Pacific heron	Chestnut teal[b]	Freckled duck	Black swan
Intermediate egret	Great cormorant	Australian shelduck	Hardhead	Plumed-whistling duck
Little egret	Pied cormorant	Grey teal	Musk duck	Black-fronted dotterel
White-faced heron		Pacific black duck	Pink-eared duck	Red-kneed dotterel
Rufous night heron		Pink-eared duck[c]	Magpie goose	Masked lapwing
Little pied cormorant			Little bittern	Brolga
Little black cormorant			Eurasian coot	Whiskered tern
Yellow-billed spoonbill			Dusky moorhen	Great crested grebe
Royal spoonbill[d]			Black-tailed native-hen	Hoary-headed grebe
Australian white ibis[d]			Purple swamphen	Australasian grebe
			Glossy ibis	Black-winged stilt
			Australian white ibis[d]	Australian pelican
			Straw-necked ibis	Silver gull[b]
			Royal spoonbill[d]	Caspian tern[b]
				Gull-billed tern
				Banded lapwing
				Red-necked avocet
				Red-capped plover

a: Preferred tree is typically river red gum.
b Breeding generally occurs in coastal settings.
c: May also occur in forks of trees.
d: Preference for live trees or among reeds, rushes and shrubs.

trees for breeding are likely to exhibit some changes (discussed above). Wetland plants that reproduce vegetatively, particularly reeds and rushes such as common reed and cumbungi, may be able to vegetatively encroach on core wetland areas that are maintained by base flows. Waterbirds utilising these habitats for breeding (Table 7.3) may exhibit some flexibility in breeding in accordance with the adaptations of vegetation (Higgins and Davies 1996; Marchant and Higgins 1990, 1993). However, this may be to the detriment of other waterbirds that utilise ephemeral or semi-permanent open waterbodies that are likely to be encroached upon by vegetatively expanding species and other species, such as river red gum, that opportunistically regenerate in these habitats.

Wetland plants that maintain seed banks that develop opportunistically in response to flooding, such as *Juncus, Cyperus, Eleocharis, Vallisneria* and some herbs and forbs, exhibit a capacity for adapting to reduced flood variability. These plants are not widely used as breeding habitat but do provide significant foraging habitat for many waterbirds, such as the black swan, hardhead, blue-billed duck, Eurasian coot, chestnut teal, grey teal, Pacific black duck, maned duck, plumed whistling-duck, Australian shelduck, black-tailed native-hen and purple swamphen (Higgins and Davies 1996; Marchant and Higgins 1990, 1993). The maintenance of foraging habitats may promote prompt breeding of waterbirds when flooding occurs, as extensive lag times to increase waterbird condition may not be required.

Reduced flood variability may favour plant species adapted to more static water conditions, such as common reed and cumbungi. Breeding of waterbirds that utilise these habitats may be enhanced. Waterbirds that prefer to breed in more static water levels, such as the great crested grebe, hoary-headed grebe, Australasian grebe, great cormorant, silver gull and black swan, may exhibit little change in breeding success (Higgins and Davies 1996; Marchant and Higgins 1990). Waterbirds that prefer to forage in areas with more static water conditions may adapt to changes to the flood regime. In particular, the fish-eating functional group of birds may adapt, as more permanent water conditions established with more persistent base flows are conducive to the establishment and maintenance of fish populations. Similarly, deep-water foragers may exhibit some resilience to reduced flood variability as their foraging habitats may increase due to the higher base flows.

When considering flood requirements for breeding, an increase in low-flow events that are likely to have shorter flood durations may favour waterbirds that require short or no lag times (Table 7.4). Similarly, waterbirds that exhibit short breeding durations of one to two months may not be greatly affected by a reduction in flood variability. Waterbirds with short or no lag times and with short breeding durations, such as whiskered tern, are well-placed for successful breeding under the projected flood conditions.

Reduced flood variability will result in less inter-flood drying in low–mid elevations of floodplain wetlands. A reduction in inter-flood drying may affect the breeding success of species that rely to some extent on inter-flood drying, such as the darter, great cormorant, little pied cormorant, little black cormorant, great egret, intermediate egret, Pacific heron, white-faced heron, rufous night heron, whiskered tern, Eurasian coot, grey teal, chestnut teal, Pacific black duck, pink-eared duck, freckled duck, maned duck, Australian shelduck, yellow-billed spoonbill, royal spoonbill and brolga (Crome 1986, 1988).

Fish

Australian native fish have water requirements that determine their distribution and recruitment success. Flooding is one of a number of factors, including water temperature, chemistry and habitat availability, which influence species and population health (see Chapter 4). Although optimal flooding and flow conditions are not well understood for most native species, the timing and duration of flows can influence the length of the spawning season, flow magnitude can determine species mobility (or migration) and flow frequency can influence species richness (Balcombe *et al.* 2006b; Puckridge *et al.* 1998). In the presence of suitable habitat, seasonal increases in water temperature, day length and flow are often cues for spawning; these normally coincide with spring and summer months in unregulated systems, and with abundant food supplies. Therefore, most species will spawn during floods if the timing is suitable, but many species are also able to recruit during periods of low flow (Balcombe *et al.* 2006; Humphries *et al.* 1999). Fecundity also varies between fish species. Ideally, flood duration should be long enough to cover the period from ovulation to metamorphosis for a number of species if the native communities are to be maintained. Declines in the duration of flooding in the Murray-Darling Basin, due to hydrological changes exacerbated by climate variability and climate change, may therefore more heavily affect native fish populations with low fecundity. Altered flow regimes can also cause declines in water quality, due to increased salinity and the build-up of agricultural chemicals.

In the Murray-Darling Basin, native fish use a variety of aquatic habitats, including channels, anabranches, floodplains, billabongs and wetlands. Habitat modification and

Table 7.4: Aspects of waterbird breeding that influence the impact of flood regime changes on breeding

	High impact			Less impact	
Species	**Short life (≤4 yr)**	**Long lag (≥6 mo)**	**Long breeding (≥4 mo)**	**Short lag (≤2 mo)**	**Short breeding (≤2 mo)**
Great crested grebe				X	X
Hoary-headed grebe		X			X
Australasian grebe				X	
Australian pelican			X	X	
Darter					
Great cormorant		X	X		
Little pied cormorant		X	X		
Little black cormorant		X	X		
Pied cormorant			X		
Great egret		X	X		
Intermediate egret		X	X		
Little egret			X		
Pacific heron		X			
White-faced heron					
Rufous night heron		X			
Silver gull					X
Whiskered tern				X	X
Caspian tern					X
Gull-billed tern			X	X	
Black swan			X	X	
Hardhead	X		X	X	
Musk duck	X		X		
Blue-billed duck	X				X
Eurasian coot				X	X
Chestnut teal			X	X	
Grey teal	X		X	X	
Australasian shoveler	X			X	
Pacific black duck	X		X	X	
Pink-eared duck	X			X	X
Freckled duck	X				
Maned duck	X		X		
Plumed whistling-duck	X		X		
Australian shelduck	X		X	X	
Black-tailed native-hen	X				X
Purple swamphen				X	
Masked lapwing				X	
Banded lapwing	X			X	
Yellow-billed spoonbill				X	X
Royal spoonbill					X
Glossy ibis		X			X
Australian white ibis		X			X
Straw-necked ibis		X			
Brolga			X	X	
Black-winged stilt	X				X
Red-necked avocet				X	X
Red-capped plover					X
Black-fronted dotterel					X
Red-kneed dotterel					

changes to flood regimes can be major influences on fish populations and species assemblages in river and wetland ecosystems. Reduced water availability and quality, and changes to the seasonality of the flow regime, are likely to have more impact on ephemeral habitats, including those on the floodplain and in adjoining wetlands, and will therefore have adverse impacts on fish that preferentially utilise those habitats. In-stream habitats are also critical because all species utilise main river channels, and channels and waterholes can provide refuge for fish in times of drought. Changes in the frequency or magnitude of large flow events and the availability of healthy floodplain habitats are likely to alter the reproductive capacity of fish reliant on increasing flows and wetland habitats, such as golden perch, silver perch and olive perchlet. These species are likely to be the most vulnerable to changes in the water regime. Other species, that use floodplain habitats only opportunistically, may exhibit a greater capacity to deal with altered flow regimes due to their ability to recruit successfully within channels.

Frogs

Considerable research is required about the water needs of frogs within the Murray-Darling Basin, but Chapter 5 does indicate that flood regime changes may affect the survival and recruitment of resident frog populations. With the exception of the *Cyclorana* species, reductions in flood frequency are likely to have a negative impact on frog populations in the Murray-Darling Basin. The water-holding frog, common green tree frog and desert tree frog may be able to breed rapidly in rain-fed pools, but the majority of other floodplain wetland frogs in the Murray-Darling Basin are reliant on extended seasonal and overbank flooding that endures for five to six months. Decreases in flood duration may markedly limit the recruitment of these frogs from tadpoles to adult life-stages.

Similarly, changes in the seasonality of flooding, which may become evident with a decrease in flood variability, may limit the survival and recruitment of some frogs. In particular, the re-release of water in the warmer summer months, which has become the norm for southern parts of the Murray-Darling Basin, may influence the common eastern froglet, eastern sign-bearing froglet, desert tree frog, eastern banjo frog and giant banjo frog, which are more active in the winter months. However, there may be some capacity for these species to adapt to flood regime changes. An increase in winter flooding in northern parts of the Murray-Darling Basin may affect the recruitment of species that prefer warmer waters for tadpole development.

Crustaceans and molluscs

There is limited information about the water needs of crustaceans and molluscs in the Murray-Darling Basin, and projecting the influence of hydrological change on these species is difficult. However, observations of population changes associated with alterations to flow regimes provide valuable information. Water resource development, particularly along the Murray and Darling rivers, has altered the natural flow regime to one with reduced flood variability and increased base flow magnitude. The result of this alteration is twofold – the stabilisation of river flows and alienation of floodplain wetlands. Chapter 6 indicates that these changes can have significant ecological impacts on crustaceans and molluscs.

The stabilisation of river flows has changed many aquatic habitats from predominantly lotic to predominantly lentic environments. This has been associated with declines in river mussels (Thoms *et al.* 1996; Walker 1985) and Murray crayfish (McCarthy 2005) which prefer lotic habitats, and increases in the abundance of yabbies (Walker 1992). The release of regular flows for irrigation use in summer may be unsuitable for McCulloch's shrimp (Richardson and Cook 2006) and the common freshwater shrimp (Richardson *et al.* 2004).

The greatest ecological impacts may be associated with decreased or limited lateral connectivity between the river and the neighbouring floodplains and wetlands. Floodplain habitats, including periodically connected billabongs and anabranches, provide primary habitat for freshwater shrimp, mussels and several snail species (Boulton and Lloyd 1991; Sheldon and Walker 1998). Reduced habitat and food availability have been linked to the decline of river snails and sculptured snails along the Murray River (Sheldon and Walker 1993, 1997). Although a range of adverse effects is likely, further research is required to more fully understand the impacts of hydrological changes on these aquatic species.

Conclusion

Given the considerable amount of water resource development that has occurred in the Murray-Darling Basin, it is essential that water be managed so that the water needs of all users are addressed. With projected reductions in water availability associated with climate change and climate variability, and increased water needs for agriculture, industry and town water supply, there is rising pressure on water resources. The Murray-Darling Basin Sustainable Yields Project indicates that most of this increased pressure is likely to be borne by the environment rather than by consumptive water users (CSIRO 2008).

The ecological significance of floodplain wetlands is well appreciated and the water needs of floodplain wetlands are becoming clearer. Considerable research is still required to fully understand the water requirements of biota, especially invertebrate species, but it is evident that flood regime changes, whether attributed to water management, climate change or climate variability, will have significant impacts on biota in floodplain wetlands. Flood regime changes may irreversibly alter the ecological value of floodplain wetlands in several bioregions.

Given the ecological value of floodplain wetlands, it is essential that consideration be given to the optimal provision of water for biota in floodplain wetlands. Water management should be directed towards supplying water to these assets to meet the specific water needs of significant or representative biota and to maintain the ecological function of floodplain wetlands in the Murray-Darling Basin. This is a complex task that requires the careful integration of science, policy and management to achieve the best use of environmental water and favourable outcomes for the environment.

References

Allan RJ (1988) El Niño Southern Oscillation influences in the Australasian region. *Progress in Physical Geography* **12**, 313–348.

Balcombe SR, Arthington AH, Foster ND, Thoms MC, Wilson GG and Bunn SE (2006) Fish assemblages of an Australian dryland river: abundance, assemblage structure and recruitment patterns in the Warrego River, Murray-Darling Basin. *Marine and Freshwater Research* **57**, 619–633.

Bernstein L, Bosch P, Canziani O, Chen Z, Christ R, Davidson O, Hare W, Huq S, Karoly D, Kattsov V, Kundzewicz Z, Liu J, Lohmann U, Manning M, Matsuno T, Menne B, Metz B, Mirza M, Nicholls N, Nurse L, Pachauri R, Palutikof J, Parry M, Qin D, Ravindranath N, Reisinger A, Ren J, Riahi K, Rosenzweig C, Rusticucci M, Schneider S, Sokona Y, Solomon S, Stott P, Stouffer R, Sugiyama T, Swart R, Tirpak D, Vogel C and Yohe G (2007) 'Climate change 2007: synthesis report'. Summary for policy-makers. Intergovernmental Panel on Climate Change: Spain.

Boulton AJ and Lloyd LN (1991) Macroinvertebrate assemblages in floodplain habitats of the lower River Murray, South Australia. *Regulated Rivers: Research and Management* **6**, 183–201.

Brock MA and Casanova MT (1997) Plant life at the edge of wetlands: ecological responses to wetting and drying patterns. In *Frontiers in Ecology: Building the Links.* (Eds NK Klomp and I Lunt) pp. 181–192. Elsevier Science: Oxford, UK.

Casanova MT and Brock MA (2000) How do depth, duration and frequency of flooding influence the establishment of wetland plant communities? *Plant Ecology* **147**, 237–250.

Crome FHJ (1986) Australian waterfowl do not necessarily breed on a rising water level. *Wildlife Research* **13**, 461–480.

Crome FHJ (1988) To drain or not to drain? Intermittent swamp drainage and waterbird breeding. *Emu* **88**, 243–248.

CSIRO (2008) 'Water availability in the Murray-Darling Basin: a report to the Australian Government from the CSIRO Murray-Darling Basin Sustainable Yields Project'. Commonwealth Scientific and Industrial Research Organisation: Canberra.

CSIRO and BOM (2007) 'Climate change in Australia'. Commonwealth Scientific and Industrial Research Organisation and Australian Bureau of Meteorology: Canberra.

Drosdowsky W (2005) The latitude of the subtropical ridge over eastern Australia: the *L* index revisited. *International Journal of Climatology* **25**, 1291–1299.

Drosdowsky W and Chambers LE (2001) Near-global sea surface temperature anomalies as predictors of Australian seasonal rainfall. *Journal of Climate* **14**, 1677–1687.

Frazier P and Page K (2006) The effect of river regulation on floodplain wetland inundation, Murrumbidgee River, Australia. *Marine and Freshwater Research* **57**, 133–141.

Frazier P, Page K and Read A (2005) Effects of flow regulation in flow regime on the Murrumbidgee River, south-eastern Australia: an assessment using a daily estimation hydrological model. *Australian Geographer* **36**, 301–314.

Hendon HH and Liebmann B (1990) The intraseasonal (30–50 day) oscillation of the Australian summer monsoon. *Journal of Atmospheric Sciences* **47**, 2909–2923.

Higgins PJ and Davies SJJF (1996) *Handbook of Australian, New Zealand and Antarctic Birds. Vol. 3: Snipe to Pigeons.* Oxford University Press: Melbourne.

Humphries P, King AJ and Koehn JD (1999) Fish, flows and flood plains: links between freshwater fishes and their environment in the Murray-Darling River system, Australia. *Environmental Biology of Fishes* **56**, 129–151.

Jolly ID (1996) The effects of river management on the hydrology and hydroecology of arid and semi-arid floodplains. In *Floodplain Processes.* (Eds MG Anderson, DE Walling and PD Bates) pp. 577–609. John Wiley and Sons: New York.

Junk WJ, Bayley PB and Sparks RE (1989) The flood pulse concept in river-floodplain system. *Canadian Fisheries and Aquatic Sciences, Special Publication* **106**, 110–127.

Kiem AS and Franks SW (2001) On the identification of ENSO-induced rainfall and runoff variability: a comparison of methods and indices. *Hydrological Sciences Journal* **46**, 715–727.

Kingsford RT (2000a) Ecological impacts of dams, water diversions and river management on floodplain wetlands in Australia. *Austral Ecology* **25**, 109–127.

Kingsford RT (2000b) Protecting rivers in arid regions or pumping them dry? *Hydrobiologia* **427**, 1–11.

Kingsford RT and Thomas RF (2004) Destruction of wetlands and waterbird populations by dams and irrigation on the Murrumbidgee River in arid Australia. *Environmental Management* **34**, 383–396.

Leslie DJ (2001) Effect of river management on colonially-nesting waterbirds in the Barmah-Millewa forest, south-eastern Australia. *Regulated Rivers: Research and Management* **17**, 21–36.

Marchant S and Higgins PJ (1990) *Handbook of Australian, New Zealand and Antarctic Birds. Vol. 1: Ratites to Ducks.* Oxford University Press: Melbourne.

Marchant S and Higgins PJ (1993) *Handbook of Australian, New Zealand and Antarctic Birds. Vol. 2: Raptors to Lapwings.* Oxford University Press: Melbourne.

McCarthy B (2005) 'Distribution of the Murray crayfish (*Euastacus armatus*) in the Mallee region of the River Murray. Report to the Mallee Catchment Management Authority and the Murray-Darling Freshwater Research Centre'. Murray-Darling Freshwater Research Centre: Mildura.

Murphy BF and Timbal B (2008) A review of recent climate variability and climate change in south-eastern Australia. *International Journal of Climatology* **28**, 859–879.

Nicholls N (1989) Sea surface temperatures and Australian winter rainfall. *Journal of Climate* **2**, 965–973.

Nicholls N (2010) Local and remote causes of the southern Australian autumn-winter rainfall decline, 1958–2007. *Climate Dynamics* **33(1)**, 63.

Nielsen D and Brock M (2009) Modified water regime and salinity as a consequence of climate change: prospects for wetlands of southern Australia. *Climatic Change* **95**, 523–533.

Power S, Casey T, Folland C, Colman A and Mehta V (1999) Inter-decadal modulation of the impact of ENSO on Australia. *Climate Dynamics* **15**, 319–324.

Puckridge JT, Sheldon F, Walker KF and Boulton AJ (1998) Flow variability and the ecology of large rivers. *Marine and Freshwater Research* **49**, 55–72.

Ralph TJ and Hesse PP (2010) Downstream hydrogeomorphic changes along the Macquarie River, southeastern Australia, leading to channel breakdown and floodplain wetlands. *Geomorphology* **118**, 48–64.

Reid MA and Brooks JJ (2000) Detecting effects of environmental water allocations in wetlands of the Murray-Darling Basin, Australia. *Regulated Rivers: Research and Management* **16**, 479–496.

Richardson AJ and Cook RA (2006) Habitat use by caridean shrimps in lowland rivers. *Marine and Freshwater Research* **57**, 695–701.

Richardson AJ, Growns JE and Cook RA (2004) Distribution and life history of caridean shrimps in regulated lowland rivers in southern Australia. *Marine and Freshwater Research* **55**, 295–308.

Roberts J and Marston F (2000) 'Water regime of wetland and floodplain plants in the Murray-Darling Basin'. CSIRO Land and Water: Canberra.

Saji NH, Goswami BN, Vinayachandran PN and Yamagata T (1999) A dipole mode in the tropical Indian Ocean. *Nature* **401**, 360–363.

Sheldon F and Walker KF (1993) Pipelines as a refuge for freshwater snails. *Regulated Rivers: Research and Management* **8**, 295–299.

Sheldon F and Walker KF (1997) Changes in biofilms induced by flow regulation could explain extinction of aquatic snails in the lower River Murray, Australia. *Hydrobiologia* **347**, 97–108.

Sheldon F and Walker KF (1998) Spatial distribution of littoral invertebrates in the lower Murray-Darling River system, Australia. *Marine and Freshwater Research* **49**, 171–182.

Sheldon F, Thoms MC, Berry O and Puckridge JT (2000) Using disaster to prevent catastrophe: referencing the impacts of flow changes in large dryland rivers. *Regulated Rivers: Research and Management* **16**, 403–420.

Thompson DWJ and Wallace JM (2000) Annular modes in the extratropical circulation. I. Month-to-month variability. *Journal of Climate* **13**, 1000–1016.

Thoms MC (2003) Floodplain-river ecosystems: lateral connections and the implications of human interference. *Geomorphology* **56**, 335–349.

Thoms MC and Sheldon F (2000) Water resource development and hydrological change in a large dryland river: the Barwon-Darling River, Australia. *Journal of Hydrology* **228**, 10–21.

Thoms MC and Walker KF (1993) Channel changes associated with two adjacent weirs on a regulated lowland alluvial river. *Regulated Rivers: Research and Management* **8**, 271–284.

Thoms M, Sheldon F, Roberts J, Harris J and Hillman T (1996) 'Scientific panel assessment of environmental flows for the Barwon-Darling River'. NSW Dept of Land and Water Conservation: Sydney.

Thoms MC, Southwell M and McGinness HM (2005) Floodplain-river ecosystems: fragmentation and water resources development. *Geomorphology* **71**, 126–138.

Ummenhofer CC, England MH, McIntosh PC, Meyers GA, Pook MJ, Risbey JS, Gupta AS and Taschetto AS (2009) What causes south-east Australia's worst droughts? *Geophysical Research Letters* **36**, L04706, doi:10.1029/2008GL036801.

van der Valk AG (1981) Succession in wetlands: a Gleasonian approach. *Ecology* **62**, 688–696.

Verdon DC and Franks SW (2006) Long-term behaviour of ENSO: interactions with the PDO over the past 400 years inferred from palaeoclimate records. *Geophysical Research Letters* **33**, L06712, doi:10.1029/2005GL025052.

Verdon DC, Wyatt AM, Kiem AS and Franks SW (2004) Multidecadal variability of rainfall and streamflow: eastern Australia. *Water Resources Research* **40**, 1–8.

Verdon DC, Franks SW and Hameed T (2006) Climate variability impacts on water resources in the Lachlan River Valley. In *30th Hydrology and Water Resources Symposium*. 4–7 December, Launceston. Engineers Australia.

Verdon-Kidd DC and Kiem AS (2009a) Nature and causes of protracted droughts in south-east Australia: comparison between the Federation, WWII, and Big Dry droughts. *Geophysical Research Letters* **36**, L22707, doi:10.1029/2009GL041067.

Verdon-Kidd DC and Kiem AS (2009b) On the relationship between large-scale climate modes and regional synoptic patterns that drive Victorian rainfall. *Hydrology and Earth System Sciences* **13**, 467–479.

Walker KF (1985) A review of the ecological effects of river regulation in Australia. *Hydrobiologia* **125**, 111–129.

Walker KF (1992) The River Murray, Australia: a semi-arid lowland river. In *The Rivers Handbook*. (Eds PA Calow and GE Petts) pp. 472–492. Blackwell Scientific: Oxford.

Chapter 8

Management of water for floodplain wetland biota

Neil Saintilan

Introduction

Environmental water management is a complex and novel task. The large freshwater wetlands which form the focus of environmental water management in the Murray-Darling Basin are hydrologically and ecologically complex. Their low gradient and dynamic geomorphology provide a challenge for hydrological modelling. The floodplains and marshes cannot be considered in isolation from the rivers and the catchments that supply them with water. Water flow through the wetlands is highly dependent on antecedent rainfall and flooding, the vigour of vegetation responses and the location and operation of structures, of which there may be hundreds. The response of wetland biota to water and the flood regime varies over a range of time-scales, and may be dependent on recruitment processes operating outside the wetland.

Even if the hydrology and ecology of the wetlands were completely understood, wetland managers operate in an environment of high climatic variability. It cannot be reliably predicted from one year to the next how much water will be available. As a result, ecological objectives from year to year require constant re-evaluation and resetting. However, as all water managers will attest, the environmental and climatic complexity pale before the complexity of managing social expectations. Water is an increasingly scarce resource in the Murray-Darling Basin, and the water requirements of the environment compete with the demands of agricultural, industrial and domestic consumption. The policy environment regulating these conflicting uses has been evolving rapidly for more than a decade.

This chapter discusses the complex scientific, policy and management environment whose primary objective is satisfying the water requirements of wetland biota. The task is interdisciplinary, not only in the scientific disciplines engaged but in the challenging task of aligning scientific research with policy and management objectives. Decision support systems are a useful platform for combining hydrological and ecological models to answer questions being asked by wetland managers. They provide the additional benefit of capturing system understanding in a way that can be tested against observation, providing a framework for the nexus between intervention and monitoring that defines adaptive environmental management.

Complexity of science

Organisms are distributed in the landscape in a way that reflects antecedent flood history (Bunn and Arthington 2002; Nilsson and Svedmark 2002). Their location in space is indicative of a hydrological regime that has to date supported all stages of their life-cycle (Lytle and Poff 2004). In a hydrologically variable system, different types of organisms respond to flow over a range of spatial and temporal scales. Kingsford *et al.* (2010) categorised these scales as micro- (tens of metres and hours), meso- (river reaches over months to years) and macro- (regional to intercontinental over decades to centuries). Groups in the first category include micro- to macrocrustaceans, bacteria and some algal communities which respond rapidly to wetting of the floodplain. The second scale is useful in describing the feeding, breeding and migration responses of most fish, amphibians and birds, as well as broad patterns in the distribution of wetland plant communities. The broadest scale must be considered when discussing waterbirds, and some aspects of the ecology of fish.

Geomorphology is often overlooked in environmental flow planning and scenario modelling, yet geomorphic change can be an important driver of wetland condition over time-scales relevant to environmental water management. The declining condition of the Ramsar-listed Southern Nature Reserve in the Macquarie Marshes is as much an outcome of channel deepening as of declining hydrological inputs (DHI 2008; Ralph 2008). Constant low flows in many systems have increased the size of flow required for overbank flooding. Where flooding is frequent, wetlands trap sediment and build the floodplain, altering the hydraulic gradient and promoting the diversion of floodwater to alternative sites (Ralph 2008). Relationships between flow and geomorphology are difficult to represent in the software environments currently used to model system hydrodynamics, yet for long time-series simulations these relationships are critical.

The challenge posed by altered hydrological regimes is to manage the redistribution of organisms to preserve biological diversity and maximise the provision of ecological goods and services, however defined. Wetlands in the Murray-Darling Basin are inherently variable, having developed in response to riverflow with strong interannual and interdecadal variability. Phases of the El Niño southern oscillation and, more broadly, the Interdecadal Pacific oscillation and the Indian Ocean dipole, influence climatic and hydrological variability over large sections of the Murray-Darling Basin during summer and winter respectively (Verdon *et al.* 2004; Verdon and Franks 2005). Our understanding of the periodicity of important climatic phases, and the triggers of change from one phase to another, is still poor.

The task of wetland management in the Murray-Darling Basin is further complicated by the threat of climate change due to greenhouse gas emissions and global warming. The CSIRO Sustainable Yields Project (CSIRO 2008a) has modelled potential hydrological impacts of climate change scenarios for all major river catchments within the Murray-Darling Basin. For example, there is likely to be a 10% increase in the period between significant inflows into the Macquarie Marshes by 2030 and the higher flood volume is likely to decrease by 16% (CSIRO 2008b). The effect of climate change exacerbates the impacts of water overallocation, and provides an additional degree of uncertainty in modelling the projected outcomes of proposed water-sharing arrangements.

Complexity of policy

Water development in the Murray-Darling Basin followed a predominantly post-war phase of dam and levee construction aimed at guaranteeing consistency of water supply for irrigated agriculture, town water supply, stock and domestic consumption and flood control. Entitlements to water in the regulated rivers were established under licences sold to water users, and in the period of development (1950s–70s) not all these entitlements were active. Further, the unusually wet conditions allowed high levels of allocation of active entitlements without obvious detriment to the environment.

It was not until the closing decades of the 20th century that an awareness developed of the environmental and social impacts of what had become a serious level of overallocation. The Murray-Darling Basin Commission established in 1997 a cap on water extraction in the Basin, at 1993–94 levels. Under the National Water Initiative (2004), state governments agreed to implement water-sharing plans which would govern the share of water allocated to the environment from dam releases and tributary flows within the Murray-Darling Basin.

This water management framework for the regulated rivers of the northern Murray-Darling Basin was largely in place by the turn of the century, and the early years of the 21st century saw the completion of most of the water-sharing plans for specific regulated catchments. The ensuing years saw the worst drought on record in the Murray-Darling Basin. In many cases this triggered the suspension of water-sharing plans, and water was managed at the discretion of the minister. The drought exposed deficiencies in all the plans, particularly the management of small to moderate flows which would previously have enhanced the resilience of significant wetlands and river reaches. The assumption that water could be allocated to the environment in wet years and to industry in dry exacerbated the impact of hyperdrought in the past decade.

In recognition of the limited prospects for early revision of plans and/or compensation of entitlement-holders, state and federal governments entered the water market to establish a reserve of strategic environmental water. This water could be used to supplement water made available under the water-sharing plans, without the compulsory reacquisition of entitlements. Funding the acquisition of entitlements from willing sellers is the main purpose of the AU$171 million Rivers Environmental Restoration Program. Following the success of that program, the Commonwealth made available an additional AU$3.4 billion for the purchase of adaptive environmental water, to be managed by the Commonwealth Environmental Water Holder.

The recent severe and prolonged drought has been one of a number of problems besetting the water reform process. Water reform has been imposed on a system designed to deliver water for extractive use, with profound structural and social impediments to systematic environmental water delivery. Examples include limitations in the size of dam release valves, precluding the volumes of flow required; inadequate gauging of flows at sites of environmental significance; low compliance with acceptable metering standards at off-take points; variability in the rigour of water-sharing plan enforcement for supplementary flows; difficulties in the regulation of water harvesting; and continued growth in allocations, particularly in the northern section of the Basin. In 2008 the federal government took greater control of the water reform process, establishing the Murray-Darling Basin Authority under the *Water Act 2007* and providing $10 billion to speed the water reform process. The authority is writing a Murray-Darling Basin Plan, with a final release planned for 2011.

Under the *Water Act 2007*, the federal government assumed overall responsibility for water planning in the Murray-Darling Basin. It will not supersede existing water-sharing plans established by the states, but the federal government will be the approving body for any plans developed after the expiry of existing plans. Expiry of most plans will occur in 2014 (2019 for Victorian plans) and it is anticipated that the federal government will be seeking conformity with sustainable diversion limits documented in the Murray-Darling Basin Plan. The task of properly accounting for the water requirements of key ecological assets is an important feature of the Murray-Darling Basin Plan and any water-sharing plan developed by the states.

Complexity of management

An important role of environmental water managers is matching the ecological requirements of species within a management area to the delivery of environmental water. The diversity of responses of individual species and their interactions makes this a task of near-infinite complexity. The wetland system comprises myriad species, each with a dynamic distribution and each of which contributes to trophic and competitive dependencies. A purely reductionist approach to system management would soon reveal fundamental limitations in knowledge, and the conflicting requirements of many constituent species. In practice, the task is simplified by the selection of a few representative species (sometimes termed 'flagship' or 'umbrella' species; Kingsford *et al.* 2010) for which the system is managed. There may be a number of reasons justifying the selection of a subset of species for active management: their water requirements could be representative of a broader suite of species, either similar in physiology or dependent on them for their basic habitat requirements; they could be iconic species for which the site is particularly significant; they might have particular significance in being rare or endangered. For example, the ecological risk assessment for Yanga National Park (Childs 2010) identified river red gum woodland and *Eleocharis* reed swamp (the two primary semi-permanent wetland vegetation habitats), the southern bell frog (an endangered species) and egrets (an iconic waterbird species) as the primary ecological assets for which water is managed in the park. Each has different water requirements. By accommodating the water regime requirements of these assets, a host of associated species should also be preserved.

The Gwydir Wetlands and Macquarie Marshes Adaptive Environmental Water Management Plans identify similar priority assets for management. For example, the Draft Gwydir Wetlands Adaptive Environmental Management Plan (DECCW 2009) identified the ecological significance of a group of broadly defined ecological communities (including marsh club-rush, water couch grassland, lignum and river cooba and coolibah/black box woodland) and species of particular significance within the wetland, including waterbirds, silver perch, woodland birds, red-bellied black snake and frogs.

Given the limited environmental water, even selecting representative species as management targets will involve environmental trade-offs. With declining inflows in the greater part of the Murray-Darling Basin linked to climate change (Murphy and Timbal 2008) it is unlikely that water availability in many major catchments will return to the conditions enjoyed in the period 1950–80. Active management of water therefore necessitates decisions about priorities for watering. This may involve the abandonment of sections of the historic floodplain, or decisions to water the floodplain for periods shorter than those that would trigger bird breeding which could not be sustained. Environmental watering strategies are by necessity filtering aspects of wetland extent and diversity. It is rare, however, that the principles of strategic conservation planning are applied in this context in a rigorous way, at the scale of the wetland mosaic or the entire Basin.

Flooding requirements in terms of timing, frequency, duration and extent may be necessary but they are not the only considerations in the active management of a species. Biotic factors of competition and predation may be equally important for a species to successfully complete a life-cycle. In some circumstances, management intervention may be required. Spencer and Wassens (2009) presented a case study of the management of the southern bell frog in the lower Murrumbidgee wetlands where the introduced common carp preys heavily on the bell frog over a range of development stages. The few wetlands with carp exclusion screens had twice the number of tadpoles and metamorphs following an environmental flow that targeted bell frogs.

There is increasing recognition of the provision of water for indigenous cultural purposes, sometimes termed 'cultural flows'. Presently, a 10 ML annual allocation is provided for cultural flows in the Macquarie catchment, and an indigenous representative sits on the Environmental Flow Reference Group. Cultural values correspond strongly to natural values in wetland conservation, and cultural flows are often used to support environmental flows targeting the maintenance of vegetation communities that are significant to Aboriginal people (Lucas, *pers. comm.* 2008). In some instances cultural flows are timed to coincide with cultural events aimed at reintroducing a younger generation of Aboriginal people to traditional uses of the wetland (Moggridge, *pers. comm.* 2009). The reconnection of wetlands and local indigenous communities is important for the long-term wellbeing of the wetlands, and it recognises the advocacy role of indigenous people that has been missing for too long.

Provision of water to the environment is a trade-off between the environmental and cultural benefits of healthy rivers and wetlands, and the economic opportunities of irrigated agriculture and other productive uses. Even small losses of agricultural capacity lead to significant social adjustment in many communities, as people leave the land and social infrastructure declines. Decisions made by governments to alter the balance between extraction and environmental use are rightly subject to intense scrutiny. In this context, the ecological requirements of rivers and wetlands need to be rigorously quantified using best available science, and water must be used with the greatest efficiency to achieve environmental outcomes. Documenting and testing the water requirements of ecological systems and system components is central to this task.

Ways forward

The tension between scientific research and environmental water management requirements often involves the reductionist tendency of hypothesis testing and the holistic requirements of system management. Managers must conceptualise the wetland system to make sense of their task, a process requiring a degree of abstraction and simplification. At the simplest level this may involve conceptual models which identify, in qualitative terms, relationships between abiotic and biotic components of the wetland. The conceptual model seeks to capture in simple, often diagrammatical, form much of the scientific understanding of the processes operating to maintain wetland form and function, and the relationships between flow and ecological response (Rogers *et al.* 2010).

However, for many systems the degree of system understanding and model representation have allowed quantitative hydrological and ecological models to be developed. A decision support system, supported by quantitative ecosystem response models, is an effective means of integrating the ecological and hydrological models required to optimise the delivery of environmental water. Decision support systems are usually computer-based systems which facilitate the comparison of alternative strategies where a range of sometimes conflicting objectives are pursued (McIntosh *et al.* 2008). Several decision support systems are used to manage environmental water in the Murray-Darling Basin, including the Murray Flow Assessment Tool (Young *et al.* 2003) developed in support of The Living Murray program, and the Water Allocation Decision Support System (Letcher 2005).

Wetland-specific decision support systems have, to date, been hindered by a lack of detailed hydrodynamic modelling (Davies and Acreman 2003). This impediment is being overcome by the application of hydrodynamic modelling to the significant wetlands of the northern Murray-Darling Basin (Saintilan *et al.* 2009). Merritt *et al.* (2010) described the IBIS decision support system, which integrates a water balance model of the Gwydir Wetlands (Powell *et al.* 2008) with models of vegetation and fish response to flow. The water balance model provides predictions of event timing, duration, depth and frequency of flooding, in daily time-steps, with the likely ecological responses modelled using discrete Bayesian network models. The approach is similar to that used in the EXCLAIM decision support system (Fu *et al.* 2009), which models the ecological implications of climate change scenarios using hydrological models of the Macquarie catchment and the water requirement information published in Chapters 2 to 6 of this volume. The model has accurately hindcast the probability of breeding success in the Macquarie Marshes, comparing model outputs with the data of Kingsford and Auld (2005).

The approach underscores the importance of good models populating the decision support system. We are still constrained by a lack of clear relationships between flow and response for many species, in particular the emergent relationships of predation and competition between species. These impediments can only be overcome by a commitment to monitoring the outcomes of environmental flows. With a substantial investment in environmental water at state and federal levels (now over $1 billion), this should not be a difficult argument. The public has a right to expect that the application of this water be appropriately audited and the environmental outcomes documented. The challenge is to use the opportunity to inform the development of better ecosystem response models, by refining our understanding of flow–ecology relationships and the hydrological behaviour of these systems. The very long dry period experienced throughout the Murray-Darling Basin this decade provides a unique opportunity to test thresholds in the life-cycles of many species and the limits of overall system resilience. The work discussed in this book is therefore ongoing.

References

Bunn SE and Arthington AH (2002) Basic principles and ecological consequences of altered flow regimes for aquatic biodiversity. *Environmental Management* **30**, 492–507.

Childs P (2010) 'An ecological risk assessment of the floodplain ecosystems of Yanga National Park, south-western NSW, Australia'. NSW Dept of Environment Climate Change and Water: Sydney.

CSIRO (2008a) 'Water availability in the Murray-Darling Basin. A report from the CSIRO to the Australian Government'. CSIRO Water for a Healthy Country: Canberra.

CSIRO (2008b) 'Water availability in the Macquarie-Castlereagh. A report to the Australian Government from the CSIRO Murray-Darling Basin Sustainable Yields Project'. CSIRO: Canberra.

Davies P and Acreman M (2003) Review of the Interim report of the SRP, the Murray Flow Assessment Tool and the SRP-REG scientific process for the Murray-Darling Basin Commission. 19 September 2003. Unpublished report to the Murray-Darling Basin Commission: Canberra.

DECCW (2009) 'Draft Adaptive Environmental Management Plan for the Gwydir Wetlands'. NSW Dept of Environment Climate Change and Water: Sydney.

DHI (2008) Hydrodynamic modelling of the Macquarie Marshes. Phase I report. Unpublished report to the NSW Dept of Environment and Climate Change: Sydney.

Fu B, Pollino C, Andrews F, Andrews A and Cuddy S (2009) EXCLAIM2: a tool for assessing climate change impacts on natural resources at a regional scale. Inaugural Annual Meeting of the Australasian Bayesian Network Modellers Society (ABNMS-09). 26–27 November. University of Melbourne: Melbourne.

Kingsford RT and Auld KM (2005) Waterbird breeding and environmental flow management in the Macquarie Marshes, arid Australia. *River Research and Applications* **21**, 187–200.

Kingsford RT, Brandis KJ, Jenkins KM, Nairn LC and Rayner TS (2010) Measuring ecosystem responses to flow across temporal and spatial scales. In *Ecosystem Response Modelling in the Murray-Darling Basin*. (Eds N Saintilan and I Overton) pp. 15–36. CSIRO Publishing: Melbourne.

Letcher RA (2005) Implementation of a water allocation decision support system in the Namoi and Gwydir valleys. In *MODSIM05: International Congress on Modelling and Simulation.* December 2005. pp. 1456–1552. Modelling and Simulation Society of Australia and New Zealand.

Lytle DH and Poff NL (2004) Adaptation to natural flow regimes. *Trends in Ecology and Evolution* **19**, 94–100.

McIntosh BS, Giupponi C, Voinov VV, Smith C, Matthews KB, Monticino M, Kolkman MJ, Crossman N, van Ittersum M, Haase D, Haase A, Mysiak J, Groot JCJ, Siber S, Verweih P, Quinn N, Waeger P, Gaber N, Hepting D, Scholten H, Sulis A, van Delden H, Gaddis E and Assaf H (2008) Bridging the gaps between design and use: developing tools to support environmental management and policy. In *Environmental Modelling, Software and Decision Support: State of the Art and New Perspectives.* (Eds AJ Jakeman, AA Voinov, AE Rizzoli and SH Chen) pp. 33–48. Elsevier: Amsterdam.

Merritt W, Powell S, Pollino C and Jakeman T (2010) IBIS: a decision support system for managers of environmental flows into wetlands. In *Ecosystem Response Modelling in the Murray-Darling Basin.* (Eds N Saintilan and I Overton) pp. 119–136. CSIRO Publishing: Melbourne.

Murphy BF and Timbal B (2008) A review of recent climate variability and climate change in south-eastern Australia. *International Journal of Climatology* **28,** 859–879.

Nilsson C and Svedmark M (2002) Basic principles and ecological consequences of changing water regimes: riparian plant communities. *Environmental Management* **30**, 468–480.

Peckham P and Molsher R (2005) Preliminary cultural values assessment of the Macquarie Marshes Nature Reserve. Unpublished report. NSW Dept of Environment, Climate Change and Water: Sydney.

Powell SJ, Letcher RA and Croke BFW (2008) Modelling floodplain inundation for environmental flows: Gwydir wetlands, Australia. *Ecological Modelling* **211,** 350–362.

Ralph TJ (2008) Channel breakdown and floodplain wetland morphodynamics in the Macquarie Marshes, south-eastern Australia. PhD thesis. Department of Physical Geography, Macquarie University.

Rogers K, Ralph T and Imgraben I (2010) Water requirements of biota, geomorphology and climate change in the Macquarie Marshes. In *Ecosystem Response Modelling in the Murray-Darling Basin.* (Eds N Saintilan and I Overton) pp. 151–170. CSIRO Publishing: Melbourne.

Saintilan N, Ling J and Wen L (2009) The development of interdisciplinary flow-ecology models for the wetlands of the northern Murray-Darling Basin. In *Ecohydrology of Surface and Groundwater Dependent Systems: Concepts, Methods and Recent Developments.* (Eds M Thoms, K Heal, E Bogh, A Chambel and V Smakhtin) pp. 86–93. IAHS Publishing: Wallingford, UK.

Spencer JA and Wassens S (2009) 'Responses of waterbirds, fish and frogs to environmental flows in the Lowbidgee wetlands in 2008–09'. Final report for the NSW Rivers Environmental Restoration Program. Rivers and Wetland Unit, NSW Dept of Environment and Climate Change: Sydney/Institute for Land, Water and Society, Charles Sturt University: Wagga Wagga.

Verdon DC and Franks SW (2005) Indian Ocean sea surface temperature variability and winter rainfall: eastern Australia. *Water Resources Research* **41,** W09413.

Verdon DC, Wyatt AM, Kiem AS and Franks SW (2004) Multidecadal variability of rainfall and streamflow: eastern Australia. *Water Resources Research* **40,** W10201.

Young WJ, Scott AC, Cuddy SM and Rennie BA (2003) 'Murray flow assessment tool: a technical description'. Client report. CSIRO Land and Water: Canberra.

Glossary

Anoxia: Without oxygen.

Antecedent flood conditions: The characteristics of prior floods, e.g. whether a series of large floods occurred prior to the current flood or whether large floods had been absent for some time.

Biofilm: The thin layer of microorganisms and detritus growing on submerged surfaces.

Body whorl: The large final coil of a gastropod shell that contains the body.

Brachiostegite: A gill cover and chamber in certain decapod crustaceans, formed by lateral expansion of the carapace.

Breeding duration: The period required for a breeding cycle to complete. In waterbirds, this commences with egg laying and is complete when young are fully fledged and independent.

Breeding lag time: The period before the commencement of breeding which allows fauna, particularly waterbirds, to build up fat reserves to increase reproductive success, establish breeding pairs and/or collect material to construct nests. Many waterbirds generally require flooding for the entire breeding duration, which incorporates the time required to build nests, lay and incubate eggs and fledge young to the point of independence from the nest.

Breeding season: The timing of breeding. This is particularly important for waterbirds with a strong seasonal breeding stimulus.

Breeding stimulus: The essential environmental conditions required to commence breeding, specifically whether breeding commences in response to flooding, season or rainfall. It may be largely driven by the conditions required for maximal generation of food.

Carapace: The protective part of the exoskeleton in crustaceans covering the cephalothorax.

Carpus: The segment joining the merus with the chela in crayfish.

Cephalothorax: The anterior section of the body in crustaceans formed by the fusion of the head and the thorax.

Chela: The large claw carried on the first limb of crayfish.

Cheliped: The claw-bearing appendage.

Columellar (fold): A twist in the columellar, the central pillar that forms the axis of the gastropod shell.

Coxae: The first segment of the leg of an arthropod.

Dabbling ducks (waterbird functional group): Waterbirds that feed by up-ending or dabbling in shallow water. This group includes some of the classic Australian waterfowl species and consists entirely of Anseriformes.

Decollate: Shedding of apical whorls in some gastropod groups.

Deep-water foragers (waterbird functional group): Waterbirds that forage from deep water, but do not exhibit a preference for fish. This functional group includes some of the classic Australian waterfowl species and is dominated by Anseriformes such the black swan, hardhead, musk duck and blue-tailed duck. It also includes the Gruiforme, the Eurasian coot.

Detritivores: Animals that consume detritus (non-living, particulate organic matter).

Dextral: Shell opening is on the right-hand side when the spire is pointing away from you and the opening (aperture of the shell) is facing upwards and towards you.

Ecohydrology: The interdisciplinary field of research that links ecological and hydrological processes.

Exopodite: The outer branch of a crustacean appendage.

Fecundity: The reproductive rate of an organism or a population or, more generally, the ability to reproduce.

Fish-eaters (waterbird functional group): Waterbirds that consume fish as their primary food source. This functional group is dominated by Pelecaniformes such as grebes, cormorants, darters and the Australian pelican, Ciconiiformes such as egrets and herons, and Charadriiformes including gulls and terns.

Flood depth: The water depth required for growth, germination and establishment of plants, or the water depth required for the establishment of breeding and reproductive success in fauna. This is considered important for flora as flooding depth may determine the proportion of exposed photosynthetic tissues. It may affect root and rhizome functions such as respiration, support, reserve storage and nutrient uptake, and may affect above-ground biomass and photosynthetic activity. This is important for fauna as specific flood depths may be required for reproduction or to support foraging during the breeding season.

Flood duration: The period of time for which biota require standing water. This is important for flora as flooding reduces soil oxygen levels, creating anoxic conditions, and may enable toxic compounds to accumulate within soils. It is also important for fauna as flooding throughout the breeding cycle may be required for reproductive success. For waterbirds, flood duration specifically incorporates the required lag time and breeding duration.

Flood frequency: The frequency of flood events at a location. This is important as biota exhibit different tolerances to flooding and drought.

Flood pulse: The inconsistent flow rates of river systems or the pulsing of river discharge over time. Prior to water resource development and water management, rivers of the Murray-Darling Basin exhibited a highly variable flood pulse.

Flood spawners (MFAT fish group): Native fish that spawn and recruit following flow rises. Major spawning occurs during periods of floodplain inundation (e.g. golden perch, silver perch).

Flood timing: The seasonality of flooding. This is important as biota often require flooding at certain times to coincide with reproductive seasons, to compensate for high air temperatures and evapotranspiration or to reduce the impacts of freezes and frosts.

Flood/flow history: The sequence of flow pulses occurring before a point in time.

Flood/flow regime: The long-term statistical pattern of river flows.

Geomorphology: The interdisciplinary field of research on landforms and the processes that create and change them over time.

Germination timing: The seasonality of plant germination. This is associated with the requirement of drawdown or flooded conditions for germination and the timing of flooding.

Glochidium: Tiny bivalved larva of hyriid freshwater mussels. Larvae are usually subtriangular with a tooth (hook) on the apex of each valve.

Gonochoric: Animals with separate sexes.

Gonopore: External genital opening.

Grasses: Any plants of the family Poaceae of Gramineae, having jointed stems, sheathing leaves and seedlike grains.

Grazing waterfowl (waterbird functional group): Herbivorous waterbirds that feed primarily by grazing from wet pasture or grasses. This group consists entirely of Anseriformes and includes the maned duck, plumed whistling-duck and Australasian shelduck.

Herbs and forbs: Annual and perennial plants whose stems do not produce woody persistent tissue. They generally die back at the end of each growing season.

Hermaphrodite: Individuals with both male and female sexual organs.

Hydroperiod: The length of time over which water is present on a floodplain wetland surface.

Hypometabolic: A state of reduced metabolic activity.

Hypoxia: Conditions of low dissolved oxygen concentrations in water (often defined in the range of 1–30% saturation).

Inter-flood dry-period: The period when there is no standing water available. This is important as many plants require an absence of flooding for seed germination. It also indicates the ability of plants to survive extended drought conditions. It may stimulate wetland productiveness and the growth of food resources that sustain fauna throughout reproduction.

Isoosmotic: Having the same concentration of solutes as the blood.

Large waders (waterbird functional group): Large long-legged birds, with a length greater than 50 cm, that wade in the water for food. This group includes Ciconiiformes such as spoonbills and ibis, and the Gruiforme, brolga.

Lateral connectivity: Linkages between the river and floodplain habitats, brought about by overbank flooding or by channels filled with water.

Low-flow regime: Flows that stay within the stream channel and prevail for most of the time.

Low-flow specialists (MFAT fish group): Native fish that spawn and recruit during low flows, within channels or floodplain habitats (e.g. rainbowfish, carp gudgeons).

Main channel generalists (MFAT fish group): Native fish that spawn and recruit during high or low flows in the channel (e.g. Australian smelt, bony bream, flathead gudgeons).

Main channel specialists (MFAT fish group): Native fish that spawn and recruit during high or low flows in the main channel. Woody debris is an important part of their habitats (e.g. Murray cod, trout cod, river blackfish).

Maintenance (plants): The water regime required to ensure growth, flowering and survival of established plants at a standard state or at heightened levels of productivity, i.e. plants not experiencing stress, which occurs under extreme conditions or outside the normal range encountered by the plant.

Merus: Jointed segment below the chela of the cheliped of crayfish.

Microhabitat: Small-scale localised patches of habitat within larger habitats with distinct hydraulic, sedimentological and water-quality characteristics. Includes snags, river edge communities and macrophyte beds.

Migration: The directed, regular or systematic movement of a group of organisms or a whole population.

Morphology: The form, structure and configuration of an organism, including the outward appearance (e.g. shape, colour, pattern) and the internal parts. In rivers and wetlands, it refers to the size, shape and arrangement of the composite landforms (e.g. channel bed and banks, floodplain, lagoons). See **Geomorphology**.

Omnivore: Animal that eats both plants and animals.

Ontogenetic stage: Part of the life-cycle, development or developmental history of an organism, from embryo to adult.

Oocyte: An immature ovum, or egg cell, involved in reproduction.

Oviduct: The tube leading from the ovary to the exterior opening (gonopore) through which eggs are passed during spawning.

Oviparity: Deposition of eggs, after which developing and hatching occur externally.

Ovoviviparity: Development and brooding of eggs within the parent's body.

Pallial: Relating to the mantle in molluscs.

Parthenogenesis: A form of asexual reproduction found in females, where embryonic development occurs without fertilisation by a male.

Patelliform: Shaped like a kneecap, limpet-like.

Pereiopods: The walking limb of a decapod crustacean. There are five pairs of walking limbs, found on the cephalothorax.

Periphyton: The assemblage of algae and other microorganisms that is attached to submerged surfaces. The term is interchangeable with biofilm.

Pleopods: The small paired abdominal appendages located beneath the abdomen in decapods, also known as swimmerets.

Potamophilous: River-loving.

Radula: A ribbon of chiton bearing rows of teeth in the buccal cavity of snails.

Rate of flood rise and fall: The rate at which flood water levels increase and drawdown occurs.

Recruitment: The process of an organism reaching a certain size or reproductive stage in its life-cycle, and a key aspect of population dynamics along with birth rate, growth rate and mortality.

Rheotaxis: A reponse to water currents. Animals that display positive rheotaxis move against the current.

Riverine: Of the river and riparian areas.

Rostrum: A long cephalic spine, an extension of the carapace, protruding forward from between the eyes of decapod crustaceans.

Sedges and rushes: Any of numerous grasslike plants of the family Cyperaceae and Juncaceae.

Semelparous: Animals that reproduce once in their life-time.

Shoreline foragers (waterbird functional group): herbivorous, carnivorous and omnivorous waterbirds that feed from the shoreline and edges of wetlands. This group includes Gruiformes such as the black-tailed native-hen and purple swamphen, and Charadriiformes such as the masked and banded lapwings.

Shrubs: Woody plants smaller than a tree, usually with multiple permanent stems branching from or near the ground.

Sinistral: A shell opening that is on the left-hand side when the spire is pointing away from you and when the opening (aperture of the shell) is facing upwards and towards you.

Small waders (waterbird functional group): Small long-legged birds, with a length less than 50 cm, that wade in the water for food. This group includes Charadriiformes such as the black-winged stilt, red-necked avocet, red-capped plover, sharp-tailed and marsh sandpipers, black-fronted and red-kneed dotterels, black-tailed godwit and common greenshank.

Spawning lag time (frogs): The interval between the commencement of calling by male frogs and the time when female frogs are ready to spawn. Lags can be caused by females needing to build up body condition prior to spawning or by delaying breeding until water temperatures are ideal for tadpole development.

Spawning: The production of large quantities of eggs by aquatic organisms (e.g. fish and frogs).

Spermatheca: Sperm sac opening to the vagina, serving as a storage chamber for sperm.

Spire: All the coils of a shell above the body whorl.

Stenotopic: Able to tolerate or adapt to only a small range of environmental conditions.

Submerged aquatic macrophytes: Plants primarily submerged beneath the water surface.

Supraorbital: Above the eye.

Survival (plants): The water regime required to enable established plants to survive, perhaps in a state of stress but not at the point of no recovery, chronic damage or cell stress. If plants are stressed, they may be able to recover to a new standard state in which growth and flowering occur. Values for survival are maximum or minimum values for flooding characteristics. For example, maximum flood depth describes the maximum depth at which plants are able to survive.

Suspension feeders: Animals that feed by removing suspended matter and food particles from water. Also known as filter-feeders.

Telson: The unpaired terminal abdominal segment of crustaceans.

Thalweg: The flow path that follows the deepest parts of the stream channel.

Trees: Plants with a permanently woody main stem or trunk, ordinarily growing to a considerable height and usually developing branches at some distance from the ground.

Trophic dynamics: The system of trophic levels and interactions which describe the relationships of organisms in a food chain or food web (based on a transfer of energy).

Trophic level: The position in a food chain or food web occupied by an organism.

Umbo: The dorsal region of maximum curvature on each valve of a bivalve shell, often called the beak.

Univoltine: Producing a single generation per year.

Veliger: Molluscan larva characterised by the velum that develops from the protoconch and comprises two large ciliated lobes. It is a free-living, planktonic larval stage.

Viviparity: Production of live offspring within the parent's body.

Water regime: The prevailing pattern of flood pulses over a period of time.

Wetland specialists (MFAT fish group): Native fish that spawn and recruit in floodplain wetlands and lakes (opportunistically, when flooding occurs) and anabranches and billabongs during in-channel flows (e.g. Australian smelt, bony bream, carp gudgeons, southern pygmy perch, hardyheads).

Index

www.ingramcontent.com/pod-product-compliance
Lightning Source LLC
LaVergne TN
LVHW070848210926
848482LV00016B/889